Pattern Theory

APPLYING MATHEMATICS

A Collection of Texts and Monographs

ADVISORY BOARD

Robert Calderbank
Jennifer Chayes
David Donoho
Weinan E
George Papanicolaou

Pattern Theory

The Stochastic Analysis of Real-World Signals

David Mumford
Agnès Desolneux

Editorial, Sales, and Customer Service Office

A K Peters, Ltd.
5 Commonwealth Road, Suite 2C
Natick, MA 01760
www.akpeters.com

Copyright © 2010 by A K Peters, Ltd.

All rights reserved. No part of the material protected by this copyright notice may be reproduced or utilized in any form, electronic or mechanical, including photocopying, recording, or by any information storage and retrieval system, without written permission from the copyright owner.

Library of Congress Cataloging-in-Publication Data

Mumford, David, 1937–
  Pattern theory : the stochastic analysis of real-world patterns / David Mumford, Agnès Desolneux.
    p. cm. – (Applying mathematics)
  Includes bibliographical references and index.
  ISBN 978-1-56881-579-4 (alk. paper)
  1. Pattern perception. 2. Pattern recognition systems. I. Desolneux, Agnès. II. Title.

Q327.M85 2010
003'.52–dc22

2009040725

Cover artwork courtesy of Luis Alvarez, Yann Gousseau, and Jean-Michel Morel; see Figure 6.18 for details.

To Ulf Grenander,

who has dared to propose mathematical models for virtually all of life, pursuing the motto on the wall of his study

Using Pattern Theory to create mathematical structures
both in the natural and the man-made world

Contents

Preface ix

Notation xi

0 What Is Pattern Theory? 1
   0.1 The Manifesto of Pattern Theory 1
   0.2 The Basic Types of Patterns 5
   0.3 Bayesian Probability Theory: Pattern Analysis
        and Pattern Synthesis 9

1 English Text and Markov Chains 17
   1.1 Basics I: Entropy and Information 21
   1.2 Measuring the n-gram Approximation with Entropy 26
   1.3 Markov Chains and the n-gram Models 29
   1.4 Words 39
   1.5 Word Boundaries via Dynamic Programming and
        Maximum Likelihood 45
   1.6 Machine Translation via Bayes' Theorem 48
   1.7 Exercises 51

2 Music and Piecewise Gaussian Models 61
   2.1 Basics III: Gaussian Distributions 62
   2.2 Basics IV: Fourier Analysis 68
   2.3 Gaussian Models for Single Musical Notes 72
   2.4 Discontinuities in One-Dimensional Signals 79
   2.5 The Geometric Model for Notes via Poisson Processes 86
   2.6 Related Models 91
   2.7 Exercises 100

3 Character Recognition and Syntactic Grouping 111
   3.1 Finding Salient Contours in Images 113
   3.2 Stochastic Models of Contours 122
   3.3 The Medial Axis for Planar Shapes 134
   3.4 Gestalt Laws and Grouping Principles 142
   3.5 Grammatical Formalisms 147
   3.6 Exercises 163

4	Image Texture, Segmentation and Gibbs Models	173
	4.1 Basics IX: Gibbs Fields	176
	4.2 $(u+v)$-Models for Image Segmentation	186
	4.3 Sampling Gibbs Fields	195
	4.4 Deterministic Algorithms to Approximate the Mode of a Gibbs Field	202
	4.5 Texture Models	214
	4.6 Synthesizing Texture via Exponential Models	221
	4.7 Texture Segmentation	228
	4.8 Exercises	234
5	Faces and Flexible Templates	249
	5.1 Modeling Lighting Variations	253
	5.2 Modeling Geometric Variations by Elasticity	259
	5.3 Basics XI: Manifolds, Lie Groups, and Lie Algebras	262
	5.4 Modeling Geometric Variations by Metrics on Diff	276
	5.5 Comparing Elastic and Riemannian Energies	285
	5.6 Empirical Data on Deformations of Faces	291
	5.7 The Full Face Model	294
	5.8 Appendix: Geodesics in Diff and Landmark Space	301
	5.9 Exercises	307
6	Natural Scenes and their Multiscale Analysis	317
	6.1 High Kurtosis in the Image Domain	318
	6.2 Scale Invariance in the Discrete and Continuous Setting	322
	6.3 The Continuous and Discrete Gaussian Pyramids	328
	6.4 Wavelets and the "Local" Structure of Images	335
	6.5 Distributions Are Needed	348
	6.6 Basics XIII: Gaussian Measures on Function Spaces	353
	6.7 The Scale-, Rotation- and Translation-Invariant Gaussian Distribution	360
	6.8 Model II: Images Made Up of Independent Objects	366
	6.9 Further Models	374
	6.10 Appendix: A Stability Property of the Discrete Gaussian Pyramid	377
	6.11 Exercises	379
Bibliography		387
Index		401

Preface

The Nature of This Book

This book is an introduction to the ideas of Ulf Grenander and the group he has led at Brown University for analyzing *all* types of signals that the world presents to us. But it is not a basic introductory graduate text that systematically expounds some area of mathematics, as one would do in a graduate course in mathematics. Nor is our goal to explain how to develop all the basic tools for analyzing speech signals or images, as one would do in a graduate course in signal processing or computer vision. In our view, what makes applied mathematics distinctive is that one starts with a collection of problems from some area of science and then seeks the appropriate mathematics for clarifying the experimental data and the underlying processes producing this data. One needs mathematical tools and, almost always, computational tools as well. The practitioner finds himself or herself engaged in a lifelong dialog: seeking models, testing them against data, identifying what is missing in the model, and refining the model. In some situations, the challenge is to find the right mathematical tools; in others, it is to find ways of efficiently calculating the consequences of the model, as well as finding the right parameters in the model. This book, then, seeks to actively bring the reader into such a dialog.

To do this, each chapter begins with some example, a class of signals whose variability and patterns we seek to model. Typically, this endeavor calls for some mathematical tools. We put these in sections labeled Basics I, II, etc., and in these sections we put on our mathematical hats and simply develop the math. In most cases, we give the basic definitions, state some results, and prove a few of them. If the reader has not seen this set of mathematical ideas at all, it is probably essential to do some background reading elsewhere for a full exposition of the topic. But if the reader has some background, this section should serve as a refresher, highlighting the definitions and results that we need. Then we compare the model with the data. Usually, new twists emerge. At some point, questions of computability emerge. Here we also have special sections labeled Algorithm I, II, etc. Each chapter ends with a set of exercises. Both of the authors have taught some or part of this material multiple times and have drawn on these classes. In fact, the book is based on lectures delivered in 1998 at the Institut Henri Poincaré (IHP) by the senior author and the notes made at the time by the junior author.

The chapters are ordered by the complexity of the signal being studied. In Chapter 1, we study text strings: discrete-valued functions in discrete time. In Chapter 2, we look at music, a real-valued function of continuous time. In Chapter 3, we examine character recognition, this being mostly about curvilinear structures in the domain of a function of two variables (namely, an image $I(x, y)$). In Chapter 4, we study the decomposition of an image into regions with distinct colors and textures, a fully two-dimensional aspect of images. In Chapter 5, we deal with faces, where diffeomorphisms of planar domains is the new player. In Chapter 6, we examine scaling effects present in natural images caused by their statistical self-similarity. For lack of time and space, we have not gone on to the logical next topic: signals with three-dimensional domains, such as various medical scans, or the reconstruction of three-dimensional scenes from one or more two-dimensional images of the scene.

There is a common thread in all these topics. Chapter 0 explains this thread, the approach to the analysis of signals that Grenander named "pattern theory." In the choice of specific algorithms and ideas, we have, of course, used the research we know best. Hence, the work of our own groups, our students and collaborators, is heavily represented. Nonetheless, we have tried to present a fair cross section, especially of computer vision research.

We want to thank many people and institutions that have helped us with the preparation of this book: the many students at the Institut Henri Poincaré, Brown University, the Université Paris Descartes, and École Normale Supérieure de Cachan, who listened and gave feedback for lectures based on the material in this book; to Jean-Michel Morel, who arranged the initial lectures; to Brown University, the Fondation des Treilles, and the Institut des Hautes Études Scientifiques, who hosted us when we worked on the book; to Tai-Sing Lee and Jason Corso for careful reading of the penultimate draft; and finally to A K Peters, for their patience over many years while we worked on the book.

Notation

$P, Q, \ldots$ denote probability distributions on a countable set, always given by uppercase letters. Their values are $P(x), Q(y), \ldots$.

$P, Q, \ldots$ as well as $\mu, \nu, \ldots$ denote probability measures on continuous spaces. Their values on measurable subsets will be $P(A), \mu(B), \ldots$. If there is a reference measure dx, their probability densities are denoted by lowercase letters: $P(dx) = p(x)dx$ or $dP/dx = p(x)$. Thus, we write

$$P(A) = \int_A dP(x) = \int_A p(x)dx.$$

$\mathcal{X}, \mathcal{Y}, \mathcal{A}, \ldots$ denote random variables from the set for the given probability distribution. Note that random variables are always indicated by this distinctive font.

$\mathbb{P}$(predicate with random variables) denotes the probability of the predicate being true. If needed, we write $\mathbb{P}_P$ to indicate the probability with respect to P. Thus,

$$P(x) = \mathbb{P}(\mathcal{X} = x) = \mathbb{P}_P(\mathcal{X} = x)$$
$$P(A) = \mathbb{P}(\mathcal{X} \in A) = \mathbb{P}_P(\mathcal{X} \in A).$$

$\mathbb{E}(f)$ denotes the expectation of the random variable or function of random variables f. If needed, we write $\mathbb{E}_P$ to indicate the expectation with respect to P. The function f, for instance, can be $\log(P(\mathcal{X}))$, which we abbreviate to $\log(P)$.

If μ is a measure on some space E, and if f is a measurable function from E to some space F, then the image measure denoted by $f_*\mu$ is the measure on F defined by $f_*\mu(U) = \mu(f^{-1}(U))$ for all measurable subsets U of F.

– 0 –
What Is Pattern Theory?

The term "pattern theory" was coined by Ulf Grenander to distinguish his approach to the analysis of patterned structures in the world from "pattern recognition." In this book, we use it in a rather broad sense to include the statistical methods used in analyzing all "signals" generated by the world, whether they be images, sounds, written text, DNA or protein strings, spike trains in neurons, or time series of prices or weather; examples from all of these appear either in Grenander's book *Elements of Pattern Theory* [94] or in the work of our colleagues, collaborators, and students on pattern theory. We believe the work in all these areas has a natural unity: common techniques and motivations. In particular, pattern theory proposes that the types of patterns (and the hidden variables needed to describe these patterns) that are found in one class of signals will often be found in the others and that their characteristic variability will be similar. Hence the stochastic models used to describe signals in one field will crop up in many other signals. The underlying idea is to find classes of stochastic models that can capture all the patterns we see in nature, so that random samples from these models have the same "look and feel" as the samples from the world itself. Then the detection of patterns in noisy and ambiguous samples can be achieved by the use of Bayes' rule, a method that can be described as "analysis by synthesis."

0.1   The Manifesto of Pattern Theory

We can express this approach to signals and their patterns in a set of five basic principles, the first three of which are as follows:

1. A wide variety of signals result from observing the world, all of which show patterns of many kinds. These patterns are caused by objects, processes, and laws present in the world but at least partially hidden from direct observation. The patterns can be used to infer information about these unobserved factors.

2. Observations are affected by many variables that are not conveniently modeled deterministically because they are too complex or too difficult to observe and often belong to other categories of events, which are irrelevant

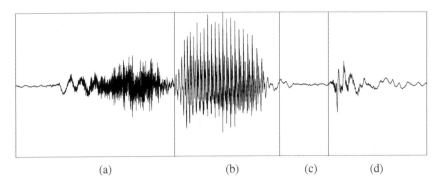

(a) (b) (c) (d)

Figure 1. A second of the raw acoustic signal during the pronunciation of the word "sheep." Note the four phones: (a) white noise during the phoneme *sh*, (b) a harmonic sound indicated by the periodic peaks (caused by the vocal chord openings) during for the vowel *ee*, (c) silence while the mouth is closed during the first part of the stop consonant *p*, (d) a burst when the mouth opens and the rest of *p* is pronounced.

to the observations of interest. To make inferences in real time or with a model of reasonable size, we must model our observations partly stochastically and partly deterministically.

3. Accurate stochastic models are needed that capture the patterns present in the signal while respecting their natural structures (i.e., symmetries, independences of parts, and marginals on key statistics). These models should be learned from the data and validated by sampling: inferences from them can be made by using Bayes' rule, provided that the models' samples resemble real signals.

As an example, a microphone or an ear responds to the pressure wave $p(t)$ transmitted by the air from a speaker's mouth. Figure 1 shows a function $p(t)$ with many obvious patterns; it divides into four distinct segments, each of which has very different character. These are four "phones" caused by changes in the configuration of the mouth and vocal cords of the speaker during the passage of air, which in turn are caused by the intention in the speaker's brain to utter a certain word. These patterns in p encode in a noisy, highly variable way the sequence of phones being pronounced and the word these phones make up. We cannot *observe* the phones or the word directly—hence, they are called *hidden variables*—but we must infer them. Early work on speech recognition attempted to make this inference deterministically by using logical rules based on binary features extracted from $p(t)$. For instance, Table 1 shows some of the features used to distinguish English consonants.

0.1. The Manifesto of Pattern Theory

	p	t	k	b	d	g	m	n	ŋ (sing)
Continuant	−	−	−	−	−	−	−	−	−
Voiced	−	−	−	+	+	+	+	+	+
Nasal	−	−	−	−	−	−	+	+	+
Labial	+	−	−	+	−	−	+	−	−
Coronal	−	+	−	−	+	−	−	+	−
Anterior	+	+	−	+	+	−	+	+	−
Strident	−	−	−	−	−	−	−	−	−

	f	θ (thing)	s	ʃ (she)	v	ð (this)	z (rose)	ʒ (beige)	tʃ (chair)	dʒ (joy)
Continuant	+	+	+	+	+	+	+	+	−	−
Voiced	−	−	−	−	+	+	+	+	−	+
Nasal	−	−	−	−	−	−	−	−	−	−
Labial	+	−	−	−	+	−	−	−	−	−
Coronal	−	+	+	+	−	+	+	+	+	+
Anterior	+	+	+	−	+	+	+	−	−	−
Strident	+	−	+	+	+	−	+	+	+	+

Table 1. Features of English consonants [3, Table 4.8].

This deterministic rule-based approach failed, and the state of the art is to use a family of precisely tuned stochastic models (hidden Markov models) and a Bayesian *maximum a posteriori* (MAP) estimator as an improvement.

The identical story played out in vision, in parsing sentences from text, in expert systems, and so forth. In all cases, the initial hope was that the deterministic laws of physics plus logical syllogisms for combining facts would give reliable methods for decoding the signals of the world in all modalities. These simple approaches seem to always fail because the signals are too variable and the hidden variables too subtly encoded. One reason for this is that there are always so many extraneous factors affecting the signal: noise in the microphone and the ear, other sounds in the room besides the speaker, variations in the geometry of the speaker's mouth, the speaker's mood, and so on. Although, in principle, one might hope to know these as well as the word being pronounced, inferring each extra factor makes the task more difficult. (In some cases, such as radioactive decay in PET scans, it is even impossible by quantum theory.) Thus stochastic models are *required*.

Now, the world is very complex and the signals it generates are likewise complex, both in their patterns and in their variability. If pattern theory makes any sense, some rules must exist whereby the patterns caused by the objects, processes, and laws in the world are not like some arbitrary recursively enumerable set. For instance, in the movie *Contact*, the main character discovers life in outer space because it broadcasts a signal encoding the primes from 2 to 101. If infants had to recognize this sort of pattern in order to learn to speak, they would never succeed. By and large, the patterns in the signals received by our senses are correctly learned by infants, at least to the level required to reconstruct the various types of objects in our world and their properties, and to communicate with adults. This is a marvelous fact, and pattern theory attempts to understand why this is so. Two further key ideas can be added to the manifesto to explain this:

4. The various objects, processes, and rules of the world produce patterns that can be described as precise *pure patterns* distorted and transformed by a limited family of *deformations*, similar across all modalities.

5. When all the stochastic factors affecting any given observation are suitably identified, they show a large amount of conditional independence.

The last point is absolutely essential to creating reasonably simple stochastic models. For example, suppose Alice and Bob are two suspects being interrogated in two separate rooms. Then, given what they both know and have planned and what they have communicated to each other, directly or indirectly, their actions are independent. See the many further examples in Pearl's very important book [174].

To apply pattern theory properly, it is essential to identify correctly the patterns present in the signal. We often have an intuitive idea of the important patterns, but the human brain does many things unconsciously and also takes many shortcuts to get things done quickly. Thus, a careful analysis of the actual data to see what they are telling us is preferable to slapping together an off-the-shelf Gaussian or log-linear model based on our guesses. Here is a very stringent test of whether a stochastic model is a good description of the world: *sample from it*. This is so obvious that one would assume everyone does this, but in actuality, this is not so. The samples from many models that are used in practice are absurd oversimplifications of real signals, and, even worse, some theories do not include the signal itself as one of its random variables (using only some derived variables), so it is not even possible to sample signals from them.[1]

In what ways is pattern theory different from the better-known field of statistical pattern recognition? Traditionally, the focus of statistical pattern recognition was the study of one or more data sets $\{\vec{x}_\alpha \in \mathbb{R}^k\}_{\alpha \in I}$ with the goals of (a) fitting

---

[1]This was, for instance, the way most traditional speech recognition systems worked: their approach was to throw away the raw speech signal in the preprocessing stage and replace it with codes designed to ignore speaker variation. In contrast, when all humans listen to speech, they are clearly aware of the idiosyncrasies of the individual speaker's voice and of any departures from normal. The idea of starting by extracting some hopefully informative features from a signal and only then classifying it is via some statistical algorithm is enshrined in classic texts such as [64].

(parametric and nonparametric) probability distributions to each data set, (b) finding optimal decision rules for classifying new data into the correct data set, and (c) separating a single data set into clusters when it appears to be a mixture. The essential issue is the "bias-variance" trade-off: to model fully the complexities of the data source but not the accidental variations of the specific data set. When Grenander first proposed pattern theory as a distinct enterprise, his approach had several very novel aspects. He put forth the following five propositions:

1. To describe the patterns in typical data sets, one should always look for appropriate hidden variables, in terms of which the patterns are more clearly described.

2. The set of variables, observed and hidden, typically forms the vertices of a graph, as in Gibbs models, and one must formulate prior probability distributions for the hidden variables as well as models for the observed variables.

3. This graph itself might be random and its variability must then be modeled.

4. One can list the different types of "deformations" patterns are subject to, thus creating the basic classes of stochastic models that can be applied.

5. These models should be used for pattern synthesis as well as analysis.

As the subject evolved, statistical pattern recognition merged with the area of neural nets, and the first two ideas were absorbed into statistical pattern recognition. These so-called graphical models are now seen as the bread and butter of the field, and discovering these hidden variables is a challenging new problem. The use of prior models has become the mainstream approach in vision and expert systems, as it has been in speech since the 1960s. The other aspects of pattern theory, however, are still quite distinctive.

0.2 The Basic Types of Patterns

Let us be more precise about the kinds of patterns and deformations referred to in point 4 above. Real-world signals show two very distinct types of patterns. We call these (1) *value patterns* and (2)*geometrical patterns*. Signals, in general, are some sort of function $f : X \to V$. The domain may be continuous (e.g., a part of space such as the retina or an interval of time) or discrete (e.g., the nodes of a graph or a discrete sample in space or time), and the range may be a vector space or binary $\{0, 1\}$ or something in the middle. In the case of value patterns, we mean that the features of this pattern are computed from the values of f or from some linear combinations of them (e.g., power in some frequency band). In the case of geometric patterns, the function f can be thought of as producing geometrical patterns in its domain (e.g., the set of its points of discontinuity). The distinction affects which extra random variables we need to describe the pattern. For value patterns, we typically add coefficients in some expansion to describe the particular signal. For geometric patterns, we add certain points or subsets of the domain or features of such subsets. Traditional statistical pattern recognition and the traditional theory of stationary processes deals only with the values of f, not the geometry of its domain. Let us be more specific and describe these patterns more explicitly by distinguishing two sorts of geometric patterns—those involving deforming the geometry and those involving a hierarchical geometric pattern:

1. *Value patterns and linear superposition.* The most standard value model creates the observed signal from the linear superposition of fixed or learned basis functions. We assume the observed signal has values in a real vector space $s : X \longrightarrow V$ and that it is expanded in terms of auxiliary functions $s_\alpha : X \longrightarrow V$ as

$$s = \Sigma_\alpha c_\alpha s_\alpha.$$

In vision, X is the set of pixels and, for a gray-level image, V is the set of real numbers. Here the coefficients $\{c_\alpha\}$ are hidden random variables, and either the functions $\{s_\alpha\}$ may be a universal basis such as sines and cosines or wavelets or some learned templates, as in Karhunen-Loeve expansions, or the $\{s_\alpha\}$ may be random. The simplest case is to allow one of them to be random and think of it as the residual (e.g., an additive noise term). An important case is that of expanding a function into its components on various scales (as in wavelet expansions), so the terms s_α are simply the components of s on scale α. See Figure 2 for an example of a face expanded into three images representing its structure on fine, medium, and coarse scales. Other variants are (1) amplitude modulation in which the low and high frequencies are combined by multiplication instead of addition, and (2) the case in which s is just a discrete sample from the full function $\Sigma c_\alpha s_\alpha$. Quite often, the goal of such an expansion, from a statistical point of view, is to make the coefficients $\{c_\alpha\}$ as independent as possible. If the coefficients are Gaussian and independent, this is called a principal components analysis (PCA) or a Karhunen-Loeve expansion; if they are non-Gaussian but independent, it is an ICA. We shall see how such expansions work very well to express the effects of lighting variation on face signals.

0.2. The Basic Types of Patterns

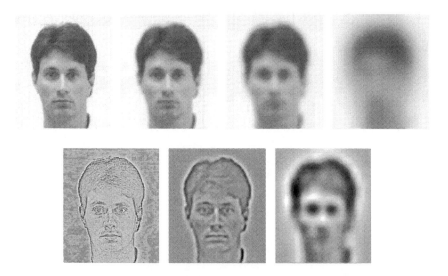

Figure 2. A multiscale decomposition of an image of the face of a well-known computer scientist. In the top-row images, the face is progressively blurred; the three bottom images are the successive differences. Adding them together, plus the blurriest version, gives back the original face. Note how the different scales contain different information: at the finest, the exact location of edges is depicted; in the middle, the local features are seen but the gray level of the skin and hair are equal; at the coarsest, the global shapes are shown. [97, Figure 1.1]

2. *Simple geometric patterns and domain warping.* Two signals generated by the same object or event in different contexts typically differ due to expansions or contractions of their domains, possibly at varying rates: phonemes may be pronounced faster or slower, and the image of a face may be distorted by varying its expression and viewing angle.[2] In speech, this is called "time warping" and in vision, this is modeled by "flexible templates." Assume that the observed signal is a random map $s : X \longrightarrow V$, as above. Then our model includes warping, which is an unobserved hidden random variable $\psi : X \longrightarrow X$ and a normalized signal s_0 such that

$$s \approx s_0 \circ \psi.$$

In other words, the warping puts s in a more standard form s_0. Here s_0 might be a fixed (i.e., nonrandom) template, or might itself be random, although one aspect of its variability has been eliminated by the warping. Note that when X is a vector space, we can describe the warping ψ as a vector of displacements $\psi(\vec{x}) - \vec{x}$ of specific points. But the components

---

[2] Also, some previously unseen parts of the face may rotate into view, and this must be modeled in another way (e.g., by three-dimensional models.)

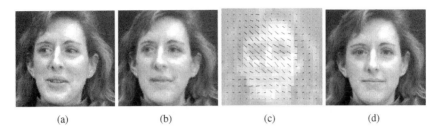

Figure 3. (a) and (d) Two images of the same woman's face. When the warping shown by the arrows in the image (c) is applied to the face in (d), image (b) is produced, which nearly matches image (a). Note, however, that no teeth are present ind (d), so the warping stretches the lips to cover the mouth. [97, Figure 4.15]

of this vector are numbers representing coordinates of points, not values of the signal (i.e., the domain of s, not the range of s as in the previous class of deformations). This is a frequent confusion in the literature. An example from the Ph.D. thesis of Peter Hallinan [97, 99] is shown in Figure 3.

3. *Hierarchical geometric patterns and parsing the signal.* A fundamental fact about real-world signals is that their statistics vary radically from point to point; that is, they do not come from a so-called stationary process. This nonstationarity is the basic statistical trace of the fact that the world has a discrete as well as a continuous behavior—it is made up of discrete events and objects. A central problem in the analysis of such signals is to tease apart these parts of its domain, so as to explicitly label the distinct objects/events affecting this signal. In speech, these distinct parts are the separate phones, phonemes, words, phrases, sentences, and finally, whole "speech acts." In vision, they are the various objects, their parts and groupings, present in the viewed scene. Note that in both cases, the objects or processes are usually embedded in each other, and hence form a hierarchy. The generic formalism for this is a grammar. Putting in this general setting, if s is a map $X \longrightarrow V$, the basic unobserved hidden random variable is a tree of subsets $X_a \subset X$, typically with labels l_a, such that for every node a with children b:

$$X_a = \bigcup_{a \to b} X_b.$$

Typically, the grammar also gives a stochastic model for an elementary signal $s_t : X_t \longrightarrow R$ for all leaves t of the tree and requires that $s\big|_{X_t} \approx s_t$. The most developed formalism for such grammars are the models called either random branching processes or probabilistic context-free grammars (PCFGs). Most situations, however, require context-sensitive grammars.

0.3. Bayesian Probability Theory: Pattern Analysis and Pattern Synthesis

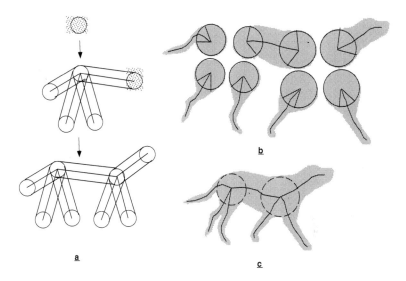

Figure 4. The grammatical decomposition of a dog into parts, from the work of Song-Chun Zhu [239, Figure 21] (Courtesy of Song-Chun Zhu). (a) The grammar generates a tree of protrusions for the limbs and a basic strip for the body. (b) Each part is given a concrete realization as a subset of the plane (with constraints so they match up). (c) These parts are assembled.

That is, the probability of a tree does not factor into terms, one for each node and its children. A parse tree of parts is very natural for the face: the parts correspond to the usual facial features—the eyes, nose, mouth, ears, eyebrows, pupils, eyelids, and so forth. An example, decomposing a dog, from the work of Zhu [239], is shown in Figure 4.

What makes the inference of the unobserved random variables in pattern theory difficult is not that any of the above models are necessarily hard to use but rather that all of them tend to coexist, and then inference becomes especially challenging. In fact, the full model of a signal may involve warping and superposition at many levels and a tree of parse trees may be needed to express the full hierarchy of parts. The world is not simple.

0.3 Bayesian Probability Theory: Pattern Analysis and Pattern Synthesis

Another key element of pattern theory is its use of Bayesian probability theory. An advantage of this Bayesian approach, compared with other vision theories, is that it requires that we first create a stochastic model and, afterwards, seek algorithms for inferring on the basis of the model. Thus, it separates the algorithms from the models. This distinguishes pattern theory from such neural network approaches such as multilayer perceptrons [104]. From our perspective, these theories try to solve two difficult tasks—both modeling and computation—at once. In a Bayesian approach, as in Hidden Markov models and Bayes nets, we first learn the models and verify them explicitly by stochastic sampling, and *then* seek algorithms for applying the models to practical problems. We believe that learning models and algorithms separately will lead to more tractable problems. Moreover, the explicit nature of the representations leads to a better understanding of the internal workings of an algorithm, and to knowing to what problems they will generalize.

We now give a brief introduction to the techniques of Bayesian probability theory. In general, we wish to infer the state of the random variable $\mathcal{S}$ describing the state of the world given some measurement, say, some observed random variable $\mathcal{I}$. Thus, the variables $\mathcal{S}$ would correspond to the hidden variables in our representations of the world (e.g., the variables representing the shape of a face), and the measurement $\mathcal{I}$ would correspond to the observed images. Within the Bayesian framework, one infers $\mathcal{S}$ by considering $P(\mathcal{S}|\mathcal{I})$, the *a posteriori* probability of the state of the world given the measurement. Note that by definition of conditional probabilities, we have

$$P(\mathcal{S}|\mathcal{I})P(\mathcal{I}) = P(\mathcal{S},\mathcal{I}) = P(\mathcal{I}|\mathcal{S})P(\mathcal{S}).$$

Dividing by $P(\mathcal{I})$, we obtain Bayes' theorem,

$$P(\mathcal{S}|\mathcal{I}) = \frac{P(\mathcal{I}|\mathcal{S})P(\mathcal{S})}{P(\mathcal{I})} = \frac{P(\mathcal{I}|\mathcal{S})P(\mathcal{S})}{\sum_{\mathcal{S}'} P(\mathcal{I}|\mathcal{S}')P(\mathcal{S}')} \propto P(\mathcal{I}|\mathcal{S})P(\mathcal{S}). \quad (1)$$

This simple theorem reexpresses $P(\mathcal{S}|\mathcal{I})$, the probability of the state given the measurement, in terms of $P(\mathcal{I}|\mathcal{S})$, the probability of observing the measurement given the state, and $P(\mathcal{S})$, the probability of the state. Each of the terms on the right-hand side (RHS) of the above equation has an intuitive interpretation.

The expression $P(\mathcal{I}|\mathcal{S})$, often termed the *likelihood function*, is a measure of the likelihood of a measurement given we know the state of the world. In this book, $\mathcal{I}$ is usually an image, and this function is also called the *imaging model*. To see this, note that given we know the state of the world (e.g., the light sources, the objects, and the reflectance properties of the surfaces of the objects), we can recreate, as an image, our particular view of the world. Yet, due to noise in our imaging system and imprecision of our models, this recreation will have an implicit degree of variability. Thus, $P(\mathcal{I}|\mathcal{S})$ probabilistically models this variability.

0.3. Bayesian Probability Theory: Pattern Analysis and Pattern Synthesis

Figure 5. A "Mooney" face [97, Figure 3.1] of a well-known psychophysicist (left), and the same image with contrast and left/right reversed, making it nearly unidentifiable (right).

The expression $P(\mathcal{S})$, referred to as the *prior model*, models our prior knowledge about the world. In vision, one often says that the prior is necessary because the reconstruction of the three-dimensional world from a two-dimensional view is not "well-posed." A striking illustration is given by "Mooney faces," images of faces in strong light that illuminate part of the face to saturation and leave the rest black (see Figure 5). These images tend at first to be confusing and then to suddenly look like a quite realistic face. But if these images are presented with the opposite contrast, they are nearly unrecognizable. We interpret this to mean that we have a lot of information on the appearance of faces in our learned prior model, and this information can be used to fill in the missing parts of the face. But if the contrast is reversed, we cannot relate our priors to this image at all.

As with the Mooney face example, in general, the image alone is not sufficient to determine the scene and, consequently, the choice of priors becomes critically important. They embody the knowledge of the patterns of the world that the visual system uses to make valid three-dimensional inferences. Some such assumptions have been proposed by workers in biological vision and include Gibson's *ecological constraints* [87] and Marr's *natural constraints* [152]. More than such general principles, however, we need probability models on representations that are sufficiently rich to model all the important patterns of the world. It is becoming increasingly clear that fully nonparametric models need to be learned to effectively model virtually all nontrivial classes of patterns (see, for example, texture modeling in Chapter 4).

What can we do with a Bayesian model of the patterned signals in the world? On the one hand, we can use it to perform probabilistic inference, such as finding the most probable estimate of the state of the world contingent on having observed a particular signal. This is called the *maximum a posteriori* (MAP) estimate of the state of the world. On the other hand, we can sample from the model—for example fixing some of the world variables S and using this distribution to construct sample signals $\mathcal{I}$ generated by various classes of objects or events. A good test of whether the prior has captured all the patterns in some class of signals is to see whether these samples are good imitations of life. From a pattern theory perspective, the analysis of the patterns in a signal and the synthesis of these signals are inseparable problems and use a common probabilistic model: computer vision should not be separated from computer graphics, nor should speech recognition be separated from speech generation.

It is helpful to also consider the patterns in signals from the perspective of information theory (see the book by Cover and Thomas [51]). This approach has its roots in the work of Barlow [13] (see also Rissanen [187]). The idea is that instead of writing out any particular perceptual signal $\mathcal{I}$ in raw form as a table of values, we seek a method of encoding $\mathcal{I}$ that minimizes its expected length in bits; that is, we take advantage of the patterns possessed by most $\mathcal{I}$ to encode them in a compressed form. We consider coding schemes that involve choosing various auxiliary variables S and then encoding the particular $\mathcal{I}$ by using these S (e.g., S might determine a specific typical signal $\mathcal{I}_S$ and we then need only to encode the deviation $\mathcal{I} - \mathcal{I}_S$). We write this:

$$\text{length}(\text{code }(\mathcal{I}, S)) = \text{length}(\text{code }(S)) + \text{length}(\text{code }(\mathcal{I} \text{ using } S)). \quad (2)$$

The mathematical problem, in the information theoretic setup, is as follows: for a given $\mathcal{I}$, find the S leading to the shortest encoding of $\mathcal{I}$. This optimal choice of S is called the *minimum description length* (MDL) estimate of the code S for a given $\mathcal{I}$:

$$\text{MDL estimate of } S = \arg\min_{S}[\text{ length}(\text{code }(S)) + \text{length}(\text{code }(\mathcal{I} \text{ using } S))]. \quad (3)$$

Moreover, one then can seek the *encoding scheme* leading to the shortest expected coding of all $\mathcal{I}$'s (given some distribution on the signals $\mathcal{I}$).

There is a close link between the Bayesian and the information-theoretic approaches given by Shannon's optimal coding theorem. This theorem states that given a class of signals $\mathcal{I}$, the coding scheme for such signals for which a random signal has the smallest expected length satisfies

$$\text{length}(\text{code }(\mathcal{I})) = -\log_2 P(\mathcal{I}), \quad (4)$$

0.3. Bayesian Probability Theory: Pattern Analysis and Pattern Synthesis

where fractional bit lengths are achieved by actually coding several $\mathcal{I}$'s at once; doing this, the left-hand side gets asymptotically close to the right-hand side when longer and longer sequences of signals are encoded at once. We may apply Shannon's theorem both to encoding $\mathcal{S}$ and to encoding $\mathcal{I}$ given $\mathcal{S}$. For these encodings, length(code($\mathcal{S}$)) = $-\log_2 p(\mathcal{S})$, and length(code($\mathcal{I}$ using $\mathcal{S}$)) = $-\log_2 p(\mathcal{I}|\mathcal{S})$, Therefore, taking $\log_2$ of Equation (1), we get Equation (2), and find that the most probable estimate of $\mathcal{S}$ is the same as the MDL estimate.

Finally, pattern theory also suggests a general framework for algorithms. Many of the early algorithms in pattern recognition were purely *bottom-up*. For example, one class of algorithms started with a signal, computed a vector of "features," or numerical quantities thought to be the essential attributes of the signal, and then compared these feature vectors with those expected for signals in various categories. This method was used to classify images of alpha-numeric characters or phonemes, for instance. Such algorithms offer no way of reversing the process, of generating typical signals. The problem these algorithms encountered was that they had no way of dealing with anything unexpected, such as a smudge on the paper partially obscuring a character, or a cough in the middle of speech. These algorithms did not say what signals were expected, only what distinguished typical signals in each category.

In contrast, a second class of algorithms works by actively reconstructing the signal being analyzed. In addition to the bottom-up stage, there is a *top-down* stage in which a signal with the detected properties is synthesized and compared to the present input signal (see Figure 6). What needs to be checked is whether the input signal agrees with the synthesized signal to within normal tolerances, or whether the residual is so great that the input has not been correctly or fully analyzed. This architecture is especially important for dealing with signals with parsed structure, in which one component of the signal partially obscures another.

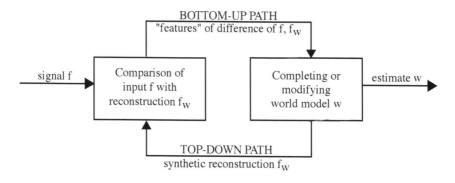

Figure 6. The fundamental architecture of pattern theory.

When this happens, the features of the two parts of the signal get confused. Only when the obscuring signal is explicitly labeled and removed can the features of the background signal be computed. We may describe this top-down stage as pattern reconstruction, distinct from the bottom-up purely pattern recognition stage.

This framework uses signal synthesis in an essential way, and this requirement for feedback gives an intriguing relation to the known properties of mammalian cortical architecture [165, 166]. Note that, although stemming from similar ideas, the idea of "analysis by synthesis" is logically separate from the Bayesian formulation and other aspects of pattern theory. Thus, the Bayesian approach might be carried out with other algorithms, and "analysis by synthesis" might be used to implement other theories.

A variant of this architecture has been introduced in tracking algorithms (by Isard and Blake [111]) and is applicable whenever sources of information become available or are introduced in stages. Instead of seeking the most probable values of the world variables $\mathcal{S}$ in one step, suppose we *sample* from the posterior distribution $P(\mathcal{S}|\mathcal{I})$. The aim is to sample sufficiently well so that no matter what later signal $\mathcal{I}'$ or later information on the world $\mathcal{S}'$ arrives, we can calculate an updated sample of $P(\mathcal{S}|\mathcal{I}, \mathcal{I}', \mathcal{S}')$ from the earlier sample.

A final word of caution: at this point in time, no algorithms have been devised and implemented that can duplicate in computers the human ability to perceive the patterns in the signals of the five senses. Pattern theory is a beautiful theoretical analysis of the problems but very much a work in progress when it comes to the challenge of creating a robot with human-like perceptual skills.

Figure 1.1. On the top, an image of an inscribed slab from the palace of Sargon II in Dur-Sharrukin, Khorsabad (alabaster, eighth century BC, Hermitage Museum, Saint Petersburg, Russia; Copyright cc-by-sa, Andrew Bossi, Flickr). Courtesy of Andrew Bossi (http://commons.wikimedia.org/wiki/File:Hermitage_Dur-Sharrukin_1.jpg). Below, an image of a palm-leaf manuscript from the library of the late K. V. Sarma in Chennai (photograph by Jenifer Mumford). For about 200 generations, mankind has been recording its history, business, and beliefs in linear strings of discrete symbols from alphabets. We will here study English text, a recent variant.

– 1 –
English Text and Markov Chains

IN THIS BOOK, we look for the patterns that occur in the signals generated by the world around us and try to develop stochastic models for them. We then use these models to detect patterns and structures in these signals, even when, as usually happens, they are ambiguous or obscured by various distortions. In this chapter, we begin with a well-known type of signal and the simplest patterns in it. The "signals" are just the strings of English characters found in any written English text. Any other language would be just as good (see Figure 1.1) if it is written with a small alphabet (e.g., Chinese would work a bit differently). Such signals are examples of *discrete one-dimensional* signals; they are made up of sequences $\{a_n\}$ whose elements lie in a finite set.

Strings of written text have, of course, layers and layers of patterns: they break up into words, sentences, and paragraphs with partial repetitions, cross-references, systematic changes in letter frequencies, and so forth. A very natural way to start to describe the patterns of some English text is to simply take random substrings of small fixed length n from a large body of text and, if enough data are available, to work out a complete table of frequencies of occurrence of each string of characters of length n. This probability table defines a stochastic model for strings of arbitrary length if we assume the Markov property that, conditional on fixing any string of $n-1$ consecutive characters, the two characters immediately before and after an occurrence of this string in English text are independent. This is called the *nth-order Markov approximation* to the full language of strings. As Shannon showed very convincingly, these models get increasingly better as n increases, and the convergence of these models to reasonable-sounding English can be measured by entropy [195]. Let's make this precise.

The general setting for our study will be signals made up of strings of letters taken from an arbitrary finite alphabet S. We will denote the size of the alphabet by $s = \#S$. Some well-known examples are the following:

- *Strings of English.* $S = \{a, b, \ldots, z, space\}$, $s = 27$ (if we ignore caps, numbers, and punctuation).

- *Strings of phonetic transcriptions of speech.* Using one version of the International Phonetic Alphabet for the phonemes used in English, we have 35 symbols:

1. 8 stop consonants {p, b, t, d, tʃ, dʒ, k, g} as in "pen, boy, top, do, chair, joy, cat, go";
2. 3 nasals {m, n, ŋ} as in "man, no, sing";
3. 9 fricatives {f, v, θ, ð, s, z, ʃ, ʒ, h} as in "fool, voice, thing, this, see, zoo, she, pleasure, ham";
4. 5 semivowels {ɹ, l, w, ʍ, j} as in "run, left, we, what, yes";
5. 10 vowels {i, ɪ, ə, ɛ, æ, ɑ, ɔ, ʌ, ʊ, u} as in "see, sit, about, bed, bad, car, law, run, put, soon."

- *Machine code.* $S = \{\mathbf{0}, \mathbf{1}\}$, $s = 2$, with which your hard disk is filled.
- *Morse code.* $S = \{\cdot, -, pause\}$, $s = 3$.
- *DNA.* $S = \{\mathbf{A, G, T, C}\}$, the 4 "codons" of life that make up the genome string, $s = 4$.
- *Proteins.* $S = $ {Leu, Ile, Val, Met, Phe, Tyr, Gly, Ala, Lys, Arg, His, Glu, Asp, Gln, Asn, Ser, Thr, Cys, Trp, Pro}, the 20 amino acids, $n = 20$.

We assume we have access to a very large set of very long signals in the alphabet S. Think of it as library of books written in the alphabet S. We want to study the simplest patterns that such signals can carry: substrings occurring more frequently or less frequently than expected. A substring of length n will be called an n-gram.[1] All this information is contained in the raw statistical tables of frequencies of n-grams. Let $\sigma = (a_1 \ldots a_n)$, $a_i \in S$, be a n-gram. Let Ω_n be the set of all such strings; there are s^n of them. We define

$$P_n(\sigma) = \text{ the frequency of } \sigma \text{ in } n\text{-grams.}$$

Then $P_n(\sigma) \geq 0$ and $\sum_\sigma P_n(\sigma) = 1$ (i.e., P_n is a probability distribution on Ω_n). As n gets larger, this table of frequencies captures more and more of the patterns in these signals. Of course, if n is too large, we won't have sufficient data to estimate P_n correctly. But, being mathematicians, we will imagine that there is some idealization of infinitely long signals so that, in principle, these probabilities can be estimated. (We will not worry about convergence here.) What we do assume is that the statistics of long signals are *stationary*. In other words, if we take any piece σ of length n out of a signal of length N, the frequency of occurrences of the resulting n-grams does not depend on where they are found in the long signal:

---

[1] This term is usually used for a string of n words, but we will use it for a string made up from a fixed finite alphabet.

I. English Text and Markov Chains

Definition 1.1. Given a finite alphabet S, a language on S is a set of probability distributions P_n on the set of strings Ω_n of length n for all $n \geq 1$ such that for all pairs of integers $n_1 < n_2$ and all i with $1 \leq i \leq n_2 - n_1 + 1$, P_{n_1} is the marginal of P_{n_2} under the map $\Omega_{n_2} \to \Omega_{n_1}$ given by $\{a_1 \cdots a_{n_2}\} \to \{a_i \cdots a_{i+n_1-1}\}$.

We can construct a full language using only the n-gram statistics for a small fixed n as follows. Notice that the frequency table P_n on n-grams gives, as marginals, frequency tables P_k on k-grams, $k < n$; hence, also conditional probabilities:[2]

$$P_n(a_n | a_1 \ldots a_{n-1}) = \frac{P_n(a_1 \ldots a_n)}{P_{n-1}(a_1 \ldots a_{n-1})} = \frac{P_n(a_1 \ldots a_n)}{\sum_{b \in S} P_n(a_1 \ldots a_{n-1} b)}.$$

The realization that these *conditional probabilities* were very informative in describing the nature of the language goes back to Markov [150]. In 1913, he published a paper analyzing the poem *Eugene Onyegin* by Pushkin and comparing it with other texts. Without a computer, he reduced the alphabet to two symbols: a vowel v and a consonant c, and he computed by hand the probabilities in this long poem:

$$P_1(v) = 0.432, \quad P_1(c) = 0.568 \quad P_2(v|c) = 0.663,$$
$$P_2(c|c) = .337, \quad P_2(v|v) = 0.128, \quad P_2(c|v) = 0.872.$$

Moreover, he showed that these numbers varied substantially in other works; hence, they captured some aspect of the author's style.

By using these conditional probabilities, we can define a simplified probability model on strings of any length $N > n$ by a sort of sliding window:

$$P_N^{(n)}(a_1 \ldots a_N) = \begin{cases} P_n(a_1 \cdots a_n) \cdot \prod_{i=n+1}^{N} P_n(a_i | a_{i-n+1} \ldots a_{i-1}) \\ \quad \text{if } P_{n-1} \text{ of all length } n-1 \text{ substrings} \\ \quad \text{of } a_1 \ldots a_N \text{ is nonzero,} \\ 0 \quad \text{otherwise.} \end{cases} \quad (1.1)$$

We can do this by starting with the n-gram distribution of any language; then $P^{(n)}$ will be called the n-gram approximation to the full language.

As described in Chapter 0, a good way to see what patterns the stochastic model given by the frequency tables P_n captures is to sample from its extension $P_N^{(n)}$ to long strings. The sampling procedure is simple: we choose $(a_1 \ldots a_n)$ randomly from the distribution P_n, then a_{n+1} randomly from the distribution

---

[2]Note that if $P_{n-1}(a_1 \ldots a_{n-1}) = 0$, then also $P_n(a_1 \ldots a_n) = 0$, and $P_n(a_n | a_1 \ldots a_{n-1})$ is undefined. To avoid special cases, it is convenient to define the probability to be zero in this case.

$P_n(b|a_2 \ldots a_n)$, then a_{n+2} from the distribution $P_n(b|a_3 \ldots a_{n+1})$, etc. This "analysis by synthesis" was first done by Shannon [194] and we reproduce his results here.

- *Random characters.*
  XFOML RXKHRJFFJUJ ZLPWCFWKCYJ FFJEYVKCQSGXYD
  QPAAMKBZAACIBZLHJQD

- *Sample from $P^{(1)}$.*
  OCRO HLI RGWR NMIELWIS EU LL NBBESEBYA TH EEI
  ALHENHTTPA OO BTTV

- *Sample from $P^{(2)}$.*
  ON IE ANTSOUTINYS ARE T INCTORE ST BE S DEAMY
  ACHIN D ILONASIVE TUCOOWE FUSO TIZIN ANDY TOBE
  SEACE CTISBE

- *Sample from $P^{(3)}$.*
  IN NO IST LAY WHEY CRATICT FROURE BERS GROCID
  PONDENOME OF DEMOSTURES OF THE REPTAGIN IS
  REGOACTIONA OF CRE

- *Sample from $P^{(4)}$.*
  THE GENERATED JOB PROVIDUAL BETTER TRAND THE
  DISPLAYED CODE ABOVERY UPONDULTS WELL THE CODERST
  IN THESTICAL IT TO HOCK BOTHE

It is amusing to see how the samples get closer and closer in some intuitive sense to reasonable English text. But we really would like to measure quantitatively how close the n-gram models $P^{(n)}$ are to the "true" probability model of length N strings. The concept of entropy will enable us to do this. The intuitive idea is that all patterns in strings increase the predictability of continuations, and hence decrease the potential information these continuations can carry. So we want to measure by a real number how predictable strings are in our models or how much new information each letter contains. Think of the two following extremes. The first one is white noise: strings are fully random, they have no patterns, they are not predictable, and they carry the most information. The second extreme is the set of periodic strings: they are fully predictable and carry no information after the basic period has been seen. To make this quantitative, we need to introduce the basic concepts of information theory.

1.1 Basics I: Entropy and Information

Entropy is a positive real number associated with an arbitrary finite probability space. It measures the information carried by a random variable, or equivalently by a probability distribution (see also Exercise 2 at the end of this chapter for an axiomatic definition of entropy).

Definition 1.2. Let $\Omega = \{a_1, \ldots, a_N\}$ and let $P : \Omega \to \mathbb{R}_+$ be a function such that $\sum P(a_i) = 1$, making Ω into a finite probability space. Or, equivalently, let $\mathcal{X}$ be a random variable with values $\{a_1, \ldots, a_N\}$ and $P(a_i) = \mathbb{P}(\mathcal{X} = a_i)$. Then the *entropy* of P or of $\mathcal{X}$ is defined by

$$H(P) = H(\mathcal{X}) = -\sum_{i=1}^{N} P(a_i) \log_2 P(a_i) = \mathbb{E}(\log_2(1/P)).$$

Notice that $H(P) \geq 0$ because $0 \leq P(a_i) \leq 1$. By convention, if $P(a_i) = 0$, then we set $P(a_i) \log_2 P(a_i) = 0$, since the limit of $x \log x$ as x goes to 0 is 0.

As was shown by Shannon [194], $H(\mathcal{X})$ can be interpreted in a quite precise way as measuring the average number of *bits* of information contained in a sample of the random variable $\mathcal{X}$. Suppose we decide to *code* the fact that we have drawn the sample a_i by the fixed bit string σ_i. A sequence of samples $a_{i_1}, \cdots, a_{i_M}$ is then encoded by the concatenation of the bit strings $\sigma_{i_1} \cdots \sigma_{i_M}$. In order to be able to decode uniquely (without the extra overhead of marking the ends of code words), we require that for all $j \neq i$, σ_j is not a prefix of σ_i. Then, if we have the code book in front of us, we can always determine where one code word ends and another begins. Such a code is called a "prefix code." We are interested in coding schemes whose average code length is as small as possible. Clearly, we want common a's to be coded by short words, uncommon by long. Shannon's fundamental result is:

Theorem 1.3.

1. *For all prefix codes σ_i, the expected length of the code for one sample $\sum_i P(a_i)|\sigma_i|$ satisfies:*

$$\sum_i P(a_i)|\sigma_i| \geq \sum_i P(a_i) \log_2(1/P(a_i)) = H(P).$$

2. *We can find a prefix code such that*

$$|\sigma_i| = \lceil \log_2(1/P(a_i)) \rceil,$$

where, for x real, $\lceil x \rceil$ denotes the integer such that $x \leq \lceil x \rceil < x + 1$. Hence, with this coding, the expected length $\sum_i P(a_i)|\sigma_i|$ is less than $H(P) + 1$.

3. *If we use* block coding *(i.e., we encode sequences of independent messages of the same type by a single code word), then the expected coding length per message tends to* $H(P)$.

This theorem justifies talking about entropy as the "number of bits of information in a sample of $\mathcal{X}$." We postpone the proof for several pages until we have defined the Kullback-Leibler distance, one of two important variants of the concept of entropy, the first of which is the notion of conditional entropy:

Definition 1.4. Let $\mathcal{X}$ be a random variable with values in $\Omega_1 = \{x_1, \ldots, x_N\}$ and probability measure $P(x_i) = \mathbb{P}(\mathcal{X} = x_i)$. Let $\Omega_2 = \{y_1, \ldots, y_M\}$ be a finite set and f a function, $f : \Omega_1 \to \Omega_2$. We define the induced probability measure P on the random variable $\mathcal{Y} = f(\mathcal{X})$ by $P(y) = \sum_{f(x)=y} P(x)$, and we define the conditional probability measures $P(x|y)$ to be $P(x)/P(y)$ when $y = f(x)$. The *conditional entropy* is then defined by

$$H(P|Q) = H(\mathcal{X}|\mathcal{Y}) = -\sum_{x,y=f(x)} P(x) \log_2 P(x|y).$$

If $\mathcal{X}$ and $\mathcal{Y}$ are two random variables for which we have a joint probability distribution (i.e., a probability measure on the space $\Omega_1 \times \Omega_2$), it is customary to extend the idea of conditional entropy, defining $H(\mathcal{X}|\mathcal{Y})$ to be $H((\mathcal{X}, \mathcal{Y})|\mathcal{Y})$.

Note that $H(\mathcal{X}|\mathcal{Y}) \geq 0$ and that (taking the general case):

$$\begin{aligned}
H(\mathcal{X}|\mathcal{Y}) &= -\sum_{x,y} P(x,y) \log_2 P(x,y|y) \\
&= -\sum_{x,y} P(x,y) \log_2 P(x,y) + \sum_{x,y} P(x,y) \log_2 P(y) \\
&= -\sum_{x} P(x,y) \log_2 P(x,y) + \sum_{y} P(y) \log_2 P(y) \\
&= H(\mathcal{X}, \mathcal{Y}) - H(\mathcal{Y}) \quad \text{or} = H(\mathcal{X}) - H(\mathcal{Y}) \text{ when } \mathcal{Y} = f(\mathcal{X}).
\end{aligned}$$

The information-theoretic meaning of $H(\mathcal{X}|\mathcal{Y})$ is thus the average number of extra bits required to describe a sample of $\mathcal{X}$ when $\mathcal{Y}$ is fixed. When $\mathcal{X}$ and $\mathcal{Y}$ are independent, then $H(\mathcal{X}, \mathcal{Y}) = H(\mathcal{X}) + H(\mathcal{Y})$ and thus $H(\mathcal{X}|\mathcal{Y})$ is simply $H(\mathcal{X})$.

The second variant of the concept of entropy is the definition of the Kullback-Leibler distance:

Definition 1.5. Let Ω be a set, not necessarily finite, with two probability measures P and Q on it. Then the Kullback-Leibler distance, or simply KL-distance,

1.1. Basics I: Entropy and Information

from P to Q, also called the divergence of Q from P, is defined by

$$D(P||Q) = \mathbb{E}_P\left(\log_2 \frac{P}{Q}\right)$$
$$= \sum_{a \in \Omega} P(a) \log_2 \frac{P(a)}{Q(a)} \quad \text{if } \Omega \text{ is finite.}$$

Although $D(P||Q)$ is *not* symmetric in P and Q, we do have the following, which justifies the use of the term of the term "distance."

Theorem 1.6. $D(P||Q) \geq 0$ and $D(P||Q) = 0$ if and only if $P = Q$.

Proof: The proof of the theorem simply uses the concavity of the log function. If $x_1, \ldots, x_n$ are positive real numbers, and if $t_1, \ldots, t_n$ are also positive real numbers such that $\sum t_i = 1$, then $\log(\sum t_i x_i) \geq \sum t_i \log x_i$, and equality holds if and only if all the x_i's are equal. More generally, we have Jensen's inequality.

If $\mathcal{X}$ is a random variable with values in $\mathbb{R}_+$, then $\log(\mathbb{E}(\mathcal{X})) \geq \mathbb{E}(\log \mathcal{X})$, with equality if and only if $\mathcal{X}$ is constant. And so we have

$$\mathbb{E}_P\left(\log_2 \frac{P}{Q}\right) = -\mathbb{E}_P\left(\log_2 \frac{Q}{P}\right) \geq -\log_2 \mathbb{E}_P\left(\frac{Q}{P}\right) = -\log_2 1 = 0,$$

and equality holds if and only if P/Q is constant, that is, if $P = Q$. □

Theorem 1.6 gives us many corollaries, such as:

Proposition 1.7. *If Ω is finite with N elements in it, and U denotes the uniform probability distribution on Ω (for each a, $U(a) = 1/N$), then if P is any probability distribution on Ω,*

$$D(P||U) = \log_2 N - H(P).$$

Hence, $0 \leq H(P) \leq \log_2 N$ and $H(P) = \log_2 N$ if and only if $P = U$.

Proposition 1.8. *For any random variables $\mathcal{X}$ and $\mathcal{Y}$, $H(\mathcal{X}|\mathcal{Y}) \leq H(\mathcal{X})$. In fact, if P is the joint distribution of $\mathcal{X}$ and $\mathcal{Y}$ and the product $P_1 P_2$ is the modified probability distribution on $\mathcal{X}$ and $\mathcal{Y}$ making them independent (where P_1 and P_2 are the marginals on $\mathcal{X}$ and $\mathcal{Y}$ given by $P_1(x) = \sum_y P(x, y)$ and $P_2(y) = \sum_x P(x, y)$), then*

$$H(\mathcal{X}) = H(\mathcal{X}|\mathcal{Y}) + D(P||P_1 P_2).$$

This comes from the simple calculation:

$$H(\mathcal{X}|\mathcal{Y}) + D(P||P_1 P_2) = -\sum_{x,y} P(x,y) \log_2(P(x,y)/P_2(y))$$
$$+ \sum_{x,y} P(x,y) \log_2(P(x,y)/P_1(x)P_2(y))$$
$$= H(\mathcal{X}).$$

More generally, the same reasoning proves that for any three random variables $\mathcal{X}, \mathcal{Y}$, and $\mathcal{Z}$,

$$H(\mathcal{X}|\mathcal{Y},\mathcal{Z}) \leq H(\mathcal{X}|\mathcal{Z}), \tag{1.2}$$

which can be interpreted to mean that adding extra information can only decrease the number of bits needed to describe $\mathcal{X}$.

We can now prove Shannon's theorem (Theorem 1.3). We first need the following lemma:

Lemma. *For any prefix code,*

$$\sum_i \frac{1}{2^{|\sigma_i|}} \leq 1.$$

Proof: Let $N = \max |\sigma_i|$, then $2^{N-|\sigma_i|}$ represents the number of bit strings of length N with prefix σ_i. Since for all i and j, σ_j is not a prefix of σ_i, these sets of strings of length N are disjoint, hence

$$\sum_i 2^{N-|\sigma_i|} \leq 2^N,$$

which proves $\sum_i 2^{-|\sigma_i|} \leq 1$.

We can now prove the first part of the fundamental theorem. Let α be the real number defined by

$$\frac{1}{\alpha} = \sum_i \frac{1}{2^{|\sigma_i|}} \leq 1.$$

We then define the probability distribution Q by $Q(a_i) = \alpha/2^{|\sigma_i|}$. The Kullback-Leibler distance from P to Q is positive due to Theorem 1.6 and is given by

$$D(P||Q) = \sum_i P(a_i) \log_2 \frac{P(a_i)}{Q(a_i)}$$
$$= \sum_i P(a_i) \log_2 P(a_i) - \sum_i P(a_i) \log_2 \frac{\alpha}{2^{|\sigma_i|}} \geq 0.$$

Thus,

$$H(P) = -\sum_i P(a_i) \log_2 P(a_i) \leq \sum_i P(a_i)|\sigma_i| - \log_2 \alpha \leq \sum_i P(a_i)|\sigma_i|,$$

and the first part of the theorem is proved.

1.1. Basics I: Entropy and Information

For the second part of the theorem, let $k_i = \lceil \log_2(1/P(a_i)) \rceil$. Then $k_i \geq \log_2(1/P(a_i))$, so that

$$\sum_i \frac{1}{2^{k_i}} \leq 1.$$

Now, we can construct a prefix code such that for all i, $|\sigma_i| = k_i$. To do this, let $K = \max k_i$ and for $1 \leq k \leq K$, let $n_k = \#\{i | k_i = k\}$. We first code the messages $\{i | k_i = K\}$ by the first K-bit code words (the order being the lexicographic one). Let the last one be $\varepsilon_1 \ldots \varepsilon_{K-1} \varepsilon_K$. Then we code the messages $\{i | k_i = K - 1\}$ by the first $K - 1$-bit code words after $\varepsilon_1 \ldots \varepsilon_{K-1}$, etc. There are enough code words to do this because

$$\sum_{k=1}^{K} n_k \frac{1}{2^k} \leq 1$$

(at each step, distinguish the case n_k as even or odd).

For the last part of the theorem, just note that if you encode with one code book a string of M independent messages $\{a_{i_1}, \cdots, a_{i_M}\}$, then the previous argument shows that the expected length is at most one more than the entropy of the space of M-tuples, which is M times the entropy $H(P)$. Thus the expected length *per message* is at most $H(P) + 1/M$. □

Let's give a very simple example: suppose $\Omega = \{a, b, c, d\}$ and that $P(a) = 0.5, P(b) = 0.25, P(c) = 0.125$, and $P(d) = 0.125$. We could use the four codes 00, 01, 10, 11 for a, b, c, d, in which case 2 would always be the coding length. But we do better if we code a by a 1-bit code (e.g., 0), b by 2 bits (e.g., 10), and c and d by 3 bits (e.g., 110 and 111), as shown in Figure 1.2. The resulting coding scheme has an expected coding length of 1.75, which saves a quarter of a bit. It is optimal since it equals $H(P)$.

The proof of Shannon's theorem also gives us an interpretation of $D(P||Q)$ in terms of information. Suppose $\mathcal{X}$ is a random variable whose true probability

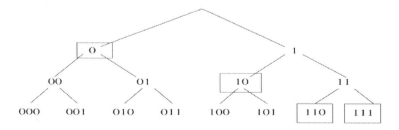

Figure 1.2. Coding four possible messages in a nonuniform way optimized to their frequencies.

distribution is P. Then if we code values of $\mathcal{X}$ using the suboptimal code associated with Q, $D(P||Q)$ is the number of extra bits this entails.

1.2 Measuring the *n*-gram Approximation with Entropy

We want to apply this machinery to languages, as defined above. As we have seen, Ω_N carries not only the true model P_N but the approximations given by the prolongations $P_N^{(1)}, P_N^{(2)}, \cdots, P_N^{(N-1)}$ defined by formula (1.1). Let's see if we can use the distances $D(P_N || P_N^{(n)})$ to see how good these approximations are.

We use the notation $a_1 \cdots a_n$ to describe *random strings* $\mathcal{S}_n$ of length n with the given probability distribution P_n or an approximation thereto. Let $f_n, g_n : \Omega_n \to \Omega_{n-1}$, be the "initial substring" and "final substring" maps; that is, $f_n(\sigma_{n-1}a) = \sigma_{n-1}$ and $g_n(a\sigma_{n-1}) = \sigma_{n-1}$. The key numbers for describing strings information-theoretically are the conditional entropies F_n:

$$F_n = H_{P_n}(a_n | a_1 \cdots a_{n-1}) = -\sum_{\sigma_n} P_n(\sigma_n) \log_2 P_n(\sigma_n | \sigma_{n-1})$$
$$= H(P_n) - H(P_{n-1}).$$

These numbers measure the predictability of the nth character given the $n-1$ preceding characters, that is, the average number of bits needed to encode the nth character in a string of length n given the $n-1$ previous ones. By formula (1.2) with $\mathcal{X} = a_n, \mathcal{Y} = a_1$, and $\mathcal{Z} = a_2 \cdots a_{n-1}$, we find

$$H(a_n | a_1 \cdots a_{n-1}) \leq H(a_n | a_2 \cdots a_{n-1}), \text{ that is, } F_n \leq F_{n-1}.$$

We let F_1 be simply $H(\mathcal{S}_1) = H(P_1)$. Using these numbers, we get the entropy of strings of length N, which represents the average number of bits needed to encode the whole string of length N:

$$H(P_N) = F_N + F_{N-1} + \cdots + F_1.$$

If, instead of the true probability distribution P_N, we use the approximation $P_N^{(n)}$, then for all $k \geq n$, the conditional entropy of σ_k with respect to $f_k(\sigma_k)$ simplifies to

$$\begin{aligned}
H_{P_k^{(n)}}(\mathcal{S}_k | f_k(\mathcal{S}_k)) &= -\sum P_k^{(n)}(\sigma_k) \log_2 P_k^{(n)}(\sigma_k | f_k(\sigma_k)) \\
&= -\sum P_k^{(n)}(\sigma_k) \log_2 P_n(a_k | a_{k-n+1} \cdots a_{k-1}) \\
&= F_n
\end{aligned}$$

1.2. Measuring the n-gram Approximation with Entropy

(using the definition of formula (1.1)), which gives us, by summing over k

$$H(P_N^{(n)}) = (N - n + 1)F_n + F_{n-1} + \cdots + F_1.$$

Finally, we can also compute the Kullback-Leibler distances in terms of the F's:

$$\begin{aligned}
D(P_N \| P_N^{(n)}) &= \sum_{\sigma_N} P_N(\sigma_N) \log_2 \left(\frac{P_N(\sigma_N)}{P_N^{(n)}(\sigma_N)} \right) \\
&= -H(P_N) - \sum_{\sigma_N} P_N(\sigma_N) \log_2 P_n(a_1 \cdots a_n) \\
&\quad - \sum_{i=n+1}^{N} \sum_{\sigma_N} P_N(\sigma_N) \log_2 P_n(a_i | a_{i-n+1} \cdots a_{i-1}) \\
&= -H(P_N) - \sum_{\sigma_N} P_N^{(n)}(\sigma_N) \log_2 P_n(a_1 \cdots a_n) \\
&\quad - \sum_{i=n+1}^{N} \sum_{\sigma_N} P_N^{(n)}(\sigma_N) \log_2 P_n(a_i | a_{i-n+1} \cdots a_{i-1}) \\
&= H(P_N^{(n)}) - H(P_N) \\
&= \sum_{k=n+1}^{N} (F_n - F_k).
\end{aligned}$$

Since the F_n are decreasing, they have a limit as n goes to infinity, which we denote by F. Combining the formulas for $H(P_N)$ and $H(P_N^{(n)})$ in terms of the F's, it follows that

$$F \leq \frac{H(P_N)}{N} \leq \frac{H(P_N^{(n)})}{N} \leq F_n + \frac{n}{N}(F_1 - F_n).$$

Letting first N go to infinity and then n, it follows that

$$\lim_{n \to \infty} \frac{H(P_N)}{N} = F.$$

Thus, F is the key number that describes the information content per symbol of long signals, called the "entropy of the source." Finally, we get

$$\frac{D(P_N \| P_N^{(n)})}{N} \leq F_n - F,$$

which shows convergence in the Kullback-Leibler sense of the nth-order models to the true model.

For some concrete results, we refer to Shannon's 1951 work, "Prediction and Entropy of Printed English" [195]. He worked on strings of English, with the reduced set of textual symbols $\{a, b, \ldots, z, space\}$, which omits capitals, numbers, and punctuation. From standard tables of frequencies, he then computed the conditional entropies,

$$F_0 = \log_2(27) \simeq 4.75; \quad F_1 = 4.03; \quad F_2 = 3.32; \quad F_3 = 3.1; \quad F_4 = 2.8,$$

and the corresponding entropies for strings of length n,

$$H_2 = 7.35; \quad H_3 = 10.45; \quad H_4 = 13.25.$$

What is especially beautiful is that Shannon devised a game, played with native English speakers equipped with frequency tables, for estimating the limit F. His game was based on reencoding a string by replacing each letter by its *rank* in the n-gram frequency tables, given the preceding $n-1$ letter. More precisely, for each letter a and each string σ of length $n-1$, we have the conditional probability of a knowing that the preceding string is σ: $P(a|\sigma)$. Then we fix σ, and for each letter a, we define $\text{rk}_\sigma(a)$ to be the rank of a when the probabilities $P(\cdot|\sigma)$ are ranked in decreasing order. For instance, $\text{rk}_\sigma(a) = 1$ when a is the most probable letter after σ. Now, for a string $\{a_k\}$, we encode it by replacing each letter a_k by its rank, knowing $(a_{k-n+1} \ldots a_{k-1})$:

$$a_k \longrightarrow \text{rk}_{(a_{k-n+1}\ldots a_{k-1})}(a_k).$$

This reencoding makes the redundancy of written text very plain: we now get huge strings of 1s ($\simeq 80\%$ are 1s). Shannon defines $Q_n(k)$ to be the frequency of k in the recoded string:

$$Q_n(k) = \sum_{\text{rk}_\sigma(a)=k} P_n(\sigma a).$$

Then he proves that the entropy $H(Q_n)$ is an upper bound for the conditional entropy F_n:

$$\forall n \geq 1, \quad F_n \leq H(Q_n).$$

Moreover,

$$\forall n \geq 1, \quad H(Q_{n+1}) \leq H(Q_n).$$

Let H^{rk} be the limit of $H(Q_n)$.

Now here's Shannon's game: ask a person to guess each successive letter in some text, knowing the whole of the text up to that letter. The person is given a dictionary and some frequency tables to help out when he or she is unsure. If

1.3. Markov Chains and the *n*-gram Models

the first guess is wrong, the person continues guessing until he or she gets the correct letter. Record now for each letter in the text the number of guesses the person needed before finding the correct letter. *This is a reencoding of the text similar to, but maybe not quite as good as, the one just described!* Let $Q^*(k)$ be the frequency with which k occurs in the subject's guessing. Shannon proved $F \leq H^{rk} \leq H(Q^*)$. By carrying this out, he got the upper bound $F \leq 1.3$.

1.3 Markov Chains and the *n*-gram Models

1.3.1 Basics II: Markov Chains

We have seen that strings of letters can be modeled as sequences of n-grams with some transition probability between the different n-grams. The right mathematical framework for such sequences is the theory of Markov chains. We now give the main definitions and properties for these Markov chains.

Definition 1.9 (Markov Chain). Let Ω be a finite set of states. A sequence $\{\mathcal{X}_0, \mathcal{X}_1, \mathcal{X}_2, \cdots\}$ of random variables taking values in Ω is said to be a Markov chain if it satisfies the Markov condition:

$$\mathbb{P}(\mathcal{X}_n = i_n | \mathcal{X}_0 = i_0, \mathcal{X}_1 = i_1, \ldots, \mathcal{X}_{n-1} = i_{n-1}) = \mathbb{P}(\mathcal{X}_n = i_n | \mathcal{X}_{n-1} = i_{n-1}),$$

for all $n \geq 1$ and all $i_0, i_1, \ldots i_n$ in Ω.
Moreover, the chain is said homogeneous if, for all $n \geq 1$ and all i, j in Ω,

$$\mathbb{P}(\mathcal{X}_n = j | \mathcal{X}_{n-1} = i) = \mathbb{P}(\mathcal{X}_1 = j | \mathcal{X}_0 = i).$$

Markov chains will always be assumed homogeneous unless otherwise specified. In this case, the transition matrix Q of the chain is defined as the $|\Omega| \times |\Omega|$ matrix of all transition probabilities $Q(i, j) = q_{i \to j} = \mathbb{P}(\mathcal{X}_1 = j | \mathcal{X}_0 = i)$.

Then, the matrix Q^n gives the law of $\mathcal{X}_n$ for the chain starting at $\mathcal{X}_0 = i$:

$$\mathbb{P}(\mathcal{X}_n = j | \mathcal{X}_0 = i) = Q^n(i, j).$$

This result can be shown by induction on n. For $n = 1$, it is the definition of Q. Assume it is true for a given n; then, for $n + 1$, we use an elementary property of the conditional probability and write

$$\begin{aligned}
\mathbb{P}(\mathcal{X}_{n+1} = j | \mathcal{X}_0 = i) &= \sum_l \mathbb{P}(\mathcal{X}_{n+1} = j | \mathcal{X}_0 = i, \mathcal{X}_n = l) \mathbb{P}(\mathcal{X}_n = l | \mathcal{X}_0 = i) \\
&= \sum_l \mathbb{P}(\mathcal{X}_{n+1} = j | \mathcal{X}_n = l) Q^n(i, l) \\
&= \sum_l Q^n(i, l) Q(l, j) = Q^{n+1}(i, j).
\end{aligned}$$

Definition 1.10. We say that the Markov chain is irreducible or primitive if, for all i, j in Ω, there exists $n \geq 0$ such that $Q^n(i, j) > 0$.

This definition means that if the chain starts at $\mathcal{X}_0 = i$, then for any j there exists an integer n such that $\mathbb{P}(\mathcal{X}_n = j | \mathcal{X}_0 = i) > 0$. In other words, all states can be "connected" through the chain. But it needn't necessarily happen that there is one n such that $Q^n(i,j) > 0$ for all j. For instance, there can be two states a and b such that: $Q(a,a) = Q(b,b) = 0$ and $Q(a,b) = Q(b,a) = 1$. Then the chain just goes back and forth between a and b; that is, $Q^n(a,a) = 0$ for all odd n, and $Q^n(a,b) = 0$ for all even n. This can be measured by defining the *period* of a state i: it is the greatest common divisor of all the n such that $Q^n(i,i) > 0$. Then, it follows easily that for any irreducible Markov chain, all states have the same period d, and there is a decomposition $\Omega = S_1 \cup \ldots \cup S_d$ such that the chain Q takes S_k to S_{k+1} with probability 1.

Definition 1.11. We say that a Markov chain is aperiodic if, for all i in Ω, the greatest common divisor of the $n \geq 1$ such that $Q^n(i,i) > 0$ is 1.

Since Ω is finite, if the Markov chain is irreducible and aperiodic, there exists an n_0 such that for all $n \geq n_0$ and for all i, j in Ω, $Q^n(i,j) > 0$.

Definition 1.12. A probability distribution Π on Ω is an equilibrium (or steady-state) probability distribution of Q iff

$$\forall j \in \Omega, \quad \sum_i \Pi(i) Q(i,j) = \Pi(j).$$

We then have the following theorem (recall that Ω is finite).

Theorem 1.13. *If Q is irreducible, then there exists a unique equilibrium probability distribution Π for Q. If, moreover, Q is aperiodic, then*

$$\forall i, j \in \Omega, \quad Q^n(i,j) \xrightarrow[n \to +\infty]{} \Pi(j).$$

Notice that in the above convergence result, the limit is independent of the starting state i of the chain. We will not give the proofs of these results here; instead, we refer the reader to [95], for instance. Before going back to the n-gram model, we end this section with some examples of Markov chains obtained from mathematical models of card shuffling.

Example (Card Shuffling). There are many different ways to shuffle cards. Some techniques are deterministic (this is, for instance, the case for "perfect shuffles," by which a deck of cards is cut in half and then perfectly interleaved). Most

1.3. Markov Chains and the n-gram Models

other techniques are random. A key point is to study the "randomization" effected by a shuffling technique, that is, to answer questions such as "what is the probability distribution on the ordering on cards after m shuffles?" All this can be translated in the mathematical framework of Markov chains. There are many published papers dealing with the mathematical properties of card shuffling, notably the widely cited papers of the mathematician (and magician) Persi Diaconis (see, for instance, [60] or [62] and all the references therein).

Let us consider a deck of n cards, labeled $0, 1, 2, \ldots, n-1$. The state space Ω_n is the set of orderings of the n cards, which has cardinality $n!$. Let G_n be the group of permutations of n elements, the symmetric group of order n. Given any ordering of the cards and a permutation on n elements, the permutation can be used to reorder the cards, and thus G_n acts on Ω_n. Ω_n is a *principal homogeneous space* under G_n: given two orderings, there is a unique permutation taking one order to the other. Card shuffling is then modeled as a probability distribution P on G_n, which induces a Markov chain on Ω_n with transition matrix $Q(x, gx) = P(g)$. In other words, shuffling cards means randomly picking up a permutation g according to the probability distribution P on G_n and rearranging the cards by g no matter what their order is now. Examples of shuffling techniques include random cut, top to random shuffle, and riffle shuffle, discussed next.

- *Random cut.* Randomly cut the deck of cards in two piles, and then put the second pile on the top of the first one. The probability distribution P in this case is

$$P(g) = \begin{cases} \frac{1}{n} & \text{if there is } k \text{ such that} \\ & g = (k, k+1, \ldots, n-1, 1, 2, \ldots, k-1), \\ 0 & \text{otherwise.} \end{cases}$$

Such a technique will never provide a good randomization of cards since it performs only translations (modulo n), that is, cyclic permutations of the cards.

- *Top to random shuffle.* Take the top card and put it back in the deck at a random position. The transition probability of such a shuffling is

$$P(g) = \begin{cases} \frac{1}{n} & \text{if } g = \text{identity, or } (2, 3, \ldots, i, 1, i+1, \ldots, n), \ 2 \leq i \leq n, \\ 0 & \text{otherwise.} \end{cases}$$

In this case, the Markov chain is irreducible and aperiodic, and its equilibrium probability distribution Π is the uniform distribution on Ω_n: for all $x \in \Omega_n$, $\Pi(x) = 1/n!$. This shuffling technique has been studied by Aldous and Diaconis [4] (and its generalization, "top m to random," by Diaconis, Fill, and Pitman [61]), and they have proven that it takes order $n \log n$ shuffles for convergence to uniformity. More precisely, their result is that the total variation distance between Q^k and Π (defined as $\frac{1}{2}\sum_{x \in \Omega_n} |Q^k(x) - \Pi(x)|$) is less than $\exp(-c)$ when $k = n(\log n + c)$.

- *Riffle shuffle.* This is a very common technique. Cut the deck of cards in two (approximately equal) piles and then interlace the two piles. The corresponding mathematical model was introduced by Gilbert and Shannon in 1955 in a technical report of Bell Laboratories. It can be described in the following way: first, randomly choose k according to the binomial distribution: $\binom{n}{k}/2^n$. Then cut the deck in two piles, one of size k and one of size $n - k$, and randomly interleave the two piles by sequentially dropping cards from the piles with a probability proportional to the size of the pile: at a given time, if the left pile has a cards and the right one has b cards, then drop a card from the left pile with probability $a/(a+b)$. Bayer and Diaconis [16] have analyzed the riffle shuffle (also called dovetail shuffle). They have proven that in this case

$$P(g) = \begin{cases} (n+1)/2^n & \text{if } g = \text{identity}, \\ 1/2^n & \text{if } g \text{ has 2 rising sequences}, \\ 0 & \text{otherwise}, \end{cases}$$

where a rising sequence is defined as a maximal subset of an arrangement of cards, consisting of successive values in increasing order. For example, if the order of six cards is 4,1,2,5,3,6, then it has two rising sequences (1,2,3 and 4,5,6), whereas the arrangement 5,4,1,2,6,3 has three rising sequences (1,2,3; 4; and 5,6). The Markov chain Q has again the uniform distribution Π on Ω_n as the equilibrium probability distribution. Bayer and Diaconis have given precise estimates of the total variation distance between Q^k and Π, and the main striking result is that for a deck of $n = 52$ cards, "seven shuffles are necessary and sufficient to approximate randomization."

1.3.2 Markov Property for the n-gram Models

We begin by giving three equivalent conditions that characterize n-gram models. At the same time, we introduce a new class of probability distributions, the exponential models, which will reappear in many places in the text. The equivalent conditions are a particular case of the so-called Gibbs-Markov equivalence, which is valid on more general graphs, and which we develop in Chapter 4.

1.3. Markov Chains and the n-gram Models

Proposition 1.14. *Let Ω_N be the space of strings $a_1 \ldots a_N$ of length N, the a_i's being taken from a finite alphabet S. Consider a probability distribution $P : \Omega_N \to \mathbb{R}_+$. Fix an integer $n \geq 1$. The following conditions are equivalent:*

1. *Conditional factorization. P has the form*

$$P(a_1 \ldots a_N) = P_0(a_1 \ldots a_{n-1}) \prod_{k=0}^{N-n} P_1(a_{k+n}|a_{k+1} \ldots a_{k+n-1}).$$

   for suitable P_0 and P_1.

2. *Exponential form.*

$$P(a_1 \ldots a_N) = \frac{1}{\mathcal{Z}} \exp\left(-\sum_{\sigma \in \Omega_n} \lambda_\sigma \cdot \#\mathrm{occ}(\sigma, a_1 \ldots a_N)\right),$$

   for suitable λ_σ's, where Ω_n is the set of strings of length n and $\#\mathrm{occ}(\sigma, a_1 \ldots a_N)$ denotes the number of occurrences of the string σ in $a_1 \ldots a_N$. Here we allow $\lambda_\sigma = \infty$ in case no strings of positive probability contain σ.

3. *Markov property. For all $I = (k+1, \ldots, k+n-1)$ of length $n-1$, let $a(I) = a_{k+1} \cdots a_{k+n-1}$; then*

$$P(a_1 \ldots a_N|a(I)) = P_1(a_1 \ldots a_k|a(I)) \cdot P_2(a_{k+n} \ldots a_N|a(I)),$$

   which means that $a(\text{before } I)$ and $a(\text{after } I)$ are conditionally independent given $a(I)$.

Proof: We first prove 1 ⇒ 2. We just write

$$\prod_{k=0}^{N-n} P(a_{k+n}|a_{k+1} \ldots a_{k+n-1})$$

$$= \exp\left(-\sum_{\sigma=(b_1 \ldots b_n)} \log \frac{1}{P(b_n|b_1 \ldots b_{n-1})} \cdot \#\mathrm{occ}(\sigma)\right),$$

and

$$P_0(a_1 \ldots a_{n-1}) = \exp\left(-\sum_{\tau=(b_1 \ldots b_{n-1})} \log \frac{1}{P_0(b_1 \ldots b_{n-1})} \cdot \delta_{\tau \text{ init.}}\right),$$

where $\delta_{\tau \text{ init.}}$ is 1 if $a_1 \ldots a_{n-1} = \tau$ and 0 otherwise. Then, finally, we notice that

$$\delta_{\tau \text{ init.}} = \#\mathrm{occ}(\tau) - \sum_b \#\mathrm{occ}(b\tau).$$

We next prove 2 ⇒ 3. We fix $I = (k+1, \ldots, k+n-1)$ and $a(I) = a_{k+1} \ldots a_{k+n-1}$. Then a string of length n cannot meet both "before I" and "after I," so that if P has the exponential characterization (2), $P(a_1 \ldots a_N | a(I))$ is simply the product of a function of $a(\text{before } I)$ and of a function of $a(\text{after } I)$.

We finally prove 3 ⇒ 1. We can write $P(a_1 \ldots a_N)$ under the form:

$$P(a_1 \ldots a_N) = P_0(a_1 \ldots a_{n-1}) P(a_n | a_1 \ldots a_{n-1}) P(a_{n+1} | a_1 \ldots a_n) \ldots P(a_N | a_1 \ldots a_{N-1}),$$

and then, since for $k \geq n+1$, a_k does not depend on a_i for $i \leq k-n$, we get

$$P(a_1 \ldots a_N) = P_0(a_1 \ldots a_{n-1}) \prod_{k=0}^{N-n} P(a_{k+n} | a_{k+1} \ldots a_{k+n-1}). \quad \square$$

We now look at the case in which the probability P on Ω_N has the Markov property with substrings of length n. As we have seen, such a model defines a family of probability models $P_M^{(n)}$ on Ω_M for every M including the given $P = P_n^{(N)}$. We call all of these models P for simplicity. Let Ω_n^* be the set of strings σ of length n with $P(\sigma) > 0$. Using the Markov property, we can convert the generation of these longer and longer strings into a Markov chain on Ω_n^* in the following way. Given a string $(a_1 \ldots a_N)$, for $1 \leq k \leq N-n+1$ let $\sigma_k = (a_k \ldots a_{k+n-1})$ be the substring of length n starting at the kth letter. Then $\sigma_1, \ldots, \sigma_{N-n+1}$ is a Markov chain with initial distribution $P_0(\sigma_1)$ and transition probabilities:

$$P(\sigma \to \tau) = P(a_1 \ldots a_n \to b_1 \ldots b_n)$$
$$= \begin{cases} P(b_n | a_1 \ldots a_n) & \text{if } b_1 = a_2, b_2 = a_3 \ldots \text{ and } b_{n-1} = a_n, \\ 0 & \text{otherwise.} \end{cases}$$

We now can use the standard theorems on Markov chains to deduce "regularities" or "ergodic conditions" on the n-gram model. In particular, an n-gram string model is irreducible and aperiodic if there exists an integer m such that for all strings σ, τ of length n there exists a string $(\sigma \rho \tau)$ of positive probability with length $|\rho| = m$. Then, for n-gram strings from such a model, there is a unique equilibrium distribution given by

$$\Pi(\tau) = \lim_{m \to \infty} P(a_m \ldots a_{m+n-1} = \tau | a_1 \ldots a_n).$$

Notice that this distribution is independent of $(a_1 \ldots a_n)$.

Given a homogeneous Markov chain $\{\mathcal{X}_0, \mathcal{X}_1, \ldots\}$, the average entropy of the first N variables can be defined by the quantity $\frac{1}{N} H(\mathcal{X}_1, \mathcal{X}_2, \ldots, \mathcal{X}_N)$. In the

1.3. Markov Chains and the n-gram Models

case of strings, this represents the entropy per symbol of text. Using the chain rule for conditional entropy, we have

$$H(\mathcal{X}_0, \mathcal{X}_1, \ldots, \mathcal{X}_N) = H(\mathcal{X}_0) + \sum_{k=1}^{N} H(\mathcal{X}_k | \mathcal{X}_0, \ldots \mathcal{X}_{k-1})$$

$$= H(\mathcal{X}_0) + \sum_{k=1}^{N} H(\mathcal{X}_k | \mathcal{X}_{k-1}).$$

If we assume that the Markov chain is irreducible and aperiodic, and if we denote by Q its transition matrix and by Π its equilibrium probability distribution, then

$$H(\mathcal{X}_k | \mathcal{X}_{k-1}) = -\sum_{i,j} P(\mathcal{X}_{k-1} = i, \mathcal{X}_k = j) \log_2 P(\mathcal{X}_k = j | \mathcal{X}_{k-1} = i)$$

$$= -\sum_{i,j} Q(i,j) P(\mathcal{X}_{k-1} = i) \log_2 Q(i,j)$$

$$\xrightarrow[k \to \infty]{} -\sum_{i,j} Q(i,j) \Pi(i) \log_2 Q(i,j).$$

By Cesaro's theorem (which states that if a sequence has a limit, then the sequence of averages has the same limit), this implies that the average entropy per symbol converges as N goes to infinity and that its limit is:

$$\lim_{N \to \infty} \frac{H(\mathcal{X}_1, \mathcal{X}_2, \ldots, \mathcal{X}_N)}{N} = -\sum_{i,j} Q(i,j) \Pi(i) \log_2 Q(i,j).$$

This limit, also called the *entropy rate* (in [51]) was introduced by Shannon [194]. We denote it by $\overline{H}(Q)$.

Let us go back to our n-gram model. If we use the n-gram distribution P_n, as we have already explained, we can convert an infinite sequence of letters a_1, $a_2 \ldots a_N \ldots$ into a Markov chain $\{\sigma_1, \sigma_2, \ldots\}$ by setting $\sigma_k = (a_k \ldots a_{k+n-1})$, and using the transition matrix Q_n, given by

$$Q_n(\sigma, \tau) = Q_n(a_1 \ldots a_n \to b_1 \ldots b_n)$$
$$= \begin{cases} \frac{P_n(a_2 \ldots a_n b_n)}{P_{n-1}(a_2 \ldots a_n)} & \text{if } b_1 = a_2, b_2 = a_3 \ldots \text{ and } b_{n-1} = a_n, \\ 0 & \text{otherwise.} \end{cases}$$

Then, we can check that an equilibrium probability distribution of this Markov chain is P_n. Indeed, if we let $(b_1 \ldots b_n)$ be a string of length n, we have

$$\sum_{a_1,\ldots,a_n} P_n(a_1\ldots a_n) Q_n(a_1\ldots a_n \to b_1\ldots b_n) =$$

$$\sum_{a_1} P_n(a_1 b_1\ldots b_{n-1}) \frac{P_n(b_1\ldots b_n)}{P_{n-1}(b_1\ldots b_{n-1})} = P_n(b_1\ldots b_n).$$

If the Markov chain is irreducible and aperiodic,[3] then P_n is the unique equilibrium probability distribution. We find that the entropy rate of this chain is equal to the conditional entropy F_n introduced in the previous section:

$$\overline{H}(Q_n) = -\sum_{a_1\ldots a_n a_{n+1}} P_n(a_1\ldots a_n) \frac{P_n(a_2\ldots a_n a_{n+1})}{P_{n-1}(a_2\ldots a_n)} \log_2 \frac{P_n(a_2\ldots a_n a_{n+1})}{P_{n-1}(a_2\ldots a_n)}$$

$$= H(P_n) - H(P_{n-1}) = F_n.$$

We have seen here that converting the n-gram model into a Markov chain allows us to use the standard theorems on Markov chains to obtain properties on the n-gram model. We will also see more properties of this type in the following section. However, many results start from the hypothesis that the Markov chain is irreducible and aperiodic. Thus, a natural question is: When is this hypothesis valid. As already stated, we can check this as follows: if there exists an integer $m > 0$ such that for all strings σ, τ of length n there exists a string $(\sigma \rho \tau)$ of positive probability with length $|\rho| = m$, then the n-gram string model is irreducible and aperiodic. It thus seems reasonable to assume that this condition is satisfied for nonzero probability n-grams of English text, at least for small n.

1.3.3 Mutual Information

We can translate the convergence of Markov chains in the language of information theory by introducing a new definition—mutual information.

Definition 1.15. Given two random variables $\mathcal{X}$ and $\mathcal{Y}$, where the possible values of $\mathcal{X}$ (respectively, $\mathcal{Y}$) are $x_1,\ldots,x_n$ (respectively, $y_1,\ldots,y_m$), let $P(x,y)$ be a joint distribution on the two variables $\mathcal{X}$ and $\mathcal{Y}$. Let

$$P_1(x) = \sum_y P(x,y) \text{ and } P_2(y) = \sum_x P(x,y)$$

---

[3] This assumption is nontrivial. For example, you might have a language in which every syllable was a single consonant followed by a single vowel with no word breaks or punctuation. Any n-gram models based on this assumption would not be aperiodic.

1.3. Markov Chains and the n-gram Models

be its marginals and
$$Q(x,y) = P_1(x)P_2(y)$$
be the distribution if $\mathcal{X}$ and $\mathcal{Y}$ were independent. Then the mutual information of $(\mathcal{X}, \mathcal{Y})$ is

$$\begin{aligned}
MI(\mathcal{X}, \mathcal{Y}) &= D(P||Q) \\
&= \sum_{x,y} P(x,y) \log_2 \frac{P(x,y)}{P_1(x)P_2(y)} \\
&= -\sum_{x,y} P(x,y) \log_2 P_1(x) - \sum_{x,y} P(x,y) \log_2 P_2(y) \\
&\quad + \sum_{x,y} P(x,y) \log_2 P(x,y) \\
&= H(\mathcal{X}) + H(\mathcal{Y}) - H(\mathcal{X}, \mathcal{Y}) = H(\mathcal{X}) - H(\mathcal{X}|\mathcal{Y}).
\end{aligned}$$

Note that $0 \leq MI(\mathcal{X}, \mathcal{Y}) \leq \min(H(\mathcal{X}), H(\mathcal{Y}))$ and that $MI(\mathcal{X}, \mathcal{Y}) = 0$ when $\mathcal{X}$ and $\mathcal{Y}$ are independent.

A convenient way to visualize the meaning of mutual information is by using the Venn diagram shown in Figure 1.3. The circle labeled $\mathcal{X}$ ihas area $H(\mathcal{X})$ and represents the bits in a sample of $\mathcal{X}$. Likewise, the circle labeled $\mathcal{Y}$ has area $H(\mathcal{Y})$ and represents the bits in a sample of $\mathcal{Y}$. A joint sample of $\mathcal{X}, \mathcal{Y}$ can be coded by giving $\mathcal{X}$ and $\mathcal{Y}$ as though they had nothing to do with other, (i.e., were independent); this would need $H(\mathcal{X}) + H(\mathcal{Y})$ bits. But knowing the value of $\mathcal{X}$ usually tells us something about the value of $\mathcal{Y}$, so the number of

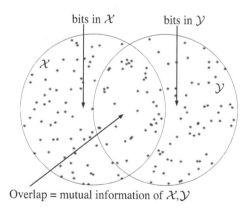

Figure 1.3. Heuristically, we represent the bits in the random variable $\mathcal{X}$ by points in the left circle; bits in $\mathcal{Y}$ as points in the right circle. The two circles together represent the information in the joint variable $(\mathcal{X}, \mathcal{Y})$ and the overlap represents the mutual information.

extra bits $H(\mathcal{Y}|\mathcal{X})$ is less than $H(\mathcal{Y})$. The diagram shows this by having the circles for $\mathcal{X}$ and $\mathcal{Y}$ overlap in a region of area $MI(\mathcal{X},\mathcal{Y})$, the number of bits of mutual information. Thus the union of the two circles has area $H(\mathcal{X},\mathcal{Y})$ and represents correctly the number of bits required to describe a sample of $\mathcal{X},\mathcal{Y}$ together.

As nice as this Venn diagram is for a way to think about information, it fails when we have three random variables $\mathcal{X},\mathcal{Y},\mathcal{Z}$! Following the usual inclusion/exclusion idea, we may be tempted to define a three-way mutual information by requiring

$$\begin{aligned}H(\mathcal{X},\mathcal{Y},\mathcal{Z}) = & H(\mathcal{X}) + H(\mathcal{Y}) + H(\mathcal{Z}) \\ & - MI(\mathcal{X},\mathcal{Y}) - MI(\mathcal{X},\mathcal{Z}) - MI(\mathcal{Y},\mathcal{Z}) \\ & + MI(\mathcal{X},\mathcal{Y},\mathcal{Z}).\end{aligned}$$

The last term should represent the area of the triple overlap or the number of bits in common to all three variables. But consider the following simple example: let $\mathcal{X},\mathcal{Y}$, and $\mathcal{Z}$ be three binary variables with $\mathcal{X}$ and $\mathcal{Y}$ independent and $\mathcal{Z}$ being 1 if $\mathcal{X} = \mathcal{Y}$, and 0 if $\mathcal{X} \neq \mathcal{Y}$. Then not only are $\mathcal{X}$ and $\mathcal{Y}$ independent, but $\mathcal{X}$ and $\mathcal{Z}$ are independent and $\mathcal{Y}$ and $\mathcal{Z}$ are independent. But clearly, knowing any two of $\mathcal{X},\mathcal{Y}$, and $\mathcal{Z}$ determines the third. Thus, $H(\mathcal{X}) = H(\mathcal{Y}) = H(\mathcal{Z}) = 1$ and $MI(\mathcal{X},\mathcal{Y}) = MI(\mathcal{X},\mathcal{Z}) = MI(\mathcal{Y},\mathcal{Z}) = 0$ but $H(\mathcal{X},\mathcal{Y},\mathcal{Z}) = 2$. Thus, the above definition makes $MI(\mathcal{X},\mathcal{Y},\mathcal{Z}) = -1$, so it cannot represent the area of the triple overlap!

Mutual information is one of many ways to measure the convergence of Markov chains, as shown by the following proposition.

Proposition 1.16. *If $\{\mathcal{X}_n\}$ is an irreducible and aperiodic Markov chain on a finite space, then for all $\varepsilon > 0$, there exists an integer m such that*

$$\forall k \; MI(\mathcal{X}_k, \mathcal{X}_{k+m}) < \varepsilon.$$

Chains with this property are called "information regular."

Proof: We fix k and let n be a positive integer, then

$$\begin{aligned}MI(\mathcal{X}_k,\mathcal{X}_{k+n}) &= \sum_{x,y} P(\mathcal{X}_k = x, \mathcal{X}_{k+n} = y) \log_2 \frac{P(\mathcal{X}_{k+n} = y | \mathcal{X}_k = x)}{P(\mathcal{X}_{k+n} = y)} \\ &= \sum_x P(\mathcal{X}_k = x) \sum_y P(\mathcal{X}_{k+n} = y | \mathcal{X}_k = x) \\ &\quad \times \log_2(P(\mathcal{X}_{k+n} = y | \mathcal{X}_k = x)) \\ &\quad - \sum_y (P(\mathcal{X}_{k+n} = y) \log_2(P(\mathcal{X}_{k+n} = y)) \\ &= H(\mathcal{X}_{k+n}) - \sum_x P(\mathcal{X}_k = x) H(\mathcal{X}_{k+n}|\mathcal{X}_k).\end{aligned}$$

1.4. Words

Let Π denote the steady-state probability distribution of the Markov chain. Then

$$\forall x, y \quad \lim_{n \to \infty} P(\mathcal{X}_{k+n} = y | \mathcal{X}_k = x) = \Pi(y),$$

and

$$\forall y \quad \lim_{n \to \infty} P(\mathcal{X}_{k+n} = y) = \Pi(y).$$

Therefore

$$\lim_{n \to \infty} H(\mathcal{X}_{k+n} | \mathcal{X}_k) = H(\Pi), \text{ and } \lim_{n \to \infty} H(\mathcal{X}_{k+n}) = H(\Pi)$$

hence $\lim_{n \to \infty} MI(\mathcal{X}_k, \mathcal{X}_{k+n}) = 0$, uniformly in k. $\square$

1.4 Words

The analysis of strings up to this point has dealt with the brute force approach to their statistics: counting substrings. For almost all human languages and for many other stochastic languages, signals are naturally divided into a sequence of *words*. This is the first higher level of structure in most stochastic languages. We have set up the analysis of English text, following Shannon, by including "space" as a character, so words are obvious: they are the units between the spaces. But this is really just a convenience for readers. In spoken languages, word boundaries are not marked; the pauses in speech are not usually the start of new words but rather are caused by the closure of the mouth just before the production of the stop consonants b,p,d,t,g,k. And in many written languages such as ancient Greek, word boundaries were not marked. Word boundaries, however, are not arbitrary breakpoints in the signal: *they are the natural places to break up the signal if we want to code it most efficiently.* We should therefore expect that they can be recovered nearly correctly by using low-level statistics.

Considering strings without markers for word boundaries, the word boundaries are the simplest example of *geometric patterns*, which we introduced in Section 0.2. The n-gram statistics are all examples of *value statistics*, (i.e., frequencies derived from the values of the signal). But finding word breaks gives a way of dividing the domain of the signal into pieces—which is a geometric pattern. The third type of pattern, domain warpings, do not occur in most languages, although colloquial Hindi, for example, has reduplicative mechanism, doubling words for strengthening or expanding meaning (e.g., "thīk thīk baṭāo") which means literally "correctly correctly show (me)', that is "state exactly."

Various approaches exist for finding the n-gram trace of word breaks, and fairly sophisticated analyses seem to be needed to get the most accurate results. But we can do surprisingly well by using a very simple method based on the concept of mutual information (as suggested by Brent [34]). Suppose you have a text from which the spaces have been removed or a phonetic transcription of speech. The idea is to look at each point in the signal and ask how much the preceding two or three characters tell us about the next two or three characters. If we are in the middle of a word, these small strings constrain each other; hence they have high mutual information. If we are between two words, these fragments are much more nearly independent.

To define these statistics, we take any n and m and write a random string $\mathcal{S}_{n+m}$ as the concatenation $\mathcal{S}'_n \mathcal{S}''_m$ of its initial substring of length n and its final substring of length m. Then we can consider how much information the first substring $\mathcal{S}'_n$ gives us about its successor $\mathcal{S}''_m$ (i.e., $MI(\mathcal{S}'_n, \mathcal{S}''_m)$). This is computable in terms of the F's:

$$\begin{aligned}
MI(\mathcal{S}'_n, \mathcal{S}''_m) &= \mathbb{E} \log_2 \left(\frac{P_{n+m}(a_1 \cdots a_{n+m})}{P_n(a_1 \cdots a_n) P_m(a_{n+1} \cdots a_{n+m})} \right) \\
&= \mathbb{E} \log_2 (P_{n+m}(a_{n+1} \cdots a_{n+m} | a_1 \cdots a_n)) \\
&\quad - \mathbb{E} \log_2 (P_m(a_1 \cdots a_m)) \\
&= \sum_{k=n+1}^{n+m} \mathbb{E} \log_2(P_k(a_k | a_1 \cdots a_{k-1})) \\
&\quad - \sum_{k=1}^{m} \mathbb{E} \log_2(P_k(a_k | a_1 \cdots a_{k-1})) \\
&= -(F_{n+1} + \cdots + F_{n+m}) + (F_1 + \cdots + F_m) \\
&= \sum_{k=1}^{m} (F_k - F_{k+n}).
\end{aligned}$$

Taking Shannon's estimates, we find that for English text, the mutual information of two adjacent characters is $F_1 - F_2 \approx 0.8$ bits. In addition, the mutual information between the first and last pair in four consecutive characters is $F_1 + F_2 - F_3 - F_4 \approx 1.36$ bits.

But we can also consider not merely the expected value but the log of the individual fractions,

$$\beta(\sigma'_n, \sigma''_m) = \log_2 \left(\frac{P_{n+m}(a_1 \cdots a_{n+m})}{P_n(a_1 \cdots a_n) P_m(a_{n+1} \cdots a_{n+m})} \right),$$

for specific strings $\sigma_{n+m} = \sigma'_n \sigma''_m$. This kind of number occurs frequently; it is the log of the *likelihood ratio* of the long string $a_1 \cdots a_{n+m}$ measured with two models, one with dependencies between the two parts considered and one with the initial and final strings independent. We call this the *binding energy* with which

1.4. Words

σ'_n and σ''_m are coupled. It also is the difference in the optimal coding length of the combined string versus that for separate coding of the two substrings. This will vary above and below its expectation $MI(\mathcal{S}'_n, \mathcal{S}''_m)$. The idea of this word-boundary–finding algorithm is to choose some small n (e.g., 2); calculate at each point in the text the mutual information between the n preceding characters and n succeeding characters, or $\beta((a_{i-n} \cdots a_{i-1}), (a_i \cdots a_{i+n-1}))$; and then put word boundaries at all i, which are local minima of β. We expect the binding energy to oscillate above and below the mutual information (e.g., 1.4 for $n = 2$).

To provide an example for this book, we downloaded from the web eight novels by Mark Twain, from which we removed all punctuation, including spaces (and numbers), eliminated capitalization and concatenated the lot into a long string with exactly 2,949,386 alphabetic characters. We next took statistics of 4-tuples from this corpus and used this to compute the binding energy for each pair of consecutive pairs $(a_{n-2}a_{n-1})$ and $(a_n a_{n+1})$ of characters. We then found all local minima of the binding energy and eliminated local minima whose value was greater than 2.5 bits as not being strong enough. At this point, we found correctly about 60% on the word boundaries. However, examination of the output shows that a large number of the mistakes were due to missing a word boundary by one character. This has a simple explanation: low binding energy indicates that the 4-tuple $(a_{n-2}a_{n-1}a_n a_{n+1})$ is not a good unit, and this may be caused by a break between a_{n-2} and a_{n-1} or between a_n and a_{n+1} as well as a break between a_{n-1} and a_n. Thus postprocessing to compare these three possible breaks was necessary. For that, we simply considered the likelihood ratios,

$$\frac{P(a_{n-3}a_{n-2})P(a_{n-1}a_n a_{n+1})}{P(a_{n-3}a_{n-2}a_{n-1})P(a_n a_{n+1})} \text{ and } \frac{P(a_{n-2}a_{n-1})P(a_n a_{n+1}a_{n+2})}{P(a_{n-2}a_{n-1}a_n)P(a_{n+1}a_{n+2})},$$

to see whether either of the breaks before or after the indicated one was better. After this step, we found 74% of the word breaks correctly. Three examples are shown in Figure 1.4, in which the text is shown without word breaks but with each new word beginning with a capital letter. The graph shows the binding energy, and the vertical hatched lines are the word breaks indicated by the binding energy algorithm. Note that some of the incorrect word breaks are syllable breaks where two words might have occurred (e.g., some-times) or where a common suffix is present (e.g., clutter-ing). Note, too, how the binding energy peaks at the letter pair qu. Clearly, finding words needs a lot more than 4-tuple statistics: the whole word lexicon should be built up (as in Ziv-Lempel coding [241]) and special care is needed for prefixes and suffixes. This occurs especially for languages such as French, with many common contractions of articles and pronouns such as le, la, se, te, me, and, even more, for agglomerative languages such as Sanskrit.

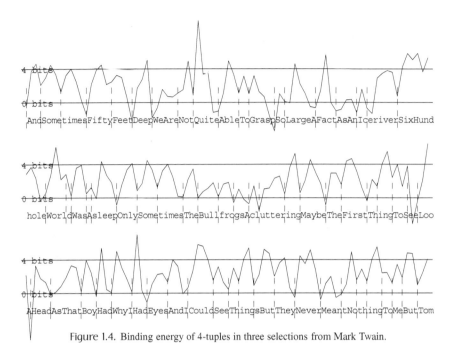

Figure 1.4. Binding energy of 4-tuples in three selections from Mark Twain.

Incidentally, for any language, one can also take the limit of $MI(\mathcal{S}_n, \mathcal{S}_m)$ as n and m go to ∞. In a doubly infinite text, this limit represents the total mutual information between the first and second half of the text if it is cut at random, which is given by

$$MI(\text{beginning}, \text{end}) = \sum_{k=1}^{\infty}(F_k - F)$$

(a sum of positive terms that may be finite or infinite). This limit may not really be well-defined in applications—where infinite texts are hard to come by![4]

Assuming we have identified the words properly, we can create a new stochastic language whose "characters" are now these words and play the same game as with the original strings of characters. That is to say, we can compute the frequency of each word and each consecutive pair of words, and so forth, and we can make Markov models of the language by using word statistics. Shannon did this for single words and for word pairs, and his work has been extended up to triples of consecutive words with some extra assumptions. The problem is that

---

[4]But consider a long-winded politician: having listened to him (or her) for what seems an infinite amount of time, having heard all his or her biases and folksy metaphors, how many bits of future speeches can be predicted? Perhaps quite a lot.

I.4. Words

even the largest data bases are not big enough to get correct word triple statistics by brute force, and more subtle techniques are called for (see, for instance, [118]). Shannon's results follow, first for single words with their correct frequencies:

REPRESENTING AND SPEEDILY IS AN GOOD APT OR COME CAN DIFFERENT NATURAL HERE HE THE A IN CAME THE TO OF TO EXPERT GRAY COME

and then with correct word pair frequencies:

THE HEAD AND IN FRONTAL ATTACK ON AN ENGLISH WRITER THAT THE CHARACTER OF THIS POINT IS THEREFORE ANOTHER METHOD FOR THE

One gets a clear sense of convergence of these models to the true stochastic language. Incidentally, single word frequencies are an excellent illustration of a rather mysterious principle called Zipf's law (named after the linguist George Zipf), according to which the frequency of the nth most common word is approximately C/n. This seems to apply to many discrete probability distributions, such as the probability that a random person lives in the nth most populous city. It holds pretty well for words; see the web site http://linkage.rockefeller.edu/wli/zipf. For English, $C \approx 0.1$, and the series had better be truncated at the total number of words, say 20,000, or it will diverge.

We can formalize the relationship between the character-based models and the word-based models. Let Ω_N be the set of strings of characters of length N without spaces and let Λ be the lexicon (i.e., the set of strings which are single words). Each word has a length, which is the number of characters in it. Let Ω_N^Λ be the set of strings of words whose total length is at least N and is less than N if the last word is deleted. We then get a map,

$$sp : \Omega_N^\Lambda \to \Omega_N$$

by spelling out the string of words as a string of characters and omitting the small spillover for the last word. If $\sigma \in \Omega_N$, then writing $\sigma = sp(\tau)$ for some $\tau \in \Omega_N^\Lambda$ is the same thing as specifying the set of word boundaries in the string σ. This is a very common situation in this book. We have the observed signal—in this case, the string of characters. This signal has extra structure that is not given to us explicitly; in this case, it is its partition into words. We seek to infer this hidden structure, which amounts to lifting the sample from the probability space of observed signals to the larger probability space of signals with extra structure.

A typical method for lifting the signal is the *maximum likelihood*, (ML) estimate. We use our best available model for the probability distribution on the larger space and find or approximate the most probable lifting of the signal from the space of observed signals to the larger space. In our case, we may put an n-gram distribution $P_N^{(n)}$ on Ω_N^Λ and seek the $\tau \in \Omega_N^\Lambda$ such that $sp(\tau) = \sigma$, which maximizes $P_N^{(n)}(\tau)$. In general, finding the ML lifting is difficult. But for this case of parsing a signal of characters into words, the method of dynamic programming is available, which we explain in the next section.

Before leaving the topic of words, we want to indicate how information theory can be used to make quantitative the coding gain achieved by using words rather than character sequences, illustrating the ideas in Section 0.3. In principle, the best possible coding of any language would be given by tabulating huge tables of n-gram statistics, say for $n = 100$, which would give a superb approximation to the true statistics of strings, expressing implicitly the many layers of patterns present in the text. However, this is totally impractical because this table would be bigger than the universe. What is needed is to reexpress the information implicit in this gigantic table by stochastic models that have only a moderate number of parameters. The first step in this direction is to replace the characters by the shorter word strings. (Later, one may want to use syntax and group the words into phrases, etc.) These more practical models invariably need hidden variables, so we have a trade-off: we need to code the hidden variables in addition to the characters, but once we do this, we get a shorter code for the characters. Let us try to quantify all of this.

We start with the two probability spaces: (1) $(\Omega, P_{\text{true}})$ of the observed signal with its true statistics, and (2) $(\Omega^\Lambda, P^\Lambda)$ the signal plus its partition into words with some approximate and manageable statistics, which are (3) connected by the spelling map $sp : \Omega^\Lambda \to \Omega$. We also assume that we have some algorithm for finding the hidden variables $A : \Omega \to \Omega^\Lambda$ such that $sp \circ A = 1_\Omega$. This may be maximum likelihood or some approximation to it. Then a character string can be encoded by partitioning it via A and using the simpler model P^Λ to code the sentence plus its hidden variables. The key number is the expected coding length, which is given by

$$\text{Expected coding length} = \sum_{\sigma \in \Omega} P_{\text{true}}(\sigma) \log_2(1/P^\Lambda(A(\sigma))).$$

We may analyze this number into three terms as follows. Let

$$\begin{aligned} \delta &= 1 - \sum_{\sigma \in \Omega} P^\Lambda(A(\sigma)) \\ &= P^\Lambda(\Omega^\Lambda - A(\Omega)) \end{aligned}$$

be the probability mass that is not used in the hidden variable model (i.e., word strings that never arise by the partitioning algorithm A). Then $Q^\Lambda = \frac{P^\Lambda \circ A}{1-\delta}$ is a

1.5. Word Boundaries via Dynamic Programming and Maximum Likelihood

probability measure on Ω, and it is easy to see that

$$\text{Expected coding length} = \log_2(1/(1-\delta))$$
$$+ \sum_{\sigma \in \Omega} P_{\text{true}}(\sigma) \log_2\left(1 \Big/ \left(\frac{P^\Lambda(A(\sigma))}{1-\delta}\right)\right)$$
$$= \log_2(1/(1-\delta)) + D(P_{\text{true}} \| Q^\Lambda) + H(P_{\text{true}}).$$

Thus we can see how close we get to optimal coding: a term arising from the inefficiencies in the coding with extra variables, and the Kullback-Leibler divergence from the true distribution to the rescaled hidden variable model.

1.5 Word Boundaries via Dynamic Programming and Maximum Likelihood

1.5.1 Algorithm I: Dynamic Programming

The dynamic programming algorithm of Bellman [19] is a very efficient algorithm to compute the minimum of a function F of n variables $x_1, \ldots, x_n$, provided this function can be decomposed as the sum of functions $f_i(x_i, x_{i+1})$. More precisely, the main result is the following theorem:

Theorem 1.17. *If $F(x_1, \ldots, x_n)$ is a real-valued function of n variables $x_i \in S_i$, S_i being a finite set, of the form*

$$F(x_1, \ldots, x_n) = f_1(x_1, x_2) + f_2(x_2, x_3) + \ldots + f_{n-1}(x_{n-1}, x_n),$$

then one can compute the global minimum of F in time $O(s^2 n)$ and space $O(sn)$, where $s = \max_i |S_i|$.

The algorithm goes like this.

Algorithm (Dynamic Programming).

1. First we initialize h_2 and Φ_2 by:

$$\forall x_2 \in S_2, \quad h_2(x_2) = \min_{x_1 \in S_1} f_1(x_1, x_2)$$
$$\forall x_2 \in S_2, \quad \Phi_2(x_2) = \operatorname*{argmin}_{x_1 \in S_1} f_1(x_1, x_2)$$

2. We now loop over the variable k. At each stage, we will have computed:

$$\forall x_k \in S_k, \quad h_k(x_k) = \min_{x_1,\ldots,x_{k-1}} [f_1(x_1,x_2) + \ldots + f_{k-1}(x_{k-1},x_k)]$$

$$\forall x_k \in S_k, \quad \Phi_k(x_k) = \operatorname*{argmin}_{x_{k-1}}(\min_{x_1,\ldots,x_{k-2}} [f_1(x_1,x_2) + \ldots + f_{k-1}(x_{k-1},x_k)]).$$

Then we define:

$$\forall x_{k+1} \in S_{k+1}, \quad h_{k+1}(x_{k+1}) = \min_{x_1,\ldots,x_k} [f_1(x_1,x_2) + \ldots + f_{k-1}(x_{k-1},x_k) + f_k(x_k,x_{k+1})]$$

$$= \min_{x_k}(h_k(x_k) + f_k(x_k,x_{k+1}))$$

$$\forall x_{k+1} \in S_{k+1}, \quad \Phi_{k+1}(x_{k+1}) = \operatorname*{argmin}_{x_k}(h_k(x_k) + f_k(x_k,x_{k+1})).$$

3. At the end, we let $h = \min_{x_n}(h_n(x_n))$ and set:

$$\overline{x}_n = \operatorname*{argmin}_{x_n}(h_n(x_n)), \quad \overline{x}_{n-1} = \Phi_n(\overline{x}_n), \cdots, \overline{x}_1 = \Phi_2(\overline{x}_2).$$

Then h is the minimum of F and $F(\overline{x}_1,\ldots,\overline{x}_n) = h$.

If we look at the complexity of the algorithm, we see that at step k, for all x_{k+1} we have to search

$$\min_{x_k}(h_k(x_k) + f_k(x_k,x_{k+1})),$$

and since there are n steps, the complexity is in $O(ns^2)$. Moreover, we have to store all the $\Phi_k(x_k)$, which means that the complexity in space is $O(sn)$.

The key thing to realize about this algorithm is that at each intermediate stage, we don't have any idea what the best value of x_k is. But, for *each* value of this x_k, we will know the best values for all the previous variables. At first acquaintance, people often find the algorithm seems nearly trivial, but when they need to apply it, it is astonishingly strong. An example is the simple path problem.

Example. Suppose that we want to find the minimum length path from A to B, knowing that we have the following graph:

1.5. Word Boundaries via Dynamic Programming and Maximum Likelihood

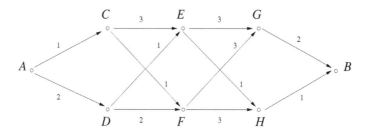

If D_X denotes the minimal distance from point A to a point $\mathcal{X}$, then we can compute in order, the minimum distances:

$$D_E = \min(d(A,C) + d(C,E), d(A,D) + d(D,E)) = 3$$
$$D_F = \min(d(A,C) + d(C,F), d(A,D) + d(D,F)) = 2$$
$$D_G = \min[D_E + d(E,G), D_F + d(F,G)] = 5$$
$$D_H = \min[D_E + d(E,H), D_F + d(F,H)] = 4$$
$$D_B = \min[D_G + d(G,B), D_H + d(H,B)] = 5$$

Finally, working backward, we find that the "best" path from A to B is A, D, E, H, B.

1.5.2 Word Boundaries Revisited

Let us illustrate how the algorithm can be used to find the maximum likelihood partition of a character string $\sigma = (a_1 \cdots a_N)$ into words. We assume the probability model on words is the first-order Markov model, that is, we assume the words λ are independent and distributed with probabilities $P(\lambda)$. Thus if $\tau = (\lambda_1 \cdots \lambda_M)$, we want to minimize

$$\log(1/P(\tau)) = -\sum_{1}^{M} \log(P(\lambda_l))$$

over all sequences τ of words that expand to the string σ. When we parse the string σ into words, each letter a_i is assigned to a specific letter in a specific word; i.e. that is to an element of the set

$$S_i = \{(\lambda, k) | \lambda \in \Lambda, 1 \leq k \leq |\lambda|, a_i = k\text{th letter in } \lambda\}.$$

We define the cost function on $S_i \times S_{i+1}$ by

$$f_i((\lambda_i, k_i), (\lambda_{i+1}, k_{i+1})) = \begin{cases} -\log(P(\lambda_{i+1})) & \text{if } k_{i+1} = 1, k_i = |\lambda_i|, \\ 0 & \text{if } \lambda_{i+1} = \lambda_i, k_{i+1} = k_i + 1, \\ \infty, & \text{otherwise}. \end{cases}$$

When we minimize the sum of the f's, we must avoid the infinite values, so the values of the x_i simply give us a partition of σ into words λ_l. and the penalty is just the sum of the corresponding $-\log(P(\lambda_l))$s, which is the negative log probability of the parse. Later we explore more applications of dynamic programming of the same nature.

1.6 Machine Translation via Bayes' Theorem

To conclude the topic of strings as signals, we introduce the use of the last major player in the book, Bayes' theorem and, at the same time, illustrate how these very low-level ideas can be used with surprising effectiveness even in situations where very high-level subtle patterns are present. Specifically, we want to describe some of the results on machine translation due to the now-disbanded IBM statistical speech and language group of Jelinek, Mercer, Brown, and Steven and Vincent Della Pietra (see [36] and [37] for a full description).

Machine translation started very naively with the idea of putting a two-language dictionary on-line and simply substituting words. This works very badly: one gets virtually unrecognizable phrases (e.g., the now classic example is the English idiom "out of sight, out of mind," translated to Chinese and back again and emerging as "the blind idiot"). The IBM group proposed putting together (i) a two-language dictionary with probabilities and (2) a simple model of probable short word sequences in the target language, and combining them by using Bayes' theorem. To be specific, let $\vec{f}$ be a French word string, and $\vec{e}$ be an English word string. Assume we observe $\vec{f}$. A French-to-English dictionary gives a probability distribution $P(\vec{e}|\vec{f})$, whose large values are tentative translations $\vec{e}$. Instead, we can use an English-to-French dictionary and a prior on $\vec{e}$ (e.g., from an n-gram model), and then use Bayes' theorem:

$$P(\vec{e}, \vec{f}) = P(\vec{e}|\vec{f})P(\vec{f}) = P(\vec{f}|\vec{e})P(\vec{e}).$$

We fix $\vec{f}$ and then seek

$$\vec{e} = \mathrm{argmax}\left[P(\vec{f}|\vec{e})P(\vec{e})\right].$$

The probability $P(\vec{f}|\vec{e})$ is given by an English-to-French dictionary, and the probability $P(\vec{e})$ is given by an n-gram prior (e.g., word triples). The resulting $\vec{e}$ is called the *maximum a posteriori* (or MAP) estimate of $\vec{e}$.

What is truly remarkable is that although the word triple model gives pseudo-English without any grammatical or semantic coherence, and although the

1.6. Machine Translation via Bayes' Theorem

English-to-French dictionary is very sloppy and has vast ambiguity, the combination of

1. low-level data in $P(\vec{e})$,

2. reasonable inclusiveness in the dictionary, especially using phrase-to-phrase links, and

3. the fact that $\vec{f}$ is syntactically and semantically correct in French

makes the MAP estimation of $\vec{e}$ quite reasonable.

Let us describe the system in more detail. The key new idea the IBM group introduced is that one needs a new random variable to describe the relation of $\vec{e}$ and $\vec{f}$: the alignment of the two sentences. Many pairs of words e_i, f_j will be simple translations one of another, but they may not be in the same order. But also some words in French or English may be simply omitted in the translation without loss of meaning, and some simple words in one sentence may expand to phrases in the other (e.g., "cheap" ↔ "bon marché," "aboriginal people" ↔ "autochtones"), and in other cases, a whole idiomatic phrase in one language may be a different idiomatic phrase in the other ("don't have any money" ↔ "sont démunis") and these may even be interleaved ("not" ↔ "ne ... pas"). All this is rather complicated! They simplify and disregard English phrases that must be treated as units, constructing instead a statistical dictionary. Thus, an alignment is a correspondence associating to each word e_j in the English sentence disjoint substrings $\vec{f}_j$ of $\vec{f}$ (the empty substring is allowed). To each English word e, the dictionary gives a list of French strings $\vec{f}(1), \ldots, \vec{f}(n)$, with their probabilities $P(\vec{f}(k)|e)$, often including the empty string, singletons, and longer strings as possible translations of the word e. Also, for the empty English string $\emptyset$, there is a table $P(f|\emptyset)$ of frequencies of French words that in some sentences may be omitted in the English translation. Then the conditional probability $P(\vec{f}, \text{alignment}|\vec{e})$ is given by the product of

- the probability that each string $\vec{f}_j$ given by the alignment is a translation of the corresponding word e_j.

- the probability that the remaining words of $\vec{f}$ are translations of the empty English string, and

- the probability of this way of interleaving the phrases $\vec{f}_j$.

The last term is an alignment term that favors keeping phrases together and roughly in the same order as $\vec{e}$.

Defining a model of this type merely starts the process. There are two further steps:

- Given $\vec{f}$, we need an algorithm for seeking $\vec{e} = \mathrm{argmax} P(\vec{f}|\vec{e}) P(\vec{e})$.

- Given lots of data $(\vec{f}(\alpha), \vec{e}(\alpha))$, can we use this to construct a statistical dictionary and an alignment model?

The IBM group did both with some heuristic simplifications. Both techniques are extensions to this case of a very powerful, very well worked out technique, originating in the analysis of speech, that works for pure linear cases (no alignments, order preserved). This is the hidden Markov model (HMM), dynamic programming, expectation-maximization (EM)-algorithm circle of ideas, whose basic idea we will explain here and further develop in the next chapters. An example of the IBM group's results is shown in Figure 1.5.

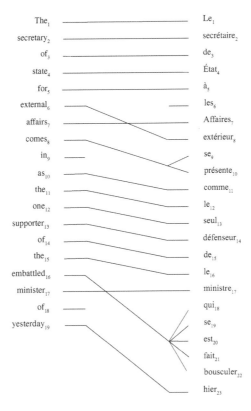

Figure 1.5. A successful French-to-English translation made by the algorithm of Brown, Stephen and Vincent Della Pietra, Mercer, and Jelinek. The input of the algorithm is the French sentence on the right (in order to perform the translation, it has been expanded: for instance aux has been replaced by à les, du by de le, etc.). The output of the algorithm is the English word string on the left. Note how the words are not aligned in the same order, nor do they match one to one.

1.7. Exercises

Here we will present a very simplified version of their model. We assume that each English word has only one French translation and that there is no rearrangement in the alignment. Then we must have a word-for-word translation.

$$\begin{array}{ccccc} f_1 & f_2 & \cdots\cdots & f_n \\ \uparrow & \uparrow & & \uparrow \\ e_1 & e_2 & \cdots\cdots & e_n \end{array}$$

We assume also a bigram model on $\vec{e}$. Then we get

$$P(\vec{e}, \vec{f}) = \prod_{i=1}^{n} P(f_i|e_i) \cdot \prod_{i=1}^{n-1} P(e_{i+1}|e_i) \cdot P(e_1),$$

where $\prod P(e_{i+1}|e_i)$ is the Markov chain of the bigram model, $P(f_i|e_i)$ is given by the word-to-word statistical dictionary, and $P(e_1)$ is the initial state probability. We can describe this model by the following graph,

$$\begin{array}{cccccccc} f_1 & & f_2 & & f_3 & & f_{n-1} & & f_n \\ | & & | & & | & & | & & | \\ e_1 & \text{---} & e_2 & \text{---} & e_3 & \text{---} & \cdots & \text{---} & e_{n-1} & \text{---} & e_n \end{array}$$

where the vertices are random variables and the edges are the factors of $P(\vec{e}, \vec{f})$. For instance,

$$\begin{array}{cccccccc} \text{il} & & \text{ouvre} & & \text{la} & & \text{porte} \\ | & & | & & | & & | \\ \text{he} & \text{---} & \text{opens} & \text{---} & \text{the} & \text{---} & \text{door} \end{array}$$

We then use this model, as above, to infer $\vec{e}$ conditional on observing $\vec{f}$. The model makes $\vec{e}$ into a Markov chain, and when $\vec{f}$ is the observed random variable, $\vec{e}$ is called a hidden Markov variable; hence, this whole model is called a hidden Markov model. . Note that if the f_i are fixed, we need to minimize

$$-\log P(\vec{e}, \vec{f}) = \sum_{i=1}^{n-1} f(e_{i+1}, e_i) + \sum_{i=1}^{n} g(e_i) + h(e_1).$$

This is the form to which dynamic programming applies, allowing us to work out rapidly the MAP English translation (although the size of the lexicon is so large that some more tricks are needed).

1.7 Exercises

1. Simulating Discrete Random Variables with MATLAB

To understand what kind of features are captured by a stochastic model, one must have a method to get samples from this model and see what they look like. This will occur again and again in this book. So, one of the main required techniques is to be able to sample from a wide variety of probability distributions. In this exercise, we study sampling from select discrete distributions, some of which will be used in the later chapters. The starting point of all the sampling methods will be sampling the uniform distribution on $[0, 1]$. MATLAB and all other computer languages have built-in algorithms for doing this.

Sampling from a binomial distribution

1. Show how to use the MATLAB function rand to get samples from a Bernoulli distribution of parameter $p \in [0, 1]$. Recall that such a distribution is a probability distribution on $\{0, 1\}$, that it is usually denoted by $b(p)$, and that it is defined by $P(\mathcal{X} = 1) = p = 1 - P(\mathcal{X} = 0)$.
2. Use this result to sample from a binomial distribution of parameters n and p. Recall that such a random variable can be obtained as the sum of n independent Bernoulli random variables with the same parameter p. Get N samples from the binomial (by using a matrix, *not* a "for" loop) and compare on the same figure the histogram of the values obtained and the binomial distribution itself.

Sampling from a Poisson distribution

1. Let $\mathcal{X}$ be a random variable following a uniform distribution on $[0, 1]$. Prove that the random variable $\mathcal{Z} = -\log \mathcal{X}$ has a density function given by e^{-x} on $\{x \geq 0\}$. (This is the exponential distribution of parameter 1).
2. Prove by induction on k that, if $\mathcal{Z}_1, \mathcal{Z}_2, \ldots, \mathcal{Z}_k$ are k independent random variables with an exponential distribution of parameter 1, then $\mathcal{Z}_1 + \mathcal{Z}_2 + \ldots + \mathcal{Z}_k$ has density function $x^{k-1} e^{-x} / (k-1)!$ on $\{x \geq 0\}$. (This is the gamma distribution of parameters $(k, 1)$.)
3. Let $\lambda > 0$, and let $(\mathcal{X}_n)_{n \geq 0}$ be a sequence of independent random variables uniformly distributed on $[0, 1]$. Then let $\mathcal{Y}$ be the random variable defined by

$$\mathcal{Y} = \min\{n \geq 0 \mid \mathcal{X}_0 \times \ldots \times \mathcal{X}_n \leq e^{-\lambda}\}.$$

Prove that the law of $\mathcal{Y}$ is the Poisson distribution P_λ of parameter λ, which means that

$$\mathbb{P}(\mathcal{Y} = k) = \frac{\lambda^k}{k!} e^{-\lambda}.$$

4. Use this result to get an algorithm for sampling from a Poisson distribution P_λ. Obtain N samples and plot on the same figure their histogram and the Poisson distribution.

The inversion method

1. The aim of the inversion method is to get samples from any distribution through the inversion of its repartition or cumulative distribution function. In the discrete framework, the method is the following one.
Let $P = (p_0, p_1, \ldots)$ be a probability distribution on $\mathbb{N}$. Let its repartition function H be defined on $\mathbb{R}_+$ by:

$$\forall k \geq 0, \forall x \in [k, k+1) \text{ then } H(x) = p_0 + p_1 + \ldots p_k.$$

Let $\mathcal{U}$ be uniform on $[0, 1]$, then prove that the random variable $\mathcal{X}$ defined by $\mathcal{X} = H^{-1}(\mathcal{U}) = \inf\{k \in \mathbb{N} \mid \mathcal{U} \leq H(k)\}$ follows the distribution P.

2. Use this result to sample from a geometric distribution of parameter p given by $\forall k \geq 1, \mathbb{P}(\mathcal{X} = k) = (1-p)^{k-1}p$. Obtain N samples in this way and plot on the same figure the histogram of the obtained values and the geometric distribution.

2. Entropy as the "Only Way to Measure Information"

In his founding paper, "A mathematical theory of communication" [194], Shannon asked how to define a quantity that will measure the information "produced" by a stochastic process. He assumed that we have a set of possible events whose probabilities of occurrence are $p_1, p_2, \ldots$. And he asked: "These probabilities are known but that is all we know concerning which event will occur. Can we find a measure of how much "choice" is involved in the selection of the event or of how uncertain we are of the outcome?" The answer to this question is that yes, under some required conditions, listed below, such a measure exists and it must be equal (up to a positive multiplicative constant) to the entropy. More precisely, the result is given as Theorem 1.18.

Theorem 1.18. *Let $H : (p_1, \ldots, p_n) \to H(p_1, p_2, \ldots, p_n)$, be a positive function defined on finite probability distributions. If H has the following properties:*

(i) *H is continuous in the p_i's.*

(ii) *If all p_i are equal, $p_i = 1/n$, then H should be a monotonic increasing function of n. (This means: "With equally likely events, there is more choice, or uncertainty, when there are more possible events.")*

(iii) *$H(p_1, p_2, \ldots, p_n) = H(p_1+p_2, p_3, \ldots, p_n) + (p_1+p_2)H(\frac{p_1}{p_1+p_2}, \frac{p_2}{p_1+p_2})$. (This means: "If you first make a choice ignoring the difference between 1 and 2, and then, if 1 or 2 came up, you choose one of them, the original H should be the weighted sum of the individual values of H in the formula above.")*

Then, there exists a positive constant C such that

$$H(p_1, p_2, \ldots, p_n) = -C \sum_i p_i \log p_i.$$

The choice of C amounts to the choice of a unit of measure.

The aim of this exercise is to prove Theorem 1.18 by following the original proof of Shannon.

1. Given n, an integer, define $A(n) = H(\frac{1}{n}, \frac{1}{n}, \ldots, \frac{1}{n})$. Prove that for all $k \geq 1$, where k is an integer,
$$A(n^k) = kA(n).$$

2. Let n and m be two integers. Let $k \geq 1$ be an integer. Prove that there exists $l \in \mathbb{N}$ such that
$$m^l \leq n^k \leq m^{l+1}.$$
Use this inequality to show that
$$\left| \frac{A(n)}{A(m)} - \frac{\log n}{\log m} \right| \leq \frac{2}{k}.$$

Conclude that there exists a constant $C > 0$ such that $A(n) = C \log n$ for all integer $n \geq 1$.

3. Assume that $(p_1, p_2, \ldots, p_n)$ is a discrete probability distribution with the property that there are integers $n_1, n_2, \ldots n_n$ such that for each i, $p_i = n_i / \sum_{j=1}^{n} n_j$. Prove that

$$C \log \sum_{i=1}^{n} n_i = H(p_1, p_2, \ldots, p_n) + C \sum_{i=1}^{n} p_i \log n_i.$$

4. End the proof of the theorem by showing that for all discrete probability distributions $(p_1, p_2, \ldots, p_n)$,

$$H(p_1, p_2, \ldots, p_n) = -C \sum_{i=1}^{n} p_i \log p_i.$$

3. A True Metric Associated with the Kullback-Leibler Distance

Fix a finite set Ω of states. Let $\mathcal{P}$ be the space of all probability distributions on Ω. Then the KL-distance $D(P\|Q)$ is not symmetric, and hence is not a true metric on $\mathcal{P}$. But there is a very elegant global metric on $\mathcal{P}$ that equals the KL-distance infinitesimally. It is a variant of what is usually called the *Bhattacharya metric*.

Enumerating the elements of Ω and letting n be the cardinality of Ω, the probability distributions are given by the vectors $(P_1, \cdots, P_n)$ such that $P_i \geq 0, \sum_i P_i = 1$. Define a bijective map from $\mathcal{P}$ to the positive octant of the $(n-1)$–sphere by mapping $\vec{P}$ to $(\sqrt{P_1}, \cdots, \sqrt{P_n})$. Show that the pullback metric of the usual global metric on the sphere (defined by the angle between any two points) is

$$d(\vec{P}, \vec{Q}) = \arccos\left(\sum_i \sqrt{P_i Q_i}\right).$$

Next compare this metric to the KL distance by considering $\vec{Q}$ near $\vec{P}$. Let $\vec{Q} = \vec{P} + t\vec{F}$, where $\sum_i F_i = 0$. Expand both $d(\vec{P}, \vec{Q})$ and $D(\vec{P}\|\vec{Q})$ in powers of t. You will find a very nice connection.

4. Rissanen's Coding of $\mathbb{N}$

The simplest way to encode all positive integers is to use their binary expansion. Formally, we have a bijection:

$$\{\text{finite strings of 0, 1 beginning with 1}\} \underset{\text{bin}}{\overset{\text{eval}}{\rightleftarrows}} \mathbb{N}.$$

The trouble with encoding positive integers like this is that it is not a prefix code. That is, if you find 101, then this might be 5 or it might be just the length 3 prefix of 101000001, or 321. We can see this also since the length $\text{bin}(n)$ of the binary expansion of n is easily checked to be $\lceil \log_2(n+1) \rceil$ and

$$\sum_n 2^{-\lceil \log_2(n+1) \rceil} \geq \tfrac{1}{2} \cdot \sum_n \tfrac{1}{n} = \infty.$$

As we have seen, for prefix codes, this sum must be at most 1. Jorma Rissanen [187, 188] came up with an elegant extension of binary expansions by making it into a prefix code.

1.7. Exercises

Essentially, you must put the binary encoding of the length of the binary coding bin(n) of n first so you know where the end is. But still, the length of this must also be given in binary even earlier, and so on. Here we describe the decoding scheme, and the exercise is to reconstruct the encoding scheme along these lines. Let eval be the map taking a binary string to the natural number it represents.

Algorithm (Rissanen's Coding—The Decoding).

1. 00 encodes 1; 010 encodes 2; 011 encodes 3; all other codes have length ≥ 4
2. Initialize σ to be the string to be decoded, and let $m = 3$.
3. Let $\sigma = \langle \sigma_1 \sigma_2 \rangle$, where σ_1 is the initial substring of length m.
4. If σ_2 begins with 0, then in fact $\sigma_2 = \langle 0 \rangle$, $n = \text{eval}(\sigma_1)$ is the number encoded and we are done.
5. Otherwise, let $m = \text{eval}(\sigma_1)$, $\sigma = \sigma_2$, and go back to step 3.

Find the encoding algorithm. Find an expression for the length $\ell^*(n)$ of the code of an integer n in terms of the function $\ell(n) = \lceil \log_2(n+1) \rceil$ and its iterates. It is well known in calculus that all the integrals:

$$\int^\infty \frac{dx}{x}, \quad \int^\infty \frac{dx}{x \log(x)}, \quad \int^\infty \frac{dx}{x \log(x) \log(\log(x))}, \text{ etc.,}$$

all diverge but more and more slowly. Using the idea behind Rissanen's coding, define a function $\log^*(x)$ so that $\int^\infty dx/(x \log^*(x))$ just barely converges.

5. Analyzing n-tuples in Some Data Bases

1. Surfing the web, find at least 1 megabyte (5+ is better) of English text. It should be ascii, not html, so you have the sequence of characters in hand. Below is some MATLAB code to do the tedious job of converting an ascii text into a string of numbers from 1 to 27 by (i) stripping out all punctuation and numerals, (ii) converting all upper- and lower-case letters to the numerals 1 to 26, (iii) converting all line returns and tabs to spaces, and (iv) collapsing all multiple spaces to single spaces represented by the number 27. The result is one long string of the numbers 1 through 27.

2. Calculate the empirical probabilities P_1 of all single letters (including space) and P_2 of all consecutive letter pairs P_2 and print these out. Check that some obvious properties hold (e.g., the conditional probability of u after a q is essentially 1). Now calculate the entropy of both distributions and of the conditional distribution of the second letter, given the first. Compare with Shannon's results. Depending on your data set, there may be differences; if so, can you see any reasons for these differences?

3. Now calculate the distributions P_3 on triples and P_4 on 4-tuples of consecutive characters. For 4-tuples, you may run into memory and speed problems, so careful MATLAB coding is essential. One method is to make a 4-tuple into a single 2-byte integer by using base 26 by coding a 4-tuple of numbers $abcd$ between 1 and 26 as

$$(a-1) + 26(b-1) + 26^2(c-1) + 26^3(d-1).$$

Again, calculate the entropies of these distributions.

4. Now play Shannon's game: make a new text, say 200 characters long, by (i) taking a random sample of four symbols from your text, and (ii) repeatedly sampling from the conditional distribution $P_4(d|abc)$ to extend the series. The results, for example, with the Bible or with IBM technical reports, should be quite suggestive.

5. Finally, process your text to perversely eliminate all spaces! Can we recover the word boundaries? To prevent problems with memory, take several random chunks, say 80 characters long, of your text, and, for each chunk, calculate the mutual information of consecutive pairs (i.e., $\log_2(P_4(abcd)/P_2(ab)P_2(cd))$). Make a figure, as in the notes, showing the graph of this function on top of the actual words (see code below) and see to what extent the local minima predict the word boundaries.

Here is the MATLAB code, to spare you annoying details, with the actual ascii coding values, which was used to read Huckleberry Finn (hfinn.txt):

```
label = fopen('hfinn.txt');
[hk,count] = fread(label,'uchar');
fclose(label);
F = double(hk);
F(F<33) = 27;                           % line feeds, tabs,
                                        %   spaces --> 27
F(F>64 & F<91) = F(F>64 & F<91)-64;     % u.c. letters --> [1,26]
F(F>96 & F<123)=F(F>96 & F<123)-96;     % l.c. letters --> [1,26]
F = F(F<28);                            % Throw out numbers,
                                        % punctuation
F2 = [0;F(1:(size(F)-1))]+F;            % Add consecutive values
G = F(F2<54);                           % The text with one space
                                        % between words

% CODE TO ELIMINATE SPACES
G = G(G<27);
% CODE TO MAKE NICE OUTPUT
% PLOTTING ARRAY MI ABOVE ACTUAL CHARCATERS
hold off, plot(MI((1:80)));
axis([0 81 -20 20]); axis off, hold on
plot([0 81], [0 0]); plot([0 81], [4 4]);
text(0,0,'0 bits');   text(0,4,'4 bits')
for i=1:80
    text(i+0.15,-2,char(96+G(i)))       % ascii codes 97,..,122
                                        % are letters
end
```

6. The Statistics of DNA Sequences

Other strings of discrete symbols besides text are interesting to analyze. Currently, perhaps the most publicized sequence is that of DNA, a string of four symbols A, C, G, and T. It is known that some parts of this sequence code for proteins and are called *exons* (they are "expressed"); other parts serve other purposes or maybe are junk and are called *introns*. A protein is also a string-like molecule made up of 20 units, the amino acids. The translation code divides the DNA at every third base and uses the 64 possible triplets in each piece to code either for one of the amino acids or for the end of the protein, called the *stop*.

1.7. Exercises

TTT	17.6	Phe	TCT	15.2	Ser	TAT	12.2	Tyr	TGT	10.6	Cys
TTC	20.3	Phe	TCC	17.7	Ser	TAC	15.3	Tyr	TGC	12.6	Cys
TTA	7.7	Leu	TCA	12.2	Ser	TAA	1.0	STOP	TGA	1.6	STOP
TTG	12.9	Leu	TCG	4.4	Ser	TAG	0.8	STOP	TGG	13.2	Trp
CTT	13.2	Leu	CCT	17.5	Pro	CAT	10.9	His	CGT	4.5	Arg
CTC	19.6	Leu	CCC	19.8	Pro	CAC	15.1	His	CGC	10.4	Arg
CTA	7.2	Leu	CCA	16.9	Pro	CAA	12.3	Gln	CGA	6.2	Arg
CTG	39.6	Leu	CCG	6.9	Pro	CAG	34.2	Gln	CGG	11.4	Arg
ATT	16.0	Ile	ACT	13.1	Thr	AAT	17.0	Asn	AGT	12.1	Ser
ATC	20.8	Ile	ACC	18.9	Thr	AAC	19.1	Asn	AGC	19.5	Ser
ATA	7.5	Ile	ACA	15.1	Thr	AAA	24.4	Lys	AGA	12.2	Arg
ATG	22.0	Met	ACG	6.1	Thr	AAG	31.9	Lys	AGG	12.0	Arg
GTT	11.0	Val	GCT	18.4	Ala	GAT	21.8	Asp	GGT	10.8	Gly
GTC	14.5	Val	GCC	27.7	Ala	GAC	25.1	Asp	GGC	22.2	Gly
GTA	7.1	Val	GCA	15.8	Ala	GAA	29.0	Glu	GGA	16.5	Gly
GTG	28.1	Val	GCG	7.4	Ala	GAG	39.6	Glu	GGG	16.5	Gly

Table 1.1. DNA codons, their frequencies out of 1000, and the corresponding proteins.

Table 1.1 (from http://www.kazusa.or.jp/codon/cgi-bin/showcodon.cgi?species=9606) shows also the frequencies out of 1000 of all triplets in all exons in the human genome. *But nature gives no indication where to start the sequence of triplets!* This gives three possible "phases," shifting the break points by 0,1, or 2.

One way to distinguish exons from introns and to find their correct starting place mod 3 is that only these sequences of triplets will have the correct statistics. A vast library of DNA sequences are available online from either GenBank (http://www.ncbi.nlm.nih.gov/Genbank/) or EMBL (http://www.ebi.ac.uk/embl/). For this exercise, download a human sequence of length 2000. We suggest base pairs 64,001–66,000 in the well-studied beta globin sequence HUMHBB (GenBank) or HSHBB (EMBL1). Find the three exons in this sequence and the starting positions mod 3. Take all subsequences of length 120 and divide each of them into 40 triplets; then calculate the likelihood ratio of this sequence occurring from the human triplet frequency table versus the uniform distribution. That is, if $(P_1, \cdots, P_{64})$ are the human gene triplet frequencies (extracted from all known exons in the human genome), and $n_i(a)$ is the number of occurrences of the ith triplet in the subsequence starting from a, calculate

$$LL(a) = \sum_i n_i(a) \log(P_i/U_i), \quad \text{where } U_i = \tfrac{1}{64} \text{ is the uniform distribution.}$$

Make the three plots $LL(3a), LL(3a+1),$ and $LL(3a+2)$. You should find broad peaks in one of these for each exon. This method doesn't give the exact beginning of each exon, but a "STOP" should indicate the end.

7. Conditional Probability and Causality

This exercise is only tangentially related to the content of this book, but it concerns a widespread mistake in interpreting conditional probabilities and, as such, connects to our topics. The following headline appeared in the *Boston Globe* a few years ago: "Teen

Exercise May Cut Breast Cancer Risk." The article reported that

$\mathbb{P}($ premenopausal breast cancer (C) | active teenage, young adult years (A)$) \approx$
$0.77 \cdot \mathbb{P}($ premenopausal breast cancer (C) | sedentary teenage, young adult years $(\sim A))$.

One reaction came from Dr. Colditz: "This really points to the benefit of sustained physical activity from adolescence through the adult years." Do you agree? What is the weak point in his deduction? If you're not sure, stop here and think about it, then read on.

Now, consider a second possibility. Suppose there is a genetic factor X, some allele or other, that has hormonal effects that predispose a young woman to more physical activities and also protects against early breast cancer:

$$\mathbb{P}(X) = \lambda,$$
$$\mathbb{P}(A \mid X) = a_1,$$
$$\mathbb{P}(A \mid \sim X) = a_2,$$
$$\mathbb{P}(C \mid X) = c_1,$$
$$\mathbb{P}(C \mid \sim X) = c_2.$$

Further, suppose that given the value of the genetic factor X, there is no connection at all between exercise and breast cancer—the two are conditionally independent. Under these assumptions, calculate the probability ratio $\mathbb{P}(C \mid A)/\mathbb{P}(C \mid \sim A)$ that had been found experimentally to be 0.77. The formula is a bit messy, but if you take an extreme case of $a_1 = 1, c_1 = 0$, you find a formula that is easy to interpret and can give arbitrarily small numbers. Show that more moderate assumptions, such as $a_1 = 2a_2, c_1 = 0.5c_2, \lambda = 0.5$, lead to a ratio less than $8/9$.

This simple example is the kernel of Judea Pearl's theory [174,175] of causal networks. Essentially, the only way to establish that causality is present is to intervene in the world. For example, take a large randomly selected group of teenage girls, convince half of them to exercise for twenty years and half to refrain from exercise, wait another twenty years and see how many in each group came down with breast cancer. If we managed to do this and it came out as before, we might reasonably conclude that there was causation.

8. Mutual Information and Weather Forecasting

This question was raised by Michael Artin: If it rains in your location 25% of the time, how do we compare a weather forecaster who says every day, "probability of rain 25%" to one who gambles by predicting with certainty "rain" or "no rain" each day? It would seem the first is quite accurate but not very informative. The correct way to measure their relative usefulness is via mutual information.

We assume the forecaster always uses probabilities from a finite set $x_1, \cdots, x_n \in [0, 1]$, such as all the multiples of 0.1. From historical data, it is easy to find (i) the frequencies $P_i = \mathbb{P}(\text{forecast} = x_i)$ of each forecast, and (ii) the conditional probabilities $Q_i = \mathbb{P}(\text{rain}|\text{forecast} = x_i)$ of it actually raining. Then $R = \sum_i P_i Q_i$ is the overall probability of rain, and the entropy $H(R)$ is the information you get each day by observing whether it rains. What, then, is the expression for the fractional number of bits of information about whether it will rain given by this weather forecaster?

Perhaps, by using the web, you can work this out in your location. As these figures tend not to be publicized, let us know what you find if you do.

1.7. Exercises

9. Guessing the Best Word via Dynamic Programming

The game here is the following: Infer the most probable sequence of letters in a word given the first letter and the last one (this is the usual Hangman game). Assume that the length k of the sequence is known. The question thus is:

What are the dots in $a_1 \cdots\cdots a_k$ to have the most probable sequence ?

We want to do this by using the n-gram model for the probability of a length k sequence of letters ($n \leq k$). Then, finding the most probable sequence given the first and last letters is a shortest path problem, and it can be solved via dynamic programming.

1. For all $m \geq 1$, let Ω_m denote the set of all length m strings of letters. For $\sigma = b_2 b_3 \ldots b_{k-1} \in \Omega_{k-2}$, the probability of $a_1 \sigma a_k$ under the n-gram model is

$$P_k^{(n)}(a_1 \sigma a_k) = $$
$$P_n(a_1 b_2 \ldots b_n) \left(\prod_{i=n+1}^{k-1} P_n(b_i | b_{i-n+1} \ldots b_{i-1}) \right) P_n(a_k | b_{k-n+1} \ldots b_{k-1}).$$

Prove that the problem of finding the σ that maximizes the above probability is the same as finding the argmin of a function F of the form $F(\lambda_1, \ldots, \lambda_{k-n+1}) = \sum_{i=1}^{k-n} f_i(\lambda_i, \lambda_{i+1})$. What are the λ_i's and the f_i's ?

2. Use the dynamic programming algorithm to infer the most probable sequence. The n-gram models can be obtained from a data base (see Exercise 5). Take a (long) word and remove all letters except the first one and the last one. Then infer the most probable strings by using different n-gram models. Is the result always the same? For $n = 2, 3, 4$, what is the longest word you can find that is correctly reconstructed in this way?

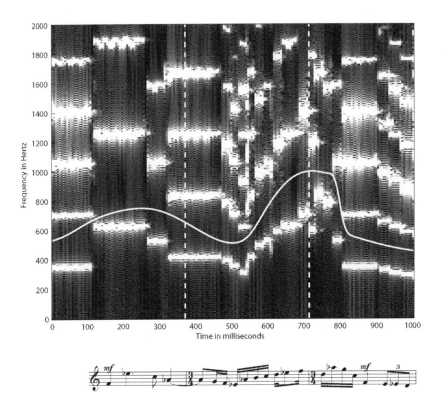

Figure 2.1. Three bars from an oboe playing *Winter711* by Jan Beran: the spectrogram (top) and the score as humans write it (bottom). The three bars are separated by vertical dotted white lines in the spectrogram and are separated as usual in the human score. The spectrogram shows the distribution of power across frequencies as time progresses. The oboe has a rich set of harmonics. Below the curving white line is the fundamental note, in which the score is easily recognized (although stretched and squeezed a bit). The pixelization of the spectrogram is an unavoidable consequence of Heisenberg's uncertainty: time and frequency cannot be simultaneously made precise. (See Color Plate I. We thank Chris Raphael, whose playing is shown in the spectrogram and who supplied us with the score.)

- 2 -

Music and Piecewise Gaussian Models

We are interested in constructing a stochastic model for the signal $s(t)$, which represents air pressure as a function of time while music is being played. This relatively simple example will provide a good introduction into the basic ideas of Pattern Theory for variables with continuous values. Except for one digression in Section 2.4, we are not concerned about the mathematical details of modeling stochastic processes of continuous time, so we always assume the sound signal has been discretely sampled. A typical piece of data might be given with a sampling interval $\Delta t = 1/8000$ seconds,[1] so that if five seconds of data are considered, we have a sequence of real numbers $s_k = s(k\Delta t), 1 \leq k \leq 40,000$, and we want a stochastic model for this finite-dimensional piece of data.

As described in Chapter 1, when words were discussed, the model needs extra "hidden" random variables that represent the patterns. As we see in Figure 2.1, the main patterns consist of what is usually called the "musical score." We need

1. the number of notes m,

2. the times $t_i = k_i \Delta t$ when new notes begin, $1 < k_1 < k_2 < \ldots < k_m < N$,

3. the frequency ω_i of the ith note in hertz (or its approximate integral period $p_i \approx 1/(\Delta t \cdot \omega_i) \in \mathbb{Z}$).

To construct this model, we eventually define a probability density $p(\vec{s}, m, \vec{t}, \vec{p})$ in *all* the variables. We can sample from this model to see whether it sounds like any music known to mankind (the simple model we will give will fail this test— but variants can improve it greatly). But more significantly, as in the French-to-English translation discussed in Chapter 1, we can use this probability distribution to reconstruct the score from an observed signal $\vec{s}_{\text{obs}}$. We recover the hidden variables $m, \vec{t},$ and $\vec{p}$ by maximizing the conditional probability

$$p(m, \vec{t}, \vec{p} \mid \vec{s}_{\text{obs}}) = \frac{p(\vec{s}_{\text{obs}}, m, \vec{t}, \vec{p})}{\sum_{m', \vec{t}', \vec{p}'} p(\vec{s}_{\text{obs}}, m', \vec{t}', \vec{p}')}.$$

---

[1]This is quite low fidelity. Compact Discs (CDs) use 44,100 samples per second.

When we use this general method, we actually have three problems:

1. the construction of the model,
2. finding an algorithm to maximize $p(m, \vec{t}, \vec{p} \mid \vec{s}_{\text{obs}})$ with respect to the variables m, $\vec{t}$, and $\vec{p}$, and
3. optimizing the parameters of the model to make it fit the data as well as possible.

We mostly treat the first two of these problems in this chapter, considering the third problem only in Section 2.6.3. Before launching into music, we need to review the basic mathematics of Gaussian distributions and of Fourier analysis. We begin with Gaussian distributions.

2.1 Basics III: Gaussian Distributions

Definition 2.1. Let $\vec{x} = (x_1, \ldots, x_n)$ denote a vector in $\mathbb{R}^n$; we then define a Gaussian distribution on $\mathbb{R}^n$ by its density

$$p(\vec{x}) = \frac{1}{Z} e^{-(\vec{x}-\vec{m})^t Q(\vec{x}-\vec{m})/2},$$

where $\vec{m} \in \mathbb{R}^n$, Q is a $n \times n$ symmetric positive definite matrix, and Z is a constant such that $\int p(\vec{x}) d\vec{x} = 1$.

Gaussian distributions are very important in probability theory, particularly because of the central limit theorem.

Theorem 2.2 (Central Limit Theorem). *If $\vec{\mathcal{X}}$ in $\mathbb{R}^n$ is any random variable with mean 0 and finite second moments, and if $\vec{\mathcal{X}}^{(1)}, \ldots, \vec{\mathcal{X}}^{(N)}$ are independent samples of $\vec{\mathcal{X}}$, then the distribution of $\frac{1}{\sqrt{N}} \sum_{k=1}^{N} \vec{\mathcal{X}}^{(k)}$ tends, as $N \to +\infty$, to a Gaussian distribution with mean 0 and the same second moments as $\vec{\mathcal{X}}$.*

Theorem 2.2 suggests the idea that the correct "default" probability models for continuous random variables are Gaussian. We rather that this idea is wrong, however, and that indeed a heuristic converse to the central limit theorem holds: any random variable in the real world that is *not an average* is also probably not Gaussian because it will have more "outliers." For example, in dimension 1, the function e^{-cx^2} goes to 0 at infinity extremely fast; hence, large values of a Gaussian random variable are extremely rare. But in the real world, distributions seldom behave this way. A good example is given by the empirical distribution of the logs of daily changes in prices (such as stock prices, commodity prices, or

2.1. Basics III: Gaussian Distributions

other financial time series, as discussed in Section 2.4.2; see Figure 2.9). Many mathematical models assume these are Gaussian distributed, but, as Mandelbrot noted many years ago [146], they are much better modeled by distributions with "heavy tails" (such as t-distributions: probability densities $p_t(x) = C_1/(C_2 + x^2)^{(t+1)/2}$). (Again, see Figure 2.9 which plots empirical data showing the *log probabilities*, which make the size of the tails much more apparent. Mandelbrot perhaps overdid it when he claimed such data have infinite variance, but we find that values such as $t = 3$ or 4 are typical.)

Proposition 2.3. *The Gaussian distribution p from Definition 2.1 has the following properties:*

1. $Z = (2\pi)^{\frac{n}{2}} (\det Q)^{-\frac{1}{2}}$.

2. $\int (\vec{x} - \vec{m}) p(\vec{x}) d\vec{x} = 0$, *which means that $\vec{m}$ is the mean of p, denoted by* $\mathbb{E}_p(\vec{x})$.

3. *If* $C_{ij} = \int (x_i - m_i)(x_j - m_j) p(\vec{x}) d\vec{x}$ *is the covariance matrix, then* $C = Q^{-1}$.

Proof: By symmetry (using the change of variables $\vec{x} - \vec{m} = \vec{m} - \vec{x}'$), we have for all i:

$$\int (x_i - m_i) p(\vec{x}) dx_1 \ldots dx_n = 0,$$

which proves Property 2. Properties 1 and 3 follow by the linear change of variables $\vec{x} = A\vec{y} + \vec{m}$, where A is such that $Q^{-1} = AA^t$. Then

$$1 = \int p(\vec{x}) d\vec{x} = \frac{|\det A|}{Z} \int e^{-(y_1^2 + \ldots + y_n^2)/2} dy_1 \ldots dy_n = \frac{(2\pi)^{n/2}}{Z\sqrt{\det Q}},$$

which proves Property 1. Moreover,

$$\begin{aligned}
C_{ij} &= \frac{1}{Z} \int (x_i - m_i)(x_j - m_j) e^{-(\vec{x}-\vec{m})^t Q(\vec{x}-\vec{m})/2} dx_1 \ldots dx_n \\
&= \frac{1}{Z} \int \sum A_{ik} y_k \sum A_{jl} y_l e^{-(y_1^2 + \ldots + y_n^2)/2} |\det A| dy_1 \ldots dy_n \\
&= \frac{|\det A|}{Z} \sum_{k,l} A_{ik} A_{jl} \left(\int y_k y_l e^{-(y_1^2 + \ldots + y_n^2)/2} dy_1 \ldots dy_n \right).
\end{aligned}$$

The integral is 0 if $k \neq l$ (by symmetry because $\int y e^{-y^2/2} = 0$), and if $k = l$ we integrate by parts, finding $\int y^2 e^{-y^2/2} = \int e^{-y^2/2} = \sqrt{2\pi}$. Thus,

$$C_{ij} = \frac{|\det A|}{Z} \sum_k A_{ik} A_{jk} (2\pi)^{n/2}.$$

Using Property 1, this becomes $C = AA^t = Q^{-1}$. □

Corollary 2.4. *A Gaussian distribution is uniquely determined by its mean and covariance. Conversely, given any vector $\vec{m}$ and any positive definite symmetric matrix C, there exists a Gaussian distribution with $\vec{m}$ and C as its mean and covariance, respectively.*

Let $p(\vec{x}) = \frac{1}{Z}e^{-(\vec{x}-\vec{m})^t Q(\vec{x}-\vec{m})/2}$ be a Gaussian distribution on $\mathbb{R}^n$. It is very helpful for our intuition to interpret and contrast in simple probability terms what $C_{ij} = 0$ and $Q_{ij} = 0$ mean:

1. Fix $i < j$, then $C_{ij} = 0$ means that "the marginal distribution on (x_i, x_j) makes x_i and x_j independent." The marginal distribution on x_i, x_j is defined by

$$p^{(i,j)}(x_i, x_j) = \int p(x_1, \ldots, x_n) dx_1 \ldots \widehat{dx_i} \ldots \widehat{dx_j} \ldots dx_n,$$

where the notation $dx_1 \ldots \widehat{dx_i} \ldots \widehat{dx_j} \ldots dx_n$ means that we integrate on all variables except x_i and x_j. Since the covariance term C_{ij} depends only on the marginal and is 0, the covariance matrix for (x_i, x_j) by themselves is diagonal and there exist constants Z_{ij}, α_i, and α_j such that the marginal distribution on x_i, x_j has the following expression

$$p^{(i,j)}(x_i, x_j) = \frac{1}{Z_{ij}} e^{-\alpha_i(x_i - m_i)^2 - \alpha_j(x_j - m_j)^2} = p^{(i)}(x_i) \cdot p^{(j)}(x_j).$$

2. Fix $i < j$, then $Q_{ij} = 0$ means that "the conditional distribution on (x_i, x_j) fixing the other variables makes them independent." For $k \neq i, j$, fix $x_k = a_k$; then since $Q_{ij} = 0$ there exist constants b_0, b_i, b_j, c_i, c_j (depending on the a_k's) such that the conditional distribution on x_i, x_j is

$$\begin{aligned} p(x_i, x_j | x_k = a_k \text{ for all } k \neq i, j) &= \text{cnst} \cdot p(a_1, .., x_i, .., x_j, .., a_n) \\ &= \frac{1}{Z} e^{-(b_0 + b_i x_i + b_j x_j + c_i x_i^2 + c_j x_j^2)} \\ &= \frac{1}{Z} e^{-c_i(x_i - m'_i)^2 - c_j(x_j - m'_j)^2} \\ &= p(x_i | x_k = a_k \text{ for all } k \neq i, j) \\ &\quad \cdot p(x_j | x_k = a_k \text{ for all } k \neq i, j). \end{aligned}$$

Remark 2.5. In some of the models that we introduce, the marginal distributions of the variables are given by simple formulas or algorithms; in others, the conditional distributions are like this, but rarely are they both.

2.1. Basics III: Gaussian Distributions

We can also extend the notion of entropy to continuous probability distributions such as the Gaussian ones. In Chapter 1, we defined the entropy of a probability distribution p on a finite (or countable) set $\Omega = \{w_i\}$ by $H(p) = \sum_i p_i \log_2(1/p_i)$, where $p_i = p(w_i)$. We now extend this definition to a probability density $p(x)$, in which case it is called the *differential entropy*.

Definition 2.6. Let $p(x)$ be a probability density on $\mathbb{R}^n$, that is,

$$dP = p(x_1, \cdots, x_n) dx_1 \cdots dx_n$$

is a probability measure. The *differential entropy* of P is defined by

$$H_d(P) \text{ or } H_d(p) = \int p(x) \log_2 \frac{1}{p(x)} dx = \mathbb{E}_P \left(\log_2(1/p) \right).$$

More generally, if Ω is any space with a reference measure $d\mu$ on it, then the differential entropy of a probability measure $dP(x) = p(x) d\mu$ on Ω absolutely continuous with respect to $d\mu$, with Radon-Nikodym derivative $p(x)$, is:

$$H_d(P) = \int p(x) \log_2 \frac{1}{p(x)} d\mu(x) = -\int \log_2 \left(\frac{dP}{d\mu} \right) dP(x).$$

Differential entropy always requires a reference measure (e.g., the Lebesgue measure for a probability measure on Euclidean space). But what does differential entropy measure? Specifying a real number exactly requires an infinite number of bits, so this isn't a useful quantity. Instead, we can compare the information in the random variable $\mathcal{X} \in \mathbb{R}^n$ with the information in a *test random variable*, namely, one that is uniformly distributed in the unit cube in $\mathbb{R}^n$. This difference, which may be positive or negative, is what differential entropy measures. To see why this is what the definition amounts to, take a large integer k and divide up $\mathbb{R}^n$ into a countable set of disjoint cubes C_α with centers P_α and sides of length 2^{-k}. Then the density function $p(x)$ is approximately constant on each C_α, so the information in a random choice of $\mathcal{X}$ is (a) the discrete α for which $\mathcal{X} \in C_\alpha$ plus (b) the residual $\mathcal{X} - P_\alpha$, which is essentially a uniform random variable in a cube of size 2^{-k}. If $p_\alpha = \int_{C_\alpha} p$, then $\sum_\alpha p_\alpha \log_2(1/p_\alpha)$ measures the information in (a), and specifying a point in the small cube C_α requires kn fewer bits than specifying a point in the unit cube in $\mathbb{R}^n$.

Indeed, specifying such points exactly would require an infinite number of bits, but specifying them up to the same decimal precision can be exactly computed as follows: Divide the unit cube of $\mathbb{R}^n$ into small cubes of side length 2^{-N} with N large. There will be 2^{nN} such cubes. The entropy of the uniform distribution on 2^{nN} values is nN; thus, this is the number of bits required to specify a point in the unit cube up to the precision 2^{-N}. Now, to compute the number of bits needed to specify a point in a cube of side length 2^{-k} (with $k < N$) up to the precision 2^{-N}, we compute the entropy of the probability distribution on 2^{nN} values, which is 0 for $2^{Nn} - 2^{(N-k)n}$ of these values and $2^{-(N-k)n}$ for the others. Its entropy is thus $\log_2(2^{(N-k)n}) = nN - kn$. Let us go back to $\mathcal{X}$. Since $p_\alpha \approx p(P_\alpha)2^{-nk}$, we get:

$$\text{(Info in } \mathcal{X}) - \text{(Info in point in unit cube)} \approx \sum p_\alpha \log_2(1/p_\alpha) - kn$$

$$\approx \int p(x) \log_2(1/p(x)) dx.$$

Alternatively, we can say that *the ideal coding of the continuous random variable $\mathcal{X}$ up to an error uniformly distributed in a small set of volume ΔV requires $H_d(g) + \log_2(1/\Delta V)$ bits.*

In the case of a finite probability space Ω with N elements, we already saw in Chapter 1 that, among all probability distributions on Ω, the uniform distribution (for all i, $p_i = 1/N$) is the one with maximal entropy. We see now that in the continuous case, the Gaussian distribution has maximal differential entropy among all distributions with given mean and variance.

Proposition 2.7. *Let $p(x)$ be a probability density on $\mathbb{R}$ with mean $\bar{x}$ and variance σ^2, and let $g(x)$ be the Gaussian distribution with same mean $\bar{x}$ and same variance σ^2:*

$$g(x) = \frac{1}{\sigma\sqrt{2\pi}} e^{-(x-\bar{x})^2/2\sigma^2};$$

then

$$H_d(g) = \log_2 \sqrt{2\pi e} + \log_2 \sigma \geq H_d(p).$$

In particular, g has maximal differential entropy among all distributions with given mean and variance.

Proof: The Kullback-Leibler distance from p to g is defined by

$$D(p\|g) = \int p(x) \log_2 \frac{p(x)}{g(x)},$$

and it is positive (due to the concavity of the log function). But, we may also compute the distance as

$$D(p\|g) = -H_d(p) + \int p(x) \left[\log_2 \sigma\sqrt{2\pi} + \frac{(x-\bar{x})^2}{2\sigma^2} \log_2(e)\right]$$

$$= -H_d(p) + \log_2 \sigma\sqrt{2\pi} + \frac{\sigma^2}{2\sigma^2} \log_2(e).$$

Since the Kullback-Leibler distance from p to g is 0 when $p = g$, we get $H_d(g) = \log_2(\sigma\sqrt{2\pi e})$, and finally

$$D(p\|g) = H_d(g) - H_d(p) \geq 0. \qquad \square$$

2.1. Basics III: Gaussian Distributions

This result can be easily extended to dimension n. Let g be the Gaussian distribution with mean $\bar{x} \in \mathbb{R}^n$ and covariance matrix C. Then the differential entropy of g is

$$H_d(g) = n \log_2 \sqrt{2\pi e} + \frac{1}{2} \log_2(\det C),$$

and it is maximal among all distributions with a given mean and covariance matrix.

Example. In Figure 2.2, we illustrate the discrete coding of a standard normal (which means Gaussian with mean 0 and variance 1) random variable with a prefix code when the range is divided into 11 bins with size $\Delta x = 0.4$, plus 2 tails. On the same figure, we plot both the density function $g(x) \cdot \Delta x$ and 2^{-l_i}, where l_i is the code length of the ith message. Inside each bar i, we give the probability p_i of that bin, and above the bar, we give the corresponding code word. In this case, the expected number of bits $\sum_i p_i l_i$ turns out to be about 3.42. Compare this to the ideal coding length found with differential entropy, which is $H_d(g) + \log_2(1/\Delta x) \approx 3.37$.

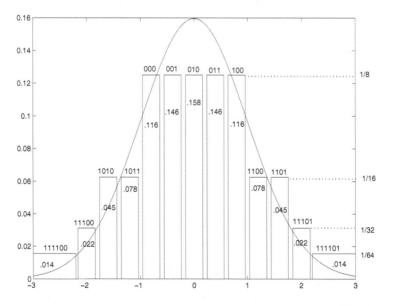

Figure 2.2. Probability density of a standard normal random variable versus $2^{-\text{codelength}}$. Inside each bar i, we give the probability p_i of that bin, and above the bar, we give the corresponding code word.

2.2 Basics IV: Fourier Analysis

Although Fourier series are probably familiar to the reader, when doing pattern theory, we need to use discrete Fourier transforms (Fourier transforms on finite abelian groups), and the transition from infinite to finite can cause much confusion. So we begin by recalling the definition of the Fourier transform in four different situations.

1. If f is a function in $L^2(\mathbb{R})$ then we can go back and forth between f and its Fourier transform $\widehat{f}$ via

$$\widehat{f}(\xi) = \int_{\mathbb{R}} e^{-2\pi i x \xi} f(x) dx; \qquad f(x) = \int_{\mathbb{R}} e^{2\pi i x \xi} \widehat{f}(\xi) d\xi.$$

In this definition, the variable x might represent a time (e.g., in seconds) and then the variable ξ represents a frequency (e.g., in hertz). *Note:* pure mathematicians never write the 2π in the Fourier transform or its inverse. This has to do with their desire to ignore any *units* attached to variables. Their convention has the result that it assigns the frequency of 314 radians per second (instead of 50 oscillations per second) to European alternating current. Also, if you follow their convention, you need to divide by 2π in the inverse Fourier transform.

2. If $(f_n) \in l^2$ is a sequence, then the Fourier transform of (f_n) is the 1-periodic function $\widehat{f}$ related to f by

$$\widehat{f}(\xi) = \sum_{n=-\infty}^{+\infty} e^{-2\pi i n \xi} f_n; \qquad f_n = \int_0^1 e^{2\pi i n \xi} \widehat{f}(\xi) d\xi.$$

3. If f is a periodic function of x with period 1, then the Fourier coefficients $\widehat{f}_n$ of f for $n \in \mathbb{Z}$ and the inversion formula are

$$\widehat{f}_n = \int_0^1 f(x) e^{-2\pi i n x} dx; \qquad f(x) = \sum_{n=-\infty}^{+\infty} \widehat{f}_n e^{2\pi i n x}.$$

4. The finite Fourier transform: if $(f_0, \ldots, f_N)$ is a finite sequence of length N, then its discrete Fourier transform is

$$\widehat{f}_m = \frac{1}{\sqrt{N}} \sum_{n=0}^{N-1} e^{-2\pi i \frac{nm}{N}} f_n; \qquad f_m = \frac{1}{\sqrt{N}} \sum_{n=0}^{N-1} e^{+2\pi i \frac{nm}{N}} \widehat{f}_n.$$

We recall here some of the main properties and a couple of examples of the Fourier transform. (In the following, we denote $f^-(x) = f(-x)$).

2.2. Basics IV: Fourier Analysis

Isometry	$\|f\|_2 = \|\hat{f}\|_2,$	$<f,g> = <\hat{f},\hat{g}>$
Product/Convolution	$\widehat{fg} = \hat{f} * \hat{g},$	$\widehat{f*g} = \hat{f} \cdot \hat{g}$
Symmetry	$\widehat{(f^-)} = \overline{\hat{f}},$	$\hat{\hat{f}} = f^-$
Translation	$\widehat{f(x-a)} = e^{-2\pi i a \xi}\hat{f},$	$\widehat{e^{2\pi i a x}f} = \hat{f}(\xi - a)$
Scaling	$\widehat{f(ax)} = \frac{1}{a}\hat{f}\left(\frac{\xi}{a}\right)$	
Derivatives	$\widehat{f'} = 2\pi i \xi \hat{f}$	$\widehat{xf(x)} = \frac{i}{2\pi}\hat{f}'$
Gaussian	$\widehat{e^{-x^2/2\sigma^2}} = \sqrt{2\pi}\sigma e^{-2\pi^2 \sigma^2 \xi^2}$	
Cauchy	$\widehat{e^{-2\pi|x|}} = \frac{1}{\pi(1+\xi^2)}$	$\widehat{\frac{1}{1+x^2}} = \pi e^{-2\pi|\xi|}$

One of the main signal processing properties of the Fourier transform is the link between the autocorrelation and the power spectrum. Recall that the autocorrelation is given by

$$f * f^-(x) = \int f(y)f(y-x)dy,$$

and thus,

$$\widehat{(f*f^-)} = |\hat{f}|^2,$$

where $|\hat{f}|^2$ is the power spectrum. Often a function has different oscillatory properties in different parts of its domain, and to describe this we need to define another variant of the Fourier transform, the *windowed Fourier transform*.

Example. We consider a signal f defined for all $x \in \mathbb{R}$, that need not die at infinity. We choose a window function w, which will generally look like Figure 2.3. Then the windowed Fourier transform of f around point a and at frequency ξ is

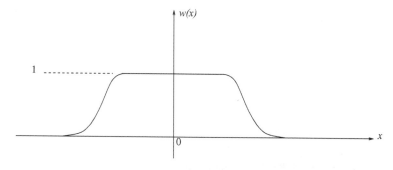

Figure 2.3. Graph of a typical window function w.

defined as $\widehat{f_a}(\xi)$, where $f_a(x) = w(x-a)f(x)$. By using the property of product/convolution conversion of the Fourier transform, we get

$$\widehat{f_a} = \widehat{w(x-a)} * \widehat{f}.$$

To work out a simple case, assume that w is a Gaussian function,

$$w(x) = \frac{1}{\sqrt{2\pi}\sigma}e^{-x^2/2\sigma^2} = g_\sigma(x).$$

With such a choice for w, the size of the window is of the order of σ. Then the Fourier transform of w is

$$\widehat{w(x-a)}(\xi) = e^{-2\pi^2\sigma^2\xi^2} \cdot e^{-2\pi i\xi a}.$$

And so the size of the "support" of $\widehat{w}$ is of the order of $1/\sigma$. So finally $\widehat{f_a} = \widehat{g_\sigma(x-a)} * \widehat{f}$ is a smoothing of $\widehat{f}$ with a kernel of width approximately $1/\sigma$.

It is important to understand the behavior of such a windowed Fourier transform as σ goes to 0 and to infinity. As $\sigma \to 0$, you get a better resolution in time (the window is small), but you get less resolution in frequency (frequencies are spread out over other frequencies). Conversely, as $\sigma \to +\infty$, you get bad resolution in time, but very good resolution in frequency. An example of this with speech is shown below in Section 2.6.2, Figure 2.12. Is it possible to have good resolution in both time and in frequency? The answer to this question is no, and the reason for this is Theorem 2.8.

Theorem 2.8 (Uncertainty Principle). *Suppose that $f \in L^2$ is a real-valued function such that $\int_{-\infty}^{+\infty} f^2(x)dx = 1$, which means therefore that $f^2 dx$ is a probability density in x. If $\overline{x} = \int xf^2 dx$ is its mean, then the standard deviation (SD) of $f^2 dx$ is defined as usual by*

$$SD(f^2 dx) = \sqrt{\int (x-\overline{x})^2 f^2 dx}.$$

Moreover, we have $\int |\widehat{f}(\xi)|^2 d\xi = 1$, which means that $|\widehat{f}(\xi)|^2 d\xi$ is also a probability density, but in the frequency variable ξ. Then

$$SD(f^2 dx) \cdot SD(|\widehat{f}(\xi)|^2 d\xi) \geq \frac{1}{2\pi}.$$

Theorem 2.8 says that $SD(f^2 dx)$ and $SD(|\widehat{f}(\xi)|^2 d\xi)$ cannot both be small, which means that you cannot localize simultaneously in time and in frequency.

2.2. Basics IV: Fourier Analysis

For a proof, see the book by H. Dym and H. McKean, *Fourier Series and Integrals* [65, pp. 116–120].

To get a better idea of how the different Fourier transforms interact, we look at an example where a simple periodic function of a real variable is sampled discretely and where the phenomenon of aliasing occurs.

Example. For some $\omega > 0$, let $f(t) = e^{2i\pi\omega t}$ be a purely periodic signal with frequency ω and period $p = 1/\omega$. Notice that $\widehat{f}(\xi) = \delta_\omega(\xi)$ is the Dirac function at ω (there is only one frequency). Let Δt be a time interval and N a large integer so that $N\Delta t \gg p$. For $0 \leq k < N$, let $f_s(k) = f(k\Delta t)$ be discrete samples of f. By using the discrete Fourier transform, we get for an integer l,

$$|\widehat{f_s}(l)|^2 = \left| \frac{1}{\sqrt{N}} \sum_{k=0}^{N-1} e^{2i\pi\omega k \Delta t} e^{-2i\pi \frac{kl}{N}} \right|^2.$$

Summing this geometric series, we get the following expression:

$$|\widehat{f_s}(l)|^2 = \frac{C}{\sin^2(\pi(\omega\Delta t - \frac{l}{N}))},$$

where C is a constant independent of l. We now distinguish two cases, depending on whether the sampling is dense or sparse:

- *Dense sampling.* $\Delta t \ll p$ or $\omega\Delta t \ll 1$. Then if l_0 is the nearest integer to $\omega\Delta t N$, we have $0 \leq l_0 < N$, and l_0 is the frequency of the signal on the discrete Fourier transform scale. The error between the true frequency ω and the frequency estimated on the discrete scale is small:

$$\left| \omega - \frac{l_0}{N\Delta t} \right| \leq \frac{1}{2N\Delta t}.$$

Moreover, the peak of $|\widehat{f_s}(l)|^2$ is found at $l = l_0$.

- *Sparse sampling.* $\Delta t > p$ or $\omega\Delta t > 1$. Then, instead of finding a peak of the discrete power near ω, you get the existence of two integers l_0 and n_0 such that $0 \leq l_0 < N$ and $\omega\Delta t \approx n_0 + \frac{l_0}{N}$. Thus, the peak of $\widehat{f_s}$ is at $l = l_0$, which means that the peak of $\widehat{f}$ has been shifted far away from the true frequency ω. Notice that in this case the signal of frequency ω has the same samples as the signal of frequency $\omega - \frac{n_0}{\Delta t}$. This is called *aliasing*.

Typical examples of both situations are shown in Figure 2.4.

2. Music and Piecewise Gaussian Models

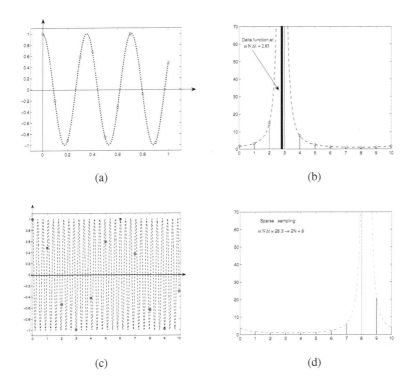

Figure 2.4. Examples of dense and sparse sampling. (a) A plot of the real part of a purely periodic signal (dotted line) $t \to \exp(2i\pi\omega t)$ with $\omega = 2.83, p \approx .35$. It is sampled at points $k\Delta t$ with $\Delta t = 0.1$, and $0 \le k < N$ with $N = 10$ in this example. (b) A plot of the sampled version of the power spectrum; since the sampling is dense ($\omega \Delta t < 1$), there is no aliasing. The peak is found at $l_0 = 3$, which is the closest integer to $\omega N \Delta t = 2.83$. (c) Everything is the same except that $\Delta t = 1$. Since the sampling is now sparse, we see aliasing. (d) The peak of the power is found at $l_0 = 8$, which is the closest integer to $\omega N \Delta t = 28.3$ modulo N.

2.3 Gaussian Models for Single Musical Notes

2.3.1 Gaussian Models for Stationary Finite Cyclic Signals

In this section, we combine the ideas from n-dimensional Gaussian distributions and from discrete Fourier transforms. Let $\vec{s} = (s_1, \ldots, s_N)$ be a periodic signal ($s_{N+1} = s_1$). We first take the Fourier transform $\hat{s}$ of $\vec{s}$. Usually we think of $\hat{s}$ as a new vector in $\mathbb{C}^N$, but instead we can now regard $\hat{s}$ simply as the coefficients of $\vec{s}$ when it is expanded in a new orthonormal basis, a rotated version of the standard unit vectors.

2.3. Gaussian Models for Single Musical Notes

The usual canonical basis of $\mathbb{C}^N$ is $\vec{e}^{(1)}, \ldots, \vec{e}^{(N)}$, where $\vec{e}^{(k)} = (0,\ldots,1,\ldots,0)$ (the 1 is at the kth place). This basis is orthonormal. But instead we can choose another orthonormal basis: $\vec{f}^{(0)}, \ldots, \vec{f}^{(N-1)}$, where $\vec{f}^{(k)}$ is defined for $0 \leq k \leq N-1$ as

$$\vec{f}^{(k)} = \frac{1}{\sqrt{N}}(1, e^{2i\pi \frac{k}{N}}, \ldots, e^{2i\pi \frac{k(N-1)}{N}}).$$

This basis is the Fourier basis. If $\vec{s}$ is the signal, in the canonical basis we have $\vec{s} = \sum_{k=1}^{N} s_k \vec{e}^{(k)}$, and in the Fourier basis (using the inverse Fourier transform), we get

$$\vec{s} = \sum_{l=0}^{N-1} \widehat{s}_l \vec{f}^{(l)}.$$

Notice that if the signal $\vec{s}$ is real, then it has the property that $\widehat{s_{N-l}} = \overline{\widehat{s}_l}$. (This is analogous to the usual equivalence for the real Fourier transform: f is real iff $\widehat{f}$ satisfies $\widehat{f}(-\xi) = \overline{\widehat{f}(\xi)}$.)

Now, let us assume that $\vec{s}$ follows a Gaussian distribution with density

$$p_Q(\vec{s}) = \frac{1}{Z} e^{-(\vec{s}-\vec{m})^t Q \overline{(\vec{s}-\vec{m})}/2}.$$

Here $\vec{s}$ may be a vector of $\mathbb{C}^n$, and in this case we have to assume that the matrix Q (which may have complex entries) is a Hermitian positive definite matrix. This simply extends real Gaussian distributions (see Definition 2.1) to complex Gaussian distributions. We do this not because we think that the real-world $\vec{s}$ are like this, but because Gaussian distributions are the simplest distributions for continuous variables and we want to see how well this works.

Definition 2.9. We say that the Gaussian distribution p_Q is stationary if it satisfies, for all integer l,

$$p_Q(T_l \vec{s}) = p_Q(\vec{s}),$$

where $(T_l \vec{s})_k = s_{k-l}$, and $k-l$ means $(k-l) \mod N$.

Using the change to the Fourier basis, we get $\widehat{Q}$ and $\widehat{m}$ such that

$$p_Q(\vec{s}) = \frac{1}{Z} e^{-(\widehat{s}-\widehat{m})^t \widehat{Q} \overline{(\widehat{s}-\widehat{m})}/2}.$$

We then have the following theorem.

Theorem 2.10. *The Gaussian distribution p_Q is stationary iff $\vec{m}$ is a constant signal and Q is a Hermitian banded matrix (meaning $Q_{i,i+j}$ depends only on j mod N, and if it is denoted by a_j then $a_j = \overline{a_{N-j}}$). Equivalently, in the Fourier basis, $\widehat{m} = (\widehat{m_0}, 0, \ldots, 0)$ and $\widehat{Q}$ is a real positive diagonal matrix.*

Proof: Since a Gaussian distribution is uniquely determined by its mean and covariance, p_Q is stationary iff $\vec{m}$ is invariant under translations (which means $\vec{m} = (m_0, \ldots, m_0)$ and so $\hat{m} = (\widehat{m_0}, 0, \ldots, 0)$), and for all i, j, l we have $Q_{i+l, j+l} = Q_{i,j}$ (which means that the matrix Q is banded). If Q is banded, then defining $a_j = Q_{i, i+j}$ (independent of j), we get $a_j = \overline{a_{N-j}}$ (because Q is also Hermitian meaning $Q^t = \overline{Q}$), and

$$(\vec{s} - \vec{m})^t Q \overline{(\vec{s} - \vec{m})} = \sum_{i,j} a_j (s_i - m_i) \overline{(s_{i+j} - m_{i+j})}.$$

In the Fourier basis, $\vec{s} = \sum_l \hat{s}_l \vec{f}^{(l)}$ and so $s_k = \sum_l \hat{s}_l f_k^{(l)}$ (where $f_k^{(l)} = e^{2i\pi \frac{kl}{N}}$ is the kth component of $\vec{f}^{(l)}$ in the canonical basis of $\mathbb{C}^n$). Then

$$\begin{aligned}
(\vec{s} - \vec{m})^t Q \overline{(\vec{s} - \vec{m})} &= \sum_{i,j} a_j \sum_{k,l} (\hat{s}_l - \widehat{m}_l)(\overline{\hat{s}_k - \widehat{m}_k}) f_i^{(l)} \overline{f_{i+j}^{(k)}} \\
&= \sum_{k,l} (\hat{s}_l - \widehat{m}_l)(\overline{\hat{s}_k - \widehat{m}_k}) \sum_j a_j e^{-2i\pi \frac{jk}{N}} <\vec{f}^{(l)}, \vec{f}^{(k)}>.
\end{aligned}$$

Since $<\vec{f}^{(l)}, \vec{f}^{(k)}> = 0$ when $k \neq l$, we finally have

$$(\vec{s} - \vec{m})^t Q \overline{(\vec{s} - \vec{m})} = \sum_l \hat{a}_l |\hat{s}_l - \widehat{m}_l|^2,$$

which means that $\hat{Q}$ is diagonal, real (because $a_j = \overline{a_{N-j}}$), and positive (because Q is). $\square$

Such a stationary distribution can be written as

$$p(\vec{s}) = \frac{1}{Z} e^{-\sum_l |\hat{s}_l|^2 / 2\sigma_l^2}.$$

Then the real and imaginary parts $\Re \hat{s}_l$ and $\Im \hat{s}_l$, respectively, are independent Gaussian random variables with mean 0 and standard deviation σ_l. Such a distribution is called *colored noise*. The particular case $Q = I_N$, $\hat{Q} = I_N$, and

$$p(\vec{s}) = \frac{1}{Z} e^{-\sum_l |\hat{s}_l|^2 / 2}$$

is called *white noise*. In general, its differential entropy is

$$\frac{N}{2} \log_2(2\pi e) + \sum_{l=0}^{l=N-1} \log_2(\sigma_l).$$

Some examples of colored noise are shown in Figure 2.5.

2.3. Gaussian Models for Single Musical Notes

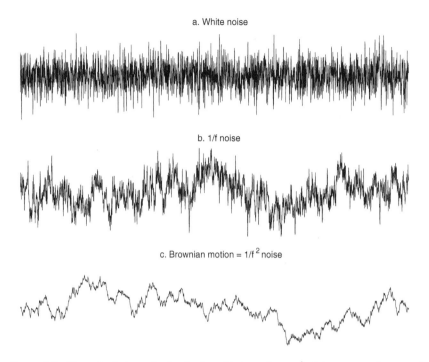

Figure 2.5. Three simulations of colored noise with the variance σ_l^2 of the power at frequency l falling off as $1/l^\alpha$, for (a) $\alpha = 0$, (b) $\alpha = 1$, and (c) $\alpha = 2$. Such noises are usually called $1/f^\alpha$ noises.

2.3.2 The Case of a Musical Note

We return to the problem of finding a stochastic model for music. We first construct a Gaussian model of a single note. Let ω be the fundamental frequency of the note being played and $p = 1/\omega$ be its period. If the signal is $s(t)$ then $s(t+p) \simeq s(t)$, which means that the signal is close to being periodic, although in real music there are always small residual variations. See Figure 2.6 for an example of such a signal taken from live music. Some deviations from perfect periodicity are shown in Graph (b). With some averaging, we can make the signal periodic and then expand it in a Fourier series with frequencies n/p. Its nth component is known as the nth harmonic. In Figure 2.6, all but three terms in the Fourier series are quite small.

How do we make a Gaussian model for this signal? We formalize the property $s(t+p) \simeq s(t)$ by assuming that the expected value of $\int (s(t+p) - s(t))^2 dt$ is quite small. We then constrain the expected total power of the signal by bounding $\int s(t)^2 dt$.

a. Six periods of a female voice singing the note sol

b. One period of the averaged detrended signal, compared to 4 samples

c. Three periods of the average signal (bold), the first harmonic (regular), and the residual (dashed)

d. Three periods of the above residual (bold), the second harmonic (regular), and the remaining residual (dashed)

Figure 2.6. (a) A small sample of vocal music; (b) the variability of individual periods due to sensor noise, background room noise, and variations in the singing; (c) and (d) the decomposition of the averaged recording into the first three harmonics (defined as the integer multiples of the fundamental frequency), which contain almost all the power.

Let us take a discrete sample of the signal s and, for simplicity, let's assume that s "wraps around" at some large integer N (i.e., $s_{N+k} = s_k$) and that p is an integer dividing N. Let $q = N/p$, the number of cycles present in the whole sample. We'll now analyze the simplest possible Gaussian model for s that gives samples that are periodic plus some small residual noise. Its density is

$$p_{a,b}(s) = \frac{1}{Z} e^{-a \sum_{k=0}^{N-1} (s(k)-s(k+p))^2/2 - b \sum_{k=0}^{N-1} s(k)^2/2} = \frac{1}{Z} e^{-\vec{s}^t Q \vec{s}/2},$$

where $a \gg b > 0$, $Q_{i,i} = b + 2a$, and $Q_{i,i+p} = -a$, for $0 \leq i \leq N-1$ and otherwise 0. Notice that Q is a positive definite quadratic form (if there is no term $b \sum_{k=0}^{N-1} s(k)^2/2$, then the quadratic form is only semi-definite). Then $p_{a,b}(s)$ is a stationary probability distribution, and thus we can diagonalize the quadratic form in the Fourier basis.

On the one hand, we have

$$\sum_k (s(k) - s(k+p))^2 = \| s - T_{-p}(s) \|^2 = \| \widehat{s} - \widehat{T_{-p}(s)} \|^2.$$

2.3. Gaussian Models for Single Musical Notes

By using the fact that $\widehat{s}(l) - \widehat{T_{-p}(s)}(l) = \widehat{s}(l)(1 - e^{2i\pi \frac{pl}{N}})$, we get

$$\sum_k (s(k) - s(k+p))^2 = \sum_l |\widehat{s}(l)|^2 |1 - e^{2i\pi \frac{pl}{N}}|^2 = 4\sum_l |\widehat{s}(l)|^2 \sin^2(\frac{\pi pl}{N}).$$

On the other hand, we have

$$\sum_k s(k)^2 = \sum_l |\widehat{s}(l)|^2,$$

so

$$p_{a,b}(s) = \frac{1}{Z} e^{-\sum_l (b+4a\sin^2(\frac{\pi pl}{N}))|\widehat{s}(l)|^2/2} = \frac{1}{Z} \Pi_l e^{-(b+4a\sin^2(\frac{\pi pl}{N}))|\widehat{s}(l)|^2/2}. \tag{2.1}$$

Then the expected power at frequency l is the mean of $|\widehat{s}(l)|^2$, which works out to be

$$\mathbb{E}(|\widehat{s}(l)|^2) = \frac{1}{b + 4a\sin^2(\frac{\pi pl}{N})}.$$

Note that this has maxima $1/b$ if l is a multiple of N/p (i.e., all frequencies that repeat in each cycle) and that all other powers are much smaller (because $a \gg b$). This is shown in Figure 2.7.

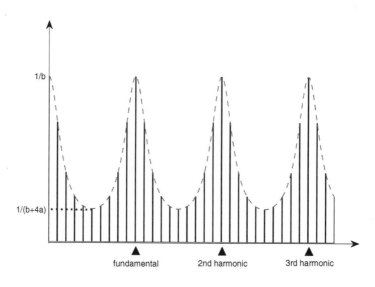

Figure 2.7. Expected power spectrum: $\mathbb{E}(|\widehat{s}(l)|^2) = 1/(b + 4a\sin^2(\frac{\pi pl}{N}))$.

It is interesting to compare the differential entropy of this music model with the white noise model given by

$$p_{a,b}(s) = \frac{1}{Z} e^{-b \sum_{k=0}^{N-1} s(k)^2/2}.$$

The difference is the number of bits saved by having this approximate periodicity. To calculate this requires calculating the determinant of Q. This determinant is analyzed for $b \ll a$ in Exercise 4 at the end of this chapter:

$$\det Q = \det \begin{pmatrix} \begin{vmatrix} b+2a & -a & \cdots & 0 & \cdots & 0 & -a \\ -a & b+2a & \cdots & 0 & \cdots & 0 & 0 \\ & \cdots & & \cdots & & \cdots & \\ 0 & 0 & \cdots & b+2a & \cdots & 0 & 0 \\ & \cdots & & \cdots & & & \\ 0 & 0 & \cdots & 0 & \cdots & b+2a & -a \\ -a & 0 & \cdots & 0 & \cdots 0 & -a & b+2a \end{vmatrix} \end{pmatrix}^p$$

$$= \left(q^2 b a^{q-1}\right)^p + O(b^{p+1}).$$

With this result, we get

$$H(\text{white noise}) - H(\text{music model}) = \left(\frac{N}{2} \log_2(2\pi e/b)\right)$$
$$- \left(\frac{N}{2} \log_2(2\pi e) + \frac{1}{2} \log_2(\det Q^{-1})\right)$$
$$\approx \frac{N-p}{2} \log_2\left(\frac{a}{b}\right) + p \log_2 q.$$

This is the information-theoretic statement of the fact that at $N - p$ frequencies, the power is reduced by about a/b.

This is not, however, an accurate model of real musical notes because the power in all harmonics (integer multiples of the fundamental frequency) is equally large. Typically, the power falls off at high frequencies; for example in Figure 2.7, there is significant power only in the fundamental, second, and third harmonics. This is reflected in another fact: notice that in our model, the Gaussian only has terms $s_k s_l$ if p divides $k - l$. Thus the restriction of the signal to each congruence class mod p (i.e., $\{s_{a+kp} | 0 \le k < N/p\}$) are independent parts of the signal. In particular, the restriction of the signal to one period is white noise. It is easy to change this and include extra parameters for the expected power of the various harmonics by using the second expression in Equation (2.1).

2.4 Discontinuities in One-Dimensional Signals

So far we have looked at single notes in music. But just as English text is broken up into words, sentences, paragraphs, and so forth, music is broken up into notes, measures, melodies, etc., and speech is broken up into phonemes, words, sentences, speech acts, etc. As discussed in Chapter 0, all natural signals tend to have a hierarchical structure of distinct parts, separated by various kinds of discontinuities, which we called hierarchical geometric patterns. In Chapter 1, we studied the statistical traces of the word boundaries in minima of log-likelihood ratios, which we called the binding energy. In this chapter, we are looking at continuous real-valued one-dimensional signals—usually functions of time and usually sampled—and here too there is a simple statistic that marks the presence of discontinuities, which we explain in this section.

First, let's look at the data. The windowed Fourier transform is a great way to "see" the discontinuities, that is, the notes in the music. This works best if we graph the log of the power in the windowed Fourier transform:

$$LP(a, \omega) = \log \left| \widehat{f_a}(\omega) \right|^2, \text{ where } f_a(x) = w(x-a)f(x).$$

This function, for five seconds of oboe playing, is shown in Figure 2.8. The temporal width of the window must be chosen carefully to resolve the changes in power in both time and frequency in the signal. We discuss this further in Section 2.6.2 (see Figure 2.12).

Figure 2.8 shows clearly that the power spectrum changes suddenly at a discrete set of times, where a new note is played. A slice of this spectrum at the fixed frequency 1220 hertz is shown in the lower graph. This particular frequency happens to be the second or third harmonic of some of the notes being played, so the power hovers at low levels and then suddenly increases when these notes are played. The power also decreases *during* some of these notes.

How do we detect when discontinuities are present and what are the simplest sort of mathematical models for them? Discontinuities are a phenomenon of functions of a real variable, so they are not really well defined when we are measuring discrete samples of a signal. One way to analyze this is to go to the continuum limit and study signals as functions of a real variable, lying in an infinite dimensional function space. The resulting random functions are called stochastic processes; these can be proven to be continuous in some cases and discontinuous in others. We introduce this theory in Chapter 3. Another way, which is based directly on the data, is to consider discontinuities as unusually large jumps in the data. Here we study this approach, based on *kurtosis*, a basic invariant of probability distributions on the real line.

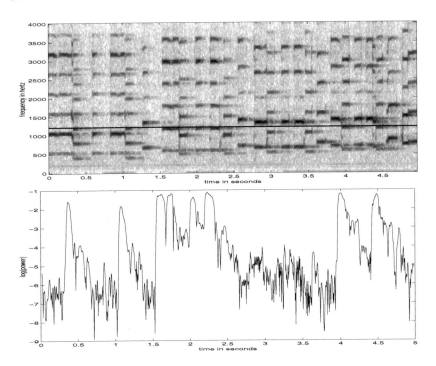

Figure 2.8. The log-power spectrum for five5 seconds of oboe, sampled at 8000 hertz, using a 32-millisecond window (top). The power at the fixed frequency 1220 hertz, corresponding to the black line on the top figure (bottom).

2.4.1 Basics V: Kurtosis

Definition 2.11. If $p(x)$ is a probability density on the real line with mean $\bar{x}$, then the *kurtosis* of p is its normalized fourth moment:

$$\kappa(p) = \frac{\int (x-\bar{x})^4\, p(x)\, dx}{\left(\int (x-\bar{x})^2\, p(x)\, dx\right)^2} = \frac{\mathbb{E}((\mathcal{X}-\bar{x})^4)}{(\mathbb{E}((\mathcal{X}-\bar{x})^2))^2},$$

where $\mathcal{X}$ is a random variable with probability density p.

Note that subtracting the mean and dividing by the square of the variance has the effect that the kurtosis of a random variable $\mathcal{X}$ and of any linear transform $a\mathcal{X}+b$ are equal. In other words, kurtosis is a dimensionless invariant of the distribution.

2.4. Discontinuities in One-Dimensional Signals

Example. We consider three examples of kurtosis.

- *Calculation of the kurtosis of a normal distribution.*

Let $p(x)\,dx$ be the normal distribution of mean $\bar{x}$ and standard deviation σ:

$$p(x) = \frac{1}{\sigma\sqrt{2\pi}} e^{-(x-\bar{x})^2/2\sigma^2}.$$

Then by change of variable $y = (x - \bar{x})/\sigma$, we get

$$\kappa = \frac{\int (x-\bar{x})^4 \, p(x)\,dx}{\left(\int (x-\bar{x})^2 \, p(x)\,dx\right)^2} = \frac{2\pi\sigma^2 \times \sigma^5}{\sigma\sqrt{2\pi} \times \sigma^6} \frac{\int y^4 e^{-y^2/2} dy}{\left(\int y^2 e^{-y^2/2} dy\right)^2}.$$

Now, by integration by parts, we have

$$\int y^4 e^{-y^2/2} dy = 3 \int y^2 e^{-y^2/2} dy = 3\sqrt{2\pi},$$

and so finally,

$$\kappa(\text{any normal distribution}) = 3.$$

- *Calculation of the kurtosis of a uniform distribution.*

Let $p(x)\,dx$ be the uniform distribution on $[-\frac{a}{2}, +\frac{a}{2}]$ (i.e., $p(x) = \frac{1}{a}$ if $x \in [-\frac{a}{2}, +\frac{a}{2}]$, and 0 else). Then we find

$$\kappa(\text{any uniform distribution}) = \frac{\int (x-\bar{x})^4 p(x) dx}{\left(\int (x-\bar{x})^2 p(x) dx\right)^2} = \frac{a^2}{a} \frac{\int_{-a/2}^{a/2} x^4 \, dx}{\left(\int_{-a/2}^{a/2} x^2 \, dx\right)^2}$$

$$= \frac{a \frac{2}{5} \times \left(\frac{a}{2}\right)^5}{\left(\frac{2}{3} \left(\frac{a}{2}\right)^3\right)^2} = \frac{9}{5}.$$

- *Calculation of the kurtosis of the Laplace distributions.*

Let $p(x)dx$ be the Laplace (or double exponential) distribution

$$p(x) = \frac{1}{Z} e^{-|x|^\alpha}.$$

Then changing variables to $y = x^\alpha$, we find

$$\int_{-\infty}^{\infty} x^{2n} e^{-|x|^\alpha} dx = \frac{2}{\alpha} \int_0^\infty y^{\frac{2n+1}{\alpha} - 1} e^{-y} dy = \frac{2}{\alpha} \Gamma\left(\frac{2n+1}{\alpha}\right).$$

This shows that the normalizing constant Z equals $\frac{2}{\alpha}\Gamma(\frac{1}{\alpha})$ and that the kurtosis is

$$\kappa(\text{Laplace distributions}) = \frac{\Gamma(5/\alpha)\Gamma(1/\alpha)}{\Gamma(3/\alpha)^2},$$

which is 6 for $\alpha = 1$, 25.2 for $\alpha = .5$, and goes to infinity as α goes to 0.

In Figure 2.9, we compare four distributions with the Gaussian with the same mean and variance: (a) a uniform distribution, (b) the Laplace distribution $p(x) = e^{-|x|}/2$, (c) the empirical distributions of log price changes of seven stocks, and (d) the log power changes of the oboe at 1220 hertz. The first (a) has kurtosis $9/5 = 1.8$, and the second (b) has kurtosis 6. Low kurtosis arises if the density function $p(x)$ has shoulders above the normal but smaller values in the middle and in the tails; high kurtosis arises if $p(x)$ has a peak near 0 and heavy tails, but lower shoulders.

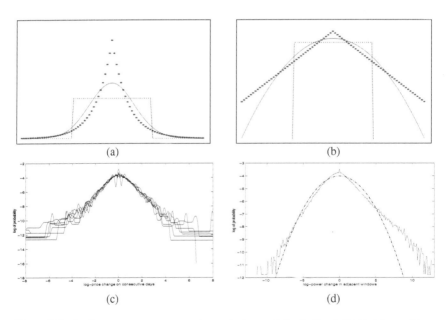

Figure 2.9. On the top row, we compare the standard normal distribution (solid line) to a uniform distribution (dotted line) and a double exponential (starred line) with matched mean and variance: (a) the probability densities; (b) the *logs* of these curves. Thus, the log of the normal gives an inverted parabola. The fact that the double exponential has heavier tails than the normal is much more evident in the log plot. On the bottom row are two plots of log histograms (they plot the log of the frequency, not the frequency) of changes in time series: (c) the log price changes of seven stocks—this is a statistic with one of the fattest tails that we have found; (d) the log power changes of an oboe and, for comparison, the log of the normal distribution with the same mean and variance (the inverted parabola). Note how the music has large tails on the positive side from the sharp note onsets.

2.4. Discontinuities in One-Dimensional Signals

To summarize, distributions with large tails and peaks at their mean but small shoulders have high kurtosis; distributions with negligible tails, big shoulders, and flat around their mean (or even small at their mean, as in bimodal distributions) have small kurtosis.

2.4.2 Measuring the Probability of Jumps via Kurtosis

Kurtosis is the lodestone of discontinuous behavior due to the following reasoning. We start with a random variable $\mathcal{X}$ that measures the change in some quantity from one measurement to the next; that is, specific values are given by $X_i = f((i+1)\Delta t) - f(i\Delta t)$. Here f could be a measurement of sound, light, or pressure (touch), or it could be a measurement of the weather or the price of something. In all these cases, in principle, one could measure them more often with a new Δt, say $\Delta t' = \Delta t/n$. Then we get changes over smaller intervals $Y_j = f((j+1)\Delta t') - f(j\Delta t')$, and clearly,

$$X_i = Y_{ni} + Y_{ni+1} + \cdots + Y_{n(i+1)-1}.$$

Now if the process is stationary (i.e., not anchored to any specific time or origin), it is reasonable to assume that, as f is varied, all the X_i have the same distribution and so do all the Y_j. Then the "zeroth" order model assumes the X_i and the Y_j are independent, and thus independent, identically distributed (iid). Of course, this is usually only an approximation because various higher-order dependencies are present. But let us make this assumption now. Then we have a decomposition of X_i into the sum of n iid variables Y_j, obtained simply by refining the sampling rate. Let's see what happens to the low-order moments of a random variable when this holds.

Proposition 2.12. *Let $\mathcal{X}$ be a real-valued random variable and assume*

$$\mathcal{X} = \mathcal{Y}_1 + \cdots + \mathcal{Y}_n$$

where the $\mathcal{Y}_i$ are iid samples of the same random variable $\mathcal{Y}$. Then

1. $\overline{\mathcal{Y}} = \overline{\mathcal{X}}/n$,

2. $\sigma^2(\mathcal{Y}) = \sigma^2(\mathcal{X})/n$,

3. $\mathbb{E}((\mathcal{Y} - \overline{\mathcal{Y}})^4) - 3\sigma^4(\mathcal{Y}) = \frac{1}{n}\left(\mathbb{E}((\mathcal{X} - \overline{\mathcal{X}})^4) - 3\sigma^4(\mathcal{X})\right)$, *hence* $\kappa(\mathcal{Y}) - 3 = n(\kappa(\mathcal{X}) - 3)$.

Proof: The first formula is clear. To handle higher moments, we use the fact that since the $\mathcal{Y}_i$ are independent, we have $\mathbb{E}\prod_k(\mathcal{Y}_{i_k} - \overline{\mathcal{Y}})^{n_k} = \prod_k \mathbb{E}(\mathcal{Y}_{i_k} - \overline{\mathcal{Y}})^{n_k}$, and this is 0 if any of the n_k are 1. Thus,

$$\mathbb{E}\left((\mathcal{X} - \overline{\mathcal{X}})^2\right) = \mathbb{E}\left(\left(\sum(\mathcal{Y}_i - \overline{\mathcal{Y}})\right)^2\right) = \sum_i \mathbb{E}\left((\mathcal{Y}_i - \overline{\mathcal{Y}})^2\right) = n\sigma^2(\mathcal{Y}),$$

$$\begin{aligned}\mathbb{E}\left((\mathcal{X} - \overline{\mathcal{X}})^4\right) &= \mathbb{E}\left(\left(\sum(\mathcal{Y}_i - \overline{\mathcal{Y}})\right)^4\right) \\ &= \sum_i \mathbb{E}\left((\mathcal{Y}_i - \overline{\mathcal{Y}})^4\right) + 6\sum_{i \neq j} \mathbb{E}\left((\mathcal{Y}_i - \overline{\mathcal{Y}})^2\right)\mathbb{E}\left((\mathcal{Y}_j - \overline{\mathcal{Y}})^2\right) \\ &= n\mathbb{E}((\mathcal{Y} - \overline{\mathcal{Y}})^4) + 3n(n-1)\sigma^4(\mathcal{Y})\end{aligned}$$

from which the second and third formulas follow. □

The so-called *cumulants* provide a simple way of relating the higher moments of $\mathcal{X}$ and $\mathcal{Y}$, but we won't need this. The essential point here is that if the kurtosis of the distribution of the long time interval changes $\mathcal{X}_i$ is larger than 3, then the kurtosis of the short time interval changes $\mathcal{Y}_j$ is even bigger and goes to infinity with n. In other words, whenever the kurtosis is bigger than that of the normal distribution, we find arbitrarily large kurtosis in the short time interval changes. This is caused by the $\mathcal{Y}_j$ having heavier and heavier tails, as is made apparent by the following theorem:

Theorem 2.13. *Let $\mathcal{X}$ be a random variable with mean 0 and variance σ^2, and assume $b = \mathbb{E}(\mathcal{X}^4) - 3\sigma^4 > 0$. Choose μ large enough so that $\mu \geq 8\sigma$ and*

$$\mathbb{E}(\mathcal{X}^4 \cdot I_{(\mu,\infty)}) \leq b/64.$$

Then if, for any n, the random variable $\mathcal{X}$ can be written as a sum $\mathcal{X} = \mathcal{Y}_1 + \cdots + \mathcal{Y}_n$ of independent identically distributed $\mathcal{Y}_i$, the following holds:

$$\mathbb{P}\left(\max_i |\mathcal{Y}_i| \geq \sqrt{\frac{b}{2\sigma^2}}\right) \geq \frac{b}{16\mu^4}.$$

Proof: The proof is elementary but a bit long. It is in four parts: first, we derive a crude bound on the tails of the distribution of $|\mathcal{Y}_i|$; second, we deduce from this a crude bound on the tails of the fourth moment of $|\mathcal{Y}_i|$; third, we have a central lemma giving a lower bound on tails, which we apply to $\mathcal{Y}_i^2$; and fourth, we put this all together.

1. Measure the tails of $\mathcal{X}$ and $\mathcal{Y}_i$ by $F(x) = \mathbb{P}(|\mathcal{X}| \geq x), F_n(x) = \mathbb{P}(|\mathcal{Y}_i| \geq x)$. Then

$$F_n(x) \leq \frac{4}{n} F(x - \sqrt{2}\sigma), \text{ if } x \geq 3\sqrt{2}\sigma.$$

2.4. Discontinuities in One-Dimensional Signals

For any i, we can write $\mathcal{X} = \mathcal{Y}_i + \mathcal{Z}_i$ where $\mathcal{Z}_i$ is the sum of all the $\mathcal{Y}$'s except $\mathcal{Y}_i$. Hence if $|\mathcal{Y}_i| \geq x$ and $|\mathcal{Z}_i| \leq \sqrt{2}\sigma$ for any i, it follows that $|\mathcal{X}| \geq x - \sqrt{2}\sigma$. But we know $\mathbb{P}(\max_i |\mathcal{Y}_i| \geq x) = 1 - (1 - F_n(x))^n$ and, by Markov's inequality, we also know $\mathbb{P}(|\mathcal{Z}_i| \geq \sqrt{2}\sigma) \leq \mathbb{P}(\mathcal{Z}^2 \geq 2\sigma^2(\mathcal{Z}_i)) \leq 1/2$. Therefore,

$$\frac{1}{2}\left(1 - (1 - F_n(x))^n\right) \leq F(x - \sqrt{2}\sigma).$$

Now, using $x \geq 3\sqrt{2}\sigma$, we check that $F(x - \sqrt{2}\sigma) \leq F(2\sqrt{2}\sigma) = \mathbb{P}(\mathcal{X}^2 \geq 8\sigma^2) \leq 1/8$ by Markov again. Using this plus the simple inequalities $\log(1+x) \leq x$ and $-2x \leq \log(1-x)$ if $0 \leq x \leq 1/2$, we get the result we want:

$$-nF_n(x) \geq n\log(1 - F_n(x)) \geq \log\left(1 - 2F(x - \sqrt{2}\sigma)\right) \geq -4F(x - \sqrt{2}\sigma).$$

2. We use a simple integration by parts fact: for any random variable $\mathcal{Z}$ with $G(z) = \mathbb{P}(\mathcal{Z} \geq z)$,

$$4\int_a^\infty z^3 G(z)dz = \mathbb{E}\left((\mathcal{Z}^4 - a^4)I_{(a,\infty)}\right).$$

If $\nu = \mu + \sqrt{2}\sigma$, it follows from the assumptions on μ that:

$$\mathbb{E}\left(\mathcal{Y}_i^4 \cdot I_{(\sqrt{2}\mu,\infty)}\right) \leq 2 \cdot \mathbb{E}\left((\mathcal{Y}_i^4 - \nu^4)I_{(\nu,\infty)}\right) = 8\int_\nu^\infty y^3 F_n(y)dy$$

$$\leq \frac{32}{n}\int_\nu^\infty x^3 F(x - \sqrt{2}\sigma)dx = \frac{32}{n}\int_\mu^\infty (x + \sqrt{2}\sigma)^3 F(x)dx$$

$$\leq \frac{64}{n}\int_\mu^\infty x^3 F(x)dx = \frac{16}{n} \cdot \mathbb{E}\left((\mathcal{X}^4 - \mu^4)I_{(\mu,\infty)}\right) \leq \frac{b}{4n}.$$

3. This is a lemma: given a positive random variable $\mathcal{Z}$, let $\lambda = \mathbb{E}(\mathcal{Z}^2)/(2\mathbb{E}(\mathcal{Z}))$; then, if $\nu \geq \lambda$ is large enough so that $\mathbb{E}(\mathcal{Z}^2 I_{(\nu,\infty)}) \leq \mathbb{E}(\mathcal{Z}^2)/4$,

$$\mathbb{E}(I_{(\lambda,\infty)}(\mathcal{Z})) \geq \frac{\mathbb{E}(\mathcal{Z}^2)}{4\nu^2}.$$

Proof of the lemma: Dividing the range of $\mathcal{Z}$ into $(0, \lambda)$, (λ, ν), and (ν, ∞), we get

$$\mathbb{E}(\mathcal{Z}^2) = \mathbb{E}(\mathcal{Z}^2 I_{(0,\lambda)}) + \mathbb{E}(\mathcal{Z}^2 I_{(\lambda,\nu)}) + \mathbb{E}(\mathcal{Z}^2 I_{(\nu,\infty)})$$

$$\leq \lambda \mathbb{E}(\mathcal{Z}) + \nu^2 \mathbb{E}(I_{(\lambda,\infty)}) + \frac{1}{4}\mathbb{E}(\mathcal{Z}^2)$$

$$= \frac{3}{4}\mathbb{E}(\mathcal{Z}^2) + \nu^2 \mathbb{E}(I_{(\lambda,\infty)}).$$

4. We apply the lemma of part 3 to $\mathcal{Z} = \mathcal{Y}_i^2$ and $\nu = \sqrt{2}\mu$. The hypothesis is satisfied because of part 2 and the fact that $\mathbb{E}(\mathcal{Y}_i^4) = b/n + 3\sigma^4/n^2$. Note that the

λ in the lemma is $\frac{b}{2\sigma^2} + \frac{3\sigma^2}{2n}$. The conclusion gives us

$$\mathbb{P}\left(\mathcal{Y}_i^2 > \frac{b}{2\sigma^2}\right) \geq \frac{b}{8n\mu^2}.$$

By using the simple estimate $1 - (1-x)^n \geq nx/2$ if $nx \leq 1$, we get the final result:

$$\mathbb{P}\left(\max_i |\mathcal{Y}_i| > \sqrt{\frac{b}{2\sigma^2}}\right) \geq 1 - \left(1 - \frac{b}{8n\mu^2}\right)^n \geq \frac{b}{16\mu^2} \qquad \square$$

The intuition behind this result is that if the kurtosis of the big changes $\mathcal{X}$ is larger than 3 and if the changes can be arbitrarily finely subdivided into independent small changes, then in the limit there is a positive probability of jumps in the measured quantity, of size at least $\sqrt{\sigma^2(\mathcal{X})(\kappa(X) - 3)/2}$. And if the process goes on forever, (i.e., if we have a sequence $\mathcal{X}_1, \ldots, \mathcal{X}_m$ of changes), then such jumps will occur almost surely as m goes to infinity.

2.5 The Geometric Model for Notes via Poisson Processes

2.5.1 Basics VI: Poisson Processes

The simplest model we can choose for the set of discontinuities is a Poisson process. These processes are precise mathematical description of what it means to throw down random points with a certain density. We denote as $\mathbb{D}$ the set of all countable discrete subsets of $\mathbb{R}$. We want to define a probability measure on $\mathbb{D}$ whose samples will be discrete sets of random points on $\mathbb{R}$ with two properties. First, they have a given density λ, meaning that the expected number of points in every interval $[a,b]$ will be $\lambda(b-a)$. Second, for any two disjoint intervals I, J, the set of points in I and the set of points in J are to be independent of each other. The key ingredient for the construction is knowing what to choose for the probability distribution on the random number $\mathcal{D}_{a,b} = |\mathcal{S} \cap [a,b]|$ of points in the interval $[a,b]$ from a random sample $\mathcal{S}$ from $\mathbb{D}$. To make things work, this must be chosen to be the Poisson distribution with mean $\lambda(b-a)$:

$$\mathbb{P}(\mathcal{D}_{a,b} = d) = e^{-\lambda(b-a)} \frac{(\lambda(b-a))^d}{d!}.$$

We construct a random $\mathcal{S} \in \mathbb{D}$ in three steps:

1. Choose, for every k in $\mathbb{Z}$, an integer d_k following the Poisson distribution with mean λ.

2.5. The Geometric Model for Notes via Poisson Processes

2. Then choose $x_1^{(k)}, \ldots, x_{d_k}^{(k)}$ independent random real numbers in $[k, k+1]$.

3. $\mathcal{S}$ is the union of all these sets: $\mathcal{S} = \{x_l^{(k)}\}$.

This construction uses a particular decomposition of the real line into intervals, namely $[k, k+1]$. Does this affect the constructed random set $\mathcal{S}$ in the sense that the "seams" of the construction can somehow be reconstructed from many samples $\mathcal{S}_k$? The answer is no, in view of Theorem 2.14, the "gluing" theorem.

Theorem 2.14. *Given a density λ and real numbers $a < b < c$, then we construct the same random set of finite points on $[a, c]$ by (a) choosing their number n from the Poisson distribution with mean $\lambda(c - a)$ and then n points uniformly on $[a, c]$, or (b) choosing k, l from Poisson distributions with means $\lambda(b - a)$ and $\lambda(c - b)$ and then choosing k points uniformly from $[a, b]$ and l points uniformly from $[b, c]$.*

Proof: Consider the trivial identity:

$$\left(e^{-\lambda(c-a)} \frac{(\lambda(c-a))^n}{n!}\right) \left(\binom{n}{k} \left(\frac{b-a}{c-a}\right)^k \left(\frac{c-b}{c-a}\right)^{n-k}\right) = \left(e^{-\lambda(b-a)} \frac{(\lambda(b-a))^k}{k!}\right) \left(e^{-\lambda(c-b)} \frac{(\lambda(c-b))^{n-k}}{(n-k)!}\right)$$

The left-hand side represents the product of the probability of choosing n points in the big interval $[a, c]$ with the first construction times the probability that k of n random points in $[a, c]$ land in $[a, b]$ and $n - k$ land in $[b, c]$. The right-hand side represents the product of the probability of choosing k points in $[a, b]$ and $n - k$ points in $[b, c]$ by using the second construction. Also, if we choose an x uniformly in $[a, c]$ and it happens to land in $[a, b]$, then it is a uniform random point in $[a, b]$. $\square$

More generally, we formalize in exactly the same way the idea of a countable random set of points $\mathcal{S}$ in a space Ω with a given expected density function μ. In the general case, let (Ω, μ) be an arbitrary measure space and $\mathbb{D}$ be the set of all countable subsets of Ω. We then have Theorem 2.15.

Theorem 2.15. *There exists a unique probability measure P on $\mathbb{D}$ such that for a random sample $\mathcal{S} \in \mathbb{D}$:*

1. *For all U measurable subsets of Ω, $\mathbb{P}(|\mathcal{S} \cap U| = n) = e^{-\lambda} \frac{\lambda^n}{n!}$, where $\lambda = \mu(U)$; hence, we have $\mathbb{E}(|\mathcal{S} \cap U|) = \lambda = \mu(U)$.*

2. *For all $k \geq 2$ and for all $U_1, U_2, \ldots, U_k$ disjoint measurable subsets of Ω, then $\mathcal{S} \cap U_1, \mathcal{S} \cap U_2 \ldots \mathcal{S} \cap U_k$ are independent.*

The construction and the proof are essentially the same as for the basic case. For more details, see the book *Poisson Processes* by J. Kingman [124].

Can we be more precise about the space on which the Poisson process lives, and thus give a simple expression for its density and calculate its differential entropy? This is easiest to do for a Poisson process on a bounded interval $[0, B]$ in $\mathbb{R}$ with density λ. Then a sample is determined by its cardinality m and the unordered set of points $\{x_1, \cdots, x_m\}$. With probability 1, the points are distinct and there is a unique order for which $0 < x_1 < \cdots < x_m < B$. Thus, the underlying space for the Poisson probability measure is the disjoint union over m, with $0 \leq m < \infty$, of the polyhedral subsets $\mathbb{D}_m$ of $\mathbb{R}^m$ defined by $0 < x_1 < \cdots < x_m < B$. Note that the cube in $\mathbb{R}^m$ given by $0 \leq x_i \leq B$ for all i is the disjoint union of the $m!$ copies of $\mathcal{S}_m$ obtained by permuting the coordinates. Thus the m-dimensional volume of $\mathbb{D}_m$ is $B^m/m!$. The probability density defining the Poisson measure is just

$$dP(m, x_1, \ldots, x_m) = e^{-\lambda B} \lambda^m dx_1 \ldots dx_m.$$

In fact, conditional on m, this is the uniform distribution on the x_i; and, integrating out the x_i, we get $P(m) = (e^{-\lambda B} \lambda^m) \cdot (B^m/m!)$, the Poisson distribution on m with mean λB. In particular, if we denote $S = \{x_1, \ldots, x_m\}$, still ordered, then the density of the probability has the simple form $Ae^{-a|S|}\Pi dx_i$, where A and a are constants (depending on λ), a formula that we use in constructing the model for music in the next sections.

The differential entropy can now be calculated as

$$-\int_{\mathbb{D}} \log_2(dP(m,\vec{x})/\Pi dx) dP(m,\vec{x})$$

$$= \sum_{m=0}^{\infty} \int_{\mathcal{S}_m} e^{-\lambda B} \lambda^m \log_2(e^{\lambda B} \lambda^{-m}) dx_1 \cdots dx_m$$

$$= \sum_{m=0}^{\infty} e^{-\lambda B} \lambda^m \int_{\mathcal{S}_m} (\lambda B \log_2(e) - m \log_2(\lambda)) dx_1 \cdots dx_m$$

$$= \lambda B \log_2(e) - \log_2(\lambda) \sum_{m=0}^{\infty} e^{-\lambda B} (\lambda B)^m m/m!$$

$$= \lambda B \log_2(e/\lambda)$$

Can we understand this intuitively? To do this, we take a small unit of measure Δx, and let's describe a sample S from the Poisson process up to accuracy Δx. Dividing $[0, B]$ into $B/\Delta x$ bins, each of them contains a point of S with probability $\lambda \Delta x$ and contains two samples with negligible probability. Thus, describing the sample S to accuracy Δx is describing $B/\Delta x$ binary variables, which are 1

2.5. The Geometric Model for Notes via Poisson Processes

with probability $\lambda \Delta x$. To describe each one, the number of bits is

$$\lambda \Delta x \log_2 \frac{1}{\lambda \Delta x} + (1 - \lambda \Delta x) \log_2 \frac{1}{(1 - \lambda \Delta x)} \approx \lambda \Delta x \left(-\log_2(\lambda \Delta x) + \log_2(e) \right).$$

Multiplying by the number of bins $B/\Delta x$, we get the differential entropy $\lambda B \log_2(e/\lambda)$ plus the expected number of samples λB times the number of bits for the required accuracy: $-\log_2(\Delta x)$.

To get a discrete model of a Poisson process, we just take $\Delta x = 1$, assume $\lambda \ll 1$, and describe each point of the sample $\mathcal{S}$ by the integral index $\mathcal{K}_l$ of the bin to which it belongs. Thus $\mathcal{S} = \{0 < \mathcal{K}_1 < \mathcal{K}_2 < \cdots < \mathcal{K}_m \leq B\}$ and

$$P(\mathcal{S}) = \lambda^m (1 - \lambda)^{B-m} = \frac{1}{Z} e^{-a|\mathcal{S}|},$$

where $m = |\mathcal{S}|$, $a = \log((1-\lambda)/\lambda)$, and $Z = (1-\lambda)^{-B}$, similar to the Poisson density worked out above.

2.5.2 The Model for Music

We construct the model for music in two stages. We recall that for this model (described in the introduction to this chapter), we need the sampled sound signal $\vec{s}$ and hidden random variables: the number of notes m, the times $\vec{t}$ where new notes begin, and the periods $\vec{p}$ of the notes.

The probability distribution $p(\vec{s}, m, \vec{t}, \vec{p})$ can be decomposed in the following way:

$$p(\vec{s}, m, \vec{t}, \vec{p}) = \prod_{l=1}^{m} p\left((\vec{s}|_{I_l}) \,\Big|\, p_l, t_l, t_{l+1} \right) \cdot p(\vec{p}, \vec{t}, m), \quad I_l = \{t \mid t_l \leq t < t_{l+1}\}.$$

In Section 2.3, we already constructed a Gaussian model for $p(s|_{[t_l,\ldots,t_{l+1}-1]} \mid p_l, t_l, t_{l+1})$:

$$p(s|_{[t_l,\ldots,t_{l+1}-1]} \mid p_l, t_l, t_{l+1}) = \frac{1}{Z} \exp\left(-a \sum_{n=t_l}^{t_{l+1}-p_l-1} (s_{n+p_l} - s_n)^2/2 - b \sum_{n=t_l}^{t_{l+1}-1} s_t^2/2 \right),$$

where $a \gg b$. Notice that this model is a value model because it constrains value patterns, and that $p(\vec{p}, \vec{t}, m)$ is a geometric model expressing how the domain $\{1, \cdots, B\}$ of the sampled sound is structured by the music. The simplest geometric model is obtained by taking the random variable $\vec{t}$ to be Poisson and each p_l to be independent of the other periods and uniformly sampled from the set of periods of all the notes the musical instrument is capable of producing (something like "atonal" music). If *per* represents this set of periods, then this gives

$$p(\vec{p}, \vec{t}, m) = \frac{1}{Z} e^{-am} 1\!\!1_{\{\vec{p} \in per^m\}},$$

where $a = \log(|per|) + \log((1-\lambda)/\lambda)$ and $Z = (1-\lambda)^B$.

The dogma of pattern theory is that we should sample from this model. It is not at all difficult to create samples using MATLAB, and in Exercise 3 at the end of this chapter, we describe how to do this. The results, however, are not very convincing—they give a totally atonal, unrhythmic "music," but we can certainly construct various models for music of increasing sophistication that sound better. Possible elaborations and constraints include:

1. a tempered scale, which means that we impose $p_l \approx$ (sampling rate)$/447 \cdot 2^{-f_k/12}$ where $f_k \in \mathbb{Z}$

2. tempo $t_{l+1} - t_l \approx a_l T_0$, where $a_l \in \mathbb{Z}$

3. gettingt a better model for harmonics (e.g., an expected power spectrum such as the one shown in Figure 2.10); but for an instrument, we may also have a resonant frequency enhancing all nearby harmonics of the note, as also shown in Figure 2.10).

Some quite sophisticated models have been studied. See, for instance, the paper of Z. Ghahramani and M. Jordan, "Factorial Hidden Markov Models" [86], which uses Bach chorales for some examples[2].

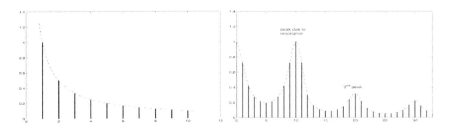

Figure 2.10. Better models for the power of harmonics: a simple decay of higher frequencies (left); resonances of the instrument enhance harmonics near specific frequencies (right).

---

[2]Another nice domain of highly structured music is marching music, which Ulf Grenander used in a student project back in the 1970s.

2.5.3 Finding the Best Possible Score via Dynamic Programming

Since music is a one-dimensional signal, we can compute by dynamic programming the best possible score or the mode of the posterior probability distribution in the hidden variables $m, \vec{p}, \vec{t}$. We might make guesses about the note boundaries and periods based on local evidence, but this is often misleading; if we look at the past and the future, we get less ambiguity. This principle is the same for speech: if we want to recover the phoneme or word pronounced at time t, we have to look at what happened both before and after t before deciding.

Our probability model for music is

$$p(\vec{s}, m, \vec{t}, \vec{p}) =$$
$$Ae^{-Cm} \prod_{k=1}^{m} \frac{1}{Z_k} e^{-a(\sum_{t=t_k}^{t_{k+1}-p_k-1}(s(t+p_k)-s(t))^2/2) - b(\sum_{t=t_k}^{t_{k+1}-1} s(t)^2/2)}.$$

Then if we fix $\vec{s} = \vec{s}_o$ and define $E(m, \vec{t}, \vec{p}) = -\log p(\vec{s}_o, m, \vec{t}, \vec{p})$, we see that it is of the form $\sum_k f(t_k, t_{k+1}, p_k)$. We consider all possible scores on $[1, t]$ including a last note, which ends at time t. The last note has a beginning time of $t' + 1 < t$ and a period p. Then for such scores, we have

$$E = E_1(\vec{s}_o|_{[0,t']}, \text{ notes up to } t') + E_2(\vec{s}_o|_{[t'+1,t]}, p) + E_3(\vec{s}_o \text{ from } t+1 \text{ on}).$$

Here, E_1 assumes the last note ends at t', E_2 assumes there is one note extending from $t'+1$ to t (so it has no other Poisson variables in it), and E_3 assumes a note begins at $t+1$.

Using the algorithm of dynamic programming (introduced in Chapter 1), we compute by induction on t the "best score" for the time interval $[0, t]$ *assuming a note ends at t*. Let $e(t') = \min E_1(\vec{s}_o|_{[0,t']}, \text{ notes up to } t')$ and assume by induction that we know $e(t')$ for all $t' < t$. We then find

$$e(t) = \min_{t' < t, p} [e(t') + E_2(\vec{s}_o|_{[t',t)}, p)],$$

and that continues the induction. Only at the end, however, do we go back and decide where the note boundaries are. In practice, we need to go only a certain amount into the future before the best choice of note boundaries becomes clear.

2.6 Related Models

2.6.1 Other Piecewise Gaussian Models

Many other types of one-dimensional signals can be fit well by piecewise Gaussian signals. We chose music as the theme of this chapter because it is an example

92   2. Music and Piecewise Gaussian Models

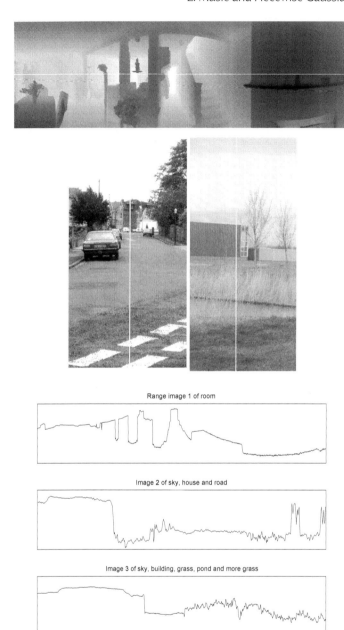

Figure 2.11. A range image of the interior of a house (top) and two usual images of the world (middle). Plots of the one-dimesnioanl slices shown as white lines in the images (bottom). The objects in each image crossed by the slices can be matched with characteristics of the corresponding functions, their values, jumps, and smoothness.

2.6. Related Models

of moderate complexity that is not difficult to implement as an exercise and that then gives good results. Another source of nice examples are one-dimensional slices of two-dimensional images of the world. Some examples are shown in Figure 2.11 and discussed in Exercise 8. These are often fit with much simpler piecewise Gaussian models (see Exercise 5).

Example (White noise with varying means and fixed variance).

The random variables are $\vec{s}$, the signal; $m, \vec{t}$, the number and values of the points of discontinuity; and the means $\vec{\mu}$ in each interval. We need to put some distribution on the means; this is known as a *hyperprior*, a prior probability distribution on the parameters of the model. We simply assume the means are independent and uniformly distributed in some interval M. Then the full model is given by

$$p(\vec{s}, m, \vec{t}, \vec{\mu}) = \frac{(1-\lambda)^{N-1}}{(\sqrt{2\pi}\sigma)^N} \cdot \prod_{l=0}^{l=m} e^{-\sum_{n=t_l}^{t_{l+1}-1}(s_n-\mu_l)^2/2\sigma^2} \cdot \frac{1}{|M|^{m+1}} \left(\frac{\lambda}{1-\lambda}\right)^m,$$

where $1 = t_0 < t_1 < \cdots < t_m < t_{m+1} = N+1$.

Minimizing this value, for fixed $\vec{s} = \vec{s}_o$, we find that μ_l is just the mean $\bar{s}_l$ of $\vec{s}_o\big|_{[t_l,t_{l+1})}$, and this gives the problem of minimizing with respect to the $\vec{t}$:

$$E(\vec{t}) = \sum_{l=0}^{l=m} \sum_{n=t_l}^{n=t_{l+1}-1} (\vec{s}_o(n) - \bar{s}_l)^2 + am,$$

where $a = 2\sigma^2 \left(\log(|M|) + \log((1-\lambda)/\lambda)\right)$. Note that the inner sums are just the variances of the data on the intervals between the $\vec{t}$. That is, if $\widehat{\sigma_l^2}$ is the variance of $\vec{s}_o\big|_{[t_l,t_{l+1})}$, then

$$E(\vec{t}) = \sum_{l=0}^{l=m} (t_{l+1} - t_l)\widehat{\sigma_l^2} + am.$$

Minimizing this value for different a gives the optimal ways of segmenting any signal into a piecewise constant function formed from the μ plus white noise of various variances. As a varies, we get segmentations with larger and smaller numbers of segments.

Example (White noise with varying means and variance).

The random variables are $\vec{s}$, the signal; $m, \vec{t}$, the number and values of the points of discontinuity; and $\vec{\mu}, \vec{\sigma}$ the means and standard deviations in each interval. Now we need a hyperprior for the variance too. We assume here that the variances are independent and uniformly distributed in the log domain (i.e., with respect to the measure $d\sigma/\sigma$). If S denotes the domain on which $\log \sigma$ is uniformly distributed, then we get the model

$$p(\vec{s}, m, \vec{t}, \vec{\mu}, \vec{\sigma}) = \frac{(1-\lambda)^{N-1}}{(2\pi)^{N/2}} \cdot \prod_{l=0}^{l=m} \frac{1}{\sigma_l^{t_{l+1}-t_l+1}} e^{-\sum_{n=t_l}^{t_{l+1}-1}(s_n-\mu_l)^2/2\sigma_l^2}$$

$$\cdot \frac{1}{(|S||M|)^{m+1}} \left(\frac{\lambda}{1-\lambda}\right)^m.$$

Note that we get one factor σ_l in the denominator for each s_n in the lth interval and one more for the density $d\sigma/\sigma$ in the prior on σ_l itself.

Minimizing this value for fixed $\vec{s} = \vec{s}_o$, we find surprising results. The means and variances both become the maximum likelihood choices for the data on the lth interval, namely the mean and variance[3] of the actual data there. So the sum in the exponent reduces to just $(t_{l+1} - t_l)/2$ and, taking out constants, the minimization reduces to that of

$$E(\vec{t}) = \sum_{l=0}^{l=m} (t_{l+1} - t_l + 1) \log(\hat{\sigma}_l) + am.$$

Here we must assume that $t_{l+1} - t_l \geq 2$ and that the data contain some generic noise or we have trouble with 0 variances.

Example (Weak string model). Here we assume that the signal in each interval between discontinuities is a random walk; that is, we assume $s_n - s_{n-1}$ are iid Gaussian when $n \neq t_l$ for any l. This wouldn't, however, make a proper probability distribution on the signal unless we also put some weak constraint on its actual values. As with music, we can do this by multiplying the probability density by $e^{-\epsilon\|\vec{s}\|^2}$ (giving us a modified random walk with restoring force). The model is

$$p(\vec{s}, m, \vec{t}) =$$
$$\frac{(1-\lambda)^{N-1}}{(2\pi)^{N/2}} \cdot \prod_{l=0}^{l=m} \sqrt{\det(Q_l)} e^{-a\sum_{n=t_l+1}^{t_{l+1}-1}(s_n-s_{n-1})^2/2 - \epsilon \sum_{n=t_l}^{t_{l+1}-1} s_n^2/2} \cdot \left(\frac{\lambda}{1-\lambda}\right)^m.$$

Here Q_l is the matrix of the quadratic form in the Gaussian,

$$Q_l = \begin{pmatrix} \epsilon+a & -a & 0 & \cdots & 0 & 0 \\ -a & \epsilon+2a & -a & \cdots & 0 & 0 \\ 0 & -a & \epsilon+2a & \cdots & 0 & 0 \\ \cdots & \cdots & \cdots & \cdots & \cdots & \cdots \\ 0 & 0 & 0 & \cdots & \epsilon+2a & -a \\ 0 & 0 & 0 & \cdots & -a & \epsilon+a \end{pmatrix}$$

---
[3]Variance in the sense of $\sum_{i=1}^{i=N}(x_i - \bar{x})^2/N$, not in the sense of the unbiased estimator of the variance $\sum_{i=1}^{i=N}(x_i - \bar{x})^2/(N-1)$.

2.6. Related Models

Minimizing the negative log of this probability gives us an optimal fit of the data to a set of Brownian walk segments separated by jumps.

Example (Weak string/noisy observation or $(u+v)$-model).
A modification of the last model is the simplest example of a popular approach to signal enhancement via denoising. One assumes that the observations $\vec{s}$ are the sum of additive Gaussian white noise $\vec{u}$ and a true signal $\vec{v}$, which itself is piecewise continuous with jumps. In this approach, the hidden variables are *both* the jumps $\vec{t}$ and the smooth signal $\vec{v}$ between the jumps. One can model $\vec{v}$ by using the previous example. If the model is used for deducing $\vec{t}$ and $\vec{v}$ with a fixed given $\vec{s} = \vec{s}_o$, then one can allow $\epsilon = 0$ and still get a proper posterior probability distribution on $\vec{v}$. The model is then

$$p(\vec{s}_o, \vec{v}, m, \vec{t}) \propto e^{-b \sum_{n=1}^{n=N}(\vec{s}_o(n)-v_n)^2/2} \cdot \prod_{l=0}^{l=m} e^{-a \sum_{n=t_l+1}^{t_{l+1}-1}(v_n-v_{n-1})^2/2} \cdot \left(\frac{\lambda}{1-\lambda}\right)^m.$$

This is the Mumford-Shah model [164] or the full weak-string model of Blake and Zisserman [24] in the one-dimensional framework. If $S = \{t_1, \cdots, t_m\}$, then the energy $E = -2\log(p)$ of this model is just

$$E(\vec{v}, S) = b \sum_{n=1}^{n=N} (\vec{s}_o(n) - v_n)^2 + a \sum_{n=2, n \notin S}^{n=N} (v_n - v_{n-1})^2 + c|S|.$$

2.6.2 Models Used in Speech Recognition

Hidden Markov models with labels. In many situations, parts of the signal that are widely separated have similar characteristics. For instance, in speech, a phoneme can be repeated many times and have similar characteristics. Then it makes better sense to let the hidden random variables be labels ℓ_n attached to each part of the signal, taking values in a set of auxiliary signal models. Then the jumps are *defined* to be the places where the labels change: $S = \{n \mid \ell_n \neq \ell_{n-1}\}$. If there are L possible labels and $m = |S|$, then there are in all $L(L-1)^m$ possible ways of assigning labels, and we have the model

$$p(\vec{s}, \vec{\ell}) = \prod_{k=0}^{k=m} p_{\ell(k)}\left(\vec{s}\,\big|_{[t_k, t_{k+1})}\right) \cdot \frac{(1-\lambda)^{N-1}}{L(L-1)^m} \left(\frac{\lambda}{1-\lambda}\right)^m, \qquad (2.2)$$

where $\ell(k)$ represents the label in the interval $[t_k, t_{k+1})$ and "dummy" discontinuities $t_0 = 1$ and $t_{m+1} = N+1$ are used to simplify the notation. Here p_ℓ is the auxiliary model attached to the label ℓ. If there are L labels in all, then at each point n in the signal, the model of Equation (2.2) assigns a probability $1 - \lambda$ of point $n+1$ having the same label and a probability $\lambda/(L-1)$ of having any of the other labels. But when labels recur at particular times, then often the order in which the labels occur near each other is not random, and a better model is to assume that some Markov chain governs the sequence of labels. If $Q(\ell \to \ell')$ is this chain, we get the more powerful class of models:

$$p(\vec{s}, \vec{\ell}) = \prod_{k=0}^{k=m} p_{\ell(k)}\left(\vec{s}\big|_{[t_k, t_{k+1}]}\right) \cdot Q_0(\ell_0) \cdot \prod_{k=1}^{k=m} Q(\ell(k-1) \to \ell(k)).$$

(One more prior is used here: a start distribution $Q_0(\ell_0)$ for the first label.)

The case of speech. In essence, the standard model of speech is an example of this model, where the speech is divided up into segments, representing the intervals of time in which a single phoneme is pronounced and the speech in each segment is evaluated by a probability model for that phoneme. Thus the $\vec{t}$'s in our model are the phoneme boundaries and the ℓ_k's are the labels of the phonemes. But like musical notes, phonemes are not stationary signals but typically have an attack and a decay. We alluded above to the fact that our model of music can be improved by incorporating this fine structure. In speech, each phoneme is divided into smaller pieces, sometimes called phones, and even within each phone, the power spectrum may vary smoothly. Thus, in Figure 2.12, we see (a) that the stop consonant "p" consists of an initial silence followed by a white noise like burst and (b) during the half second in which the vowel "ee" is spoken, the fundamental (vocal chord) frequency decreases by about 25% while the formants rise slowly and then fall suddenly. To complicate matters further, the pronunciation of each phoneme is often affected by the preceding and following phoneme (for the simple reason that the mouth must transition from one configuration to another at each phoneme boundary).

For many years, speech recognition software made a big departure from the "manifesto" of pattern theory by using what engineers call *vector quantization* (see, for instance [184]). Instead of having a probability distribution for the signal $p(\vec{s}|\vec{t}, \vec{\ell})$, they first take the local power spectrum of $\vec{s}$ via the windowed Fourier transform and then convert it to a 1- or 2-byte "code word" $\vec{c}$ using the "code book." This code book is nothing but a division of all possible power spectra into 256 or 256^2 bins. Then, finally, they develop a probability model $p(\vec{c}|\vec{t}, \vec{p})$ for the sequence of code words rather than the signal itself. Since $\vec{c}$ is discrete, our Gaussian model for $\vec{s}$ has been replaced by a probability table for the finite number of code words. As a result, *you cannot sample speech from this model*; you can only sample strings of code words. Thus, you cannot see which patterns in speech have been captured and which not. More recently, models have been used that incorporate the actual phoneme sounds by using Gaussian distributions

2.6. Related Models

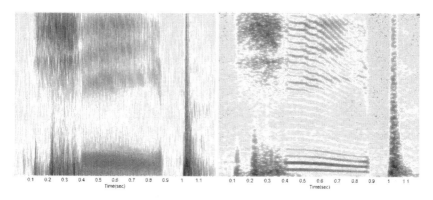

Figure 2.12. Spectral analysis of the word "sheep" (ʃ i p in IPA): On the left, the log-power from a windowed Fourier transform with a window size of 6 msec, giving fine temporal resolution; on the right the window size is 40 msec, giving fine frequency resolution instead. This is a good illustration of Theorem 2.8, which states that we cannot have fine temporal and frequency resolution at the same time. The initial 0.4 sec is the fricative sh, colored noise with power in low and high frequencies. In the middle 0.4–0.9 sec is the vowel ee. It is "voiced"—the vocal chords are vibrating—which creates bursts seen in the thin vertical lines on the left or as the multiple bands of harmonics on the right. Finally, at 0.9–1.1 sec, we have the stop consonant p with a mouth closure and a burst. The horizontal bands seen most clearly on the left during the vowel are the formants—frequencies at which the mouth cavity resonates, reinforcing specific harmonics of the burst. The vertical scales of both spectograms are 0 to 4000 Hertz. (See Color Plate II.)

$p_\ell(\vec{s})$ and mixtures of Gaussians, weighted sums of finite sets of Gaussians. The variables in these models are the linear functions of the local log-power spectra of the signal, a point in some Euclidean feature space such as $\mathbb{R}^{40}$, allowing the phases to be entirely random. A recent survey of the state of the art is presented in [78].

Perhaps the biggest difficulty in speech recognition is that there are such large variations in the way different speakers pronounce the same phoneme. The patterns in speech are not only what the person says but also the way he or she says it, which means that the observed signal $\vec{s}$ is affected by two hidden random variables: first, the identity of the speaker, and second, $\vec{\ell}$, the phoneme/word sequence. (Face recognition presents the same problem, where the two hidden random variables are the "illumination" and the "person you are looking at.") This was one of the reasons for using a code book: it was designed explicitly to try to eliminate speaker variation. But if we take this approach, we are precluded from developing a decent stochastic model for speaker variation. It is better to explicitly model the speakers—the variations in their vocal tracts, speed of articulation, and so forth. One approach does this by applying a linear transformation to feature space, depending on the speaker. Nonetheless, the authors' personal experiences with computer phone-number retrieval using verbal commands suggests there is a long way to go.

A final problem with speech recognition is that these models have a really huge number of unknown coefficients (on the order of thousands or tens of thousands). But there is a remarkable algorithm, the EM algorithm, which can be used to "train" the model, which is discussed next.

2.6.3 Algorithm II: Learning Parameters via EM

All of the models discussed in this chapter have the following form: a one-dimensional signal is to be split into segments on each of which a stochastic model is to be fit. These stochastic models for the segments always have parameters: period and power, and maybe a model for power in various harmonics, for musical notes; mean and variance for image slices; or the parameters in the models p_ℓ for each label in the HMM just discussed. Moreover, in the HMM we also had parameters for the transition probabilities between labels. If these parameters are known, we can find the best possible segmentation via dynamic programming. But a fundamental question in fitting these models to data is choosing the parameters. There is a "chicken or egg" problem here: knowing the segments, we can fit the parameters, and knowing the parameters, we can fit the segments. But we need to do both. The basic tool for this is the expectation-maximization algorithm (EM), or simply the *EM algorithm*. It was discovered in full generality by Dempster, Laird, and Rubin [55] following earlier work in the case of speech recognition by Baum and collaborators [14].

The abstract setup for EM is as follows: We are given observations for a set of random variables $\{\mathcal{X}_i\}$. We have a model $P(\{\mathcal{X}_i\}, \{\mathcal{Y}_j\} | \{\lambda_\alpha\})$ that depends on certain hidden random variables $\{\mathcal{Y}_j\}$ and also on certain parameters $\{\lambda_\alpha\}$. Given the observations $\mathcal{X}_i = \widehat{x}_i$, we seek the values of the parameters that maximize the probability of the observations:

$$F(\{\lambda_\alpha\}) = \sum_{\{y_j\}} P(\{\widehat{x}_i\}, \{y_j\} | \{\lambda_\alpha\}).$$

Before describing the algorithm in general, let us start with a simple example, which will give the essential idea.

Example. Suppose $\{\mathcal{X}_i\}$ are iid samples from a mixture distribution on $\mathbb{R}$ made up of 50 percent uniform distribution on $[-C, C]$ and 50 percent Gaussian with standard deviation 1 but unknown mean μ:

$$p(x|\mu) = \frac{1}{2\sqrt{2\pi}} e^{-(x-\mu)^2/2} + \frac{1}{4C} \chi_{[-C,C]}(x).$$

We want to maximize with respect to μ:

$$\prod_i \left(\frac{1}{2\sqrt{2\pi}} e^{-(\widehat{x}_i - \mu)^2/2} + \frac{1}{4C} \chi_{[-C,C]}(\widehat{x}_i) \right).$$

2.6. Related Models

The idea is to consider as hidden variables whether the samples are drawn from the uniform or the Gaussian component. We let $\ell_i = g$ if $\widehat{x}_i$ is drawn from the Gaussian, and $\ell_i = u$ if it is drawn from the uniform. Then we can write

$$\frac{1}{2\sqrt{2\pi}} e^{-(\widehat{x}_i - \mu)^2 / 2} + \frac{1}{4C} \chi_{[-C,C]}(\widehat{x}_i) = \sum_{\{\ell_i\}} P(\widehat{x}_i, \ell_i),$$

$$P(x, g) = \frac{1}{2\sqrt{2\pi}} e^{-(x-\mu)^2/2}, \quad P(x, u) = \frac{1}{4C} \chi_{[-C,C]}(x).$$

The EM algorithm, in this simple case, is the following quite intuitive procedure. We proceed through a sequence of guesses μ_n for the correct value of μ. For each guess μ_n, we consider the likelihood that $\widehat{x}_i$ comes from the Gaussian:

$$L_i^{(n)} = \frac{\frac{1}{2\sqrt{2\pi}} e^{-(\widehat{x}_i - \mu_n)^2 / 2}}{\frac{1}{2\sqrt{2\pi}} e^{-(\widehat{x}_i - \mu_n)^2 / 2} + \frac{1}{4C} \chi_{[-C,C]}(\widehat{x}_i)}$$

and define the new μ as a weighted average of the $\widehat{x}_i$ with these likelihoods as weights:

$$\mu_{n+1} = \frac{\sum_i L_i^{(n)} \widehat{x}_i}{\sum_i L_i^{(n)}}.$$

Dempster, Laird, and Rubin [55] saw how to recast this to apply to the general situation. Abbreviating $\{\widehat{x}_i\}$, $\{y_j\}$, and $\{\lambda_\alpha\}$ to $\widehat{x}, y$, and λ, their key idea is to define the following weighted log-likelihood:

$$Q(\lambda_{\text{new}}, \lambda_{\text{old}}) = \sum_y \log(P(\widehat{x}, y | \lambda_{\text{new}})) \cdot P(y | \widehat{x}, \lambda_{\text{old}}).$$

Then the EM algorithm is the following.

Algorithm (EM Algorithm).
Iterate the following two steps:

1. Expectation step: compute the log-likelihood

$$Q(\lambda_{\text{new}}, \lambda_{\text{old}}) = \sum_y \log(P(\widehat{x}, y | \lambda_{\text{new}})) \cdot P(y | \widehat{x}, \lambda_{\text{old}}).$$

2. Maximization step: find a λ_{new} that maximizes Q.

There are two questions about EM: why does this work and when can these steps be carried out effectively? The reason EM often works is shown by Theorem 2.16.

Theorem 2.16. $P(\widehat{x}|\lambda_{\text{new}}) = \sum_y P(\widehat{x}, y|\lambda_{\text{new}}) \geq P(\widehat{x}|\lambda_{\text{old}}) = \sum_y P(\widehat{x}, y|\lambda_{\text{old}})$ *with equality only if* $\lambda_{\text{new}} = \lambda_{\text{old}}$ *and then in this case* $\lambda = \lambda_{\text{new}} = \lambda_{\text{old}}$ *is a critical point of* $P(\widehat{x}|\lambda) = \sum_y P(\widehat{x}, y|\lambda)$.

Proof: Since $Q(\lambda_{\text{new}}, \lambda_{\text{old}}) \geq Q(\lambda_{\text{old}}, \lambda_{\text{old}})$, we have

$$0 \leq \sum_y \log \frac{P(\widehat{x}, y|\lambda_{\text{new}})}{P(\widehat{x}, y|\lambda_{\text{old}})} \cdot P(y|\widehat{x}, \lambda_{\text{old}})$$

$$= \sum_y \log \frac{P(y|\widehat{x}, \lambda_{\text{new}})}{P(y|\widehat{x}, \lambda_{\text{old}})} \cdot P(y|\widehat{x}, \lambda_{\text{old}}) + \log \frac{P(\widehat{x}|\lambda_{\text{new}})}{P(\widehat{x}|\lambda_{\text{old}})}.$$

But the first term is minus the KL distance from $P(y|\widehat{x}, \lambda_{\text{new}})$ to $P(y|\widehat{x}, \lambda_{\text{old}})$, so it is nonpositive. Thus, the second term is positive, and this proves the inequality. If one has equality, the KL distance between the λ_{old} and λ_{new} models is zero, so they are equal. Moreover, differentiating $Q(\lambda, \lambda_{\text{old}})$ with respect to λ, we find that

$$\left.\frac{\partial}{\partial \lambda} Q(\lambda, \lambda_{\text{old}})\right|_{\lambda = \lambda_{\text{old}}} = \left.\frac{\partial}{\partial \lambda} \log(P(\widehat{x}|\lambda))\right|_{\lambda = \lambda_{\text{old}}}.$$

Thus, if $\lambda_{\text{old}} = \lambda_{\text{new}}$, it is a maximum of $Q(\lambda, \lambda_{\text{old}})$, and thus both terms above equal 0, showing that is it also a critical point of $P(\widehat{x}|\lambda)$. □

Of course, this does not guarantee that EM will find the *global maximum* of $P(\widehat{x}|\lambda)$. There are even known examples in which the fixed point $\lambda_{\text{old}} = \lambda_{\text{new}}$ is a local minimum of $P(\widehat{x}|\lambda)$ (see [154])! However, in most cases, EM converges to a local maximum in which it often gets stuck. It is always essential to start the algorithm with the best available parameters.

The good situations for which EM can be carried out those for which the probability model is the exponential of a sum of terms of the form $f_k(\lambda) \cdot g_k(\widehat{x}, y)$, where the expectation of each term $g_k(\widehat{x}, y)$ for the probabilities $P(y|\widehat{x}, \lambda)$ can be computed. This holds for the last example and also for the HMM situations described in Section 2.6.2.

2.7 Exercises

1. Simulating Continuous Random Variables with MATLAB

Box-Muller algorithm

1. Let $\mathcal{U}$ and $\mathcal{V}$ be two random variables independent and uniformly distributed on $[0, 1]$. Prove that the random variables $\mathcal{X}$ and $\mathcal{Y}$ defined by

$$\mathcal{X} = \sqrt{-2\ln(\mathcal{U})} \cos(2\pi \mathcal{V}) \text{ and } \mathcal{Y} = \sqrt{-2\ln(\mathcal{U})} \sin(2\pi \mathcal{V})$$

2.7. Exercises

are independent and follow the Gaussian distribution with mean 0 and variance 1 (denoted by $\mathcal{N}(0, 1)$ and also called the normal distribution). This result is closely connected with the well-known method for evaluating $\int_{-\infty}^{+\infty} e^{-x^2/2} dx$ by squaring it and passing it onto polar coordinates. Can you see why?

2. Use this result (called the Box-Muller algorithm) and the MATLAB function rand to sample standard normal random variables: take N samples and plot on the same figure their histogram and the density function of the standard normal distribution.

Simulation of Gaussian vectors

1. Prove that if $\mathcal{X} = (\mathcal{X}_1, \ldots, \mathcal{X}_k)$ is a Gaussian vector with mean 0 and covariance matrix equal to the identity, and if $A \in \mathcal{M}_k(\mathbb{R})$ is such that $AA^t = C$, then $\mathcal{Y} = A\mathcal{X} + \mu$ is a Gaussian vector with mean 0 and covariance C. How can this be used in MATLAB for sampling Gaussian vectors?

2. Toy example: get N samples of a Gaussian vector with mean μ and covariance C, where C=[5 3 2 ; 3 4 1 ; 2 1 2] and mu=[-2 1 4]. (*Hint*: use the MATLAB function chol). Check the obtained results from the MATLAB functions mean and cov.

The inversion method

1. Prove the following result:
   The cumulative distribution function (cdf) of a random variable $\mathcal{X}$ is defined by $F(x) = \mathbb{P}(\mathcal{X} \leq x)$. If $\mathcal{U}$ is a random variable uniformly distributed on $[0, 1]$, then $\mathcal{X} = F^{-1}(\mathcal{U})$ is a random variable with cdf F.

2. Use this result (called the inversion method) to sample from an exponential distribution of parameter λ. (Recall that its density function is $\lambda \exp(-\lambda x)$ for $x \geq 0$ and 0 otherwise.) Take N samples and compare on the same figure the histogram of the obtained values and the density function.

3. Do part 2 for a Cauchy distribution (the density function is $1/\pi(1 + x^2)$).

Checking the law of large numbers

1. Get N independent samples $X_1, \ldots, X_N$ of a distribution of your choice (e.g., Bernoulli, normal, exponential). Then plot on the same figure the function $n \mapsto (X_1 + \ldots + X_n)/n$ and the value of $\mathbb{E}(\mathcal{X})$. What do you see?

2. Do part 1 for a Cauchy distribution, and check experimentally that the law of large numbers fails in this case.

The rejection method

The idea of the rejection method is to use two density functions: the first one f is the one you want to sample, and the second one g is a law that is easy to sample (for instance, a uniform one). Prove the following result:

Let f and g be two density functions on $\mathbb{R}^n$. Assume that there is a constant c such that: $\forall x \in \mathbb{R}^n$, $cg(x) \geq f(x)$. Let $\mathcal{X}$ be a random variable with density function g, and let $\mathcal{U}$ be a random variable with uniform distribution on $[0, 1]$, independent of $\mathcal{X}$. Let E be the event "$c\mathcal{U}g(\mathcal{X}) < f(\mathcal{X})$." Then the conditional law of $\mathcal{X}$ given E has density function f. The algorithm for sampling the law f from g is then the following:

```
Repeat
    sample X from the density g
    sample U uniformly on [0, 1]
Until (cUg(X) < f(X)).
```

What is the law of the number of runs of this loop?

2. Correlation and Mutual Information

Suppose $(\mathcal{X}, \mathcal{Y}) \in \mathbb{R}^2$ is a two-dimensional real-valued random variable. Then we can consider the normalized correlation,

$$\text{Cor}(\mathcal{X}, \mathcal{Y}) = \frac{\mathbb{E}((\mathcal{X} - \bar{x}) \cdot (\mathcal{Y} - \bar{y}))}{\sigma_x \cdot \sigma_y},$$

where $\bar{x} = \text{mean}(\mathcal{X})$, $\bar{y} = \text{mean}(\mathcal{Y})$, $\sigma_x = \text{SD}(\mathcal{X})$, and $\sigma_y = \text{SD}(\mathcal{Y})$, and their mutual information,

$$MI(\mathcal{X}, \mathcal{Y}) = \mathbb{E} \log_2 \left(\frac{p(x, y)}{p(x) \cdot p(y)} \right),$$

where $p(x, y) = \text{prob. density}(x, y), p(x) = \text{prob. density}(x), p(y) = \text{prob. density}(y)$.

1. Recall the proof that the normalized correlation is always between 1 and -1. (*Hint*: this follows from Cauchy's inequality.)
2. Show that if $(\mathcal{X}, \mathcal{Y})$ is Gaussian, then

    $$MI(\mathcal{X}, \mathcal{Y}) = -\tfrac{1}{2} \log_2(1 - \text{Cor}^2(\mathcal{X}, \mathcal{Y})).$$

3. Find an example where $\text{Cor}(\mathcal{X}\mathcal{Y}) = 0$ but $MI(\mathcal{X}, \mathcal{Y}) = \infty$. (*Hint*: consider the case $\mathcal{X} = \pm \mathcal{Y}$.)
4. Find an example where $\text{Cor}(\mathcal{X}, \mathcal{Y}) \approx 1$ but $MI(\mathcal{X}, \mathcal{Y}) \approx 0$. (*Hint*: consider cases where the support of $p(x, y)$ is concentrated very near $(0, 0) \cup (1, 1)$.)

3. Synthesizing Music

The easiest way to synthesize random samples from stationary Gaussian processes is by randomly choosing Fourier coefficients from a Gaussian distribution with mean zero and whatever variance you want and then taking the Fourier transform. The Fourier coefficients are independent, and you can use the Box-Muller algorithm or, in MATLAB, the built-in routine x = randn(N,1) and multiply by the desired standard deviation. The steps for synthesizing music follow.

1. Synthesize colored noise with a *power law power spectrum*. In other words, take the real and imaginary parts of the Fourier coefficients $\hat{s}_k, 0 \le k < N$ to be random normal variables with mean 0 and standard deviation $1/\min(k, N-k)^\lambda$, but with

2.7. Exercises

$\hat{s}_0 = 0, \hat{s}_{N-k} = \bar{\hat{s}}_k$.[4] Note that low frequencies come from the Fourier coefficients with index either near zero or near N. You can take the size of the vector N to be 1024. Do this for $\lambda = 0, 0.5, 1$, and 2 and plot the results. What do you see? We return to such power laws in Chapter 6 and connect this at that point to Mandlebrot's fractal landscapes in Exercise 6.

2. Synthesize random single notes of music by taking the Gaussian distribution to be $\frac{1}{Z}e^{Q(\vec{s})}$ where $Q(\vec{s}) = \Sigma_{k=0}^{N-1}(s(k+p) - s(k))^2 + cs(k)^2$. Here, $s(k)$ is assumed to "wrap around" above N. Recall that we worked out the corresponding power in Section 2.3.2. You may try $N = 8192, p = 32$, and various quite large c's to look for samples that resemble music. Listen to what you have produced via MATLAB's command sound, whose default is 8192 Hz, making your note have frequency 256 Hz.

3. Some notes that are better approximations to various instruments can be synthesized by choosing a better rule for the power in various frequencies. Try experimenting so only a few harmonics are present. A realistic note is also amplitude modulated: it may have an attack where the power rises and then a slow decline before the next note. This can be achieved by multiplying the actual sound by a suitable function.

4. Finally, let's introduce multiple notes. Take a longer period, say 2^{15} samples, and now divide it into intervals by a handful of Poisson points and put notes (as above) in each interval. You can synthesize the actual model described in Section 2.5.2, but this will sound like some weird composer. But now you are in a position to improve the model in many ways, for example by using measures, bars, scales, and keys.

4. Some Determinants

In the analysis of cyclic signals, we need some determinants. For any $q \geq 2$, define the symmetric circulant matrix A_q by

$$A_q = \begin{pmatrix} 2 & -1 & 0 & \cdots & 0 & 0 & -1 \\ -1 & 2 & -1 & \cdots & 0 & 0 & 0 \\ 0 & -1 & 2 & \cdots & 0 & 0 & 0 \\ \cdots & & & \cdots & & & \cdots \\ 0 & 0 & 0 & \cdots & -1 & 2 & -1 \\ -1 & 0 & 0 & \cdots & 0 & -1 & 2 \end{pmatrix}$$

Let $\zeta = e^{2\pi i/q}$. Then show that

$$e_k = \begin{pmatrix} 1 \\ \zeta^k \\ \zeta^{2k} \\ \vdots \\ \zeta^{(q-1)k} \end{pmatrix}$$

---

[4] Alternately, ignore the condition $\hat{s}_{N-k} = \bar{\hat{s}}_k$ and take the real part after the Fourier transform. This requires twice the computation but with one-dimensional Fourier transforms and fast computers, it makes very little difference.

are the eigenvectors of A_q for $0 \leq k \leq q-1$ with eigenvalues

$$\lambda_k = -(\zeta^k - 2 + \zeta^{-k}) = 4\sin^2(\pi k/q).$$

Deduce that

$$\det(bI + aA_q) = a^q \prod_{k=0}^{q-1} \left(b/a + 4\sin^2(\pi k/q)\right).$$

If $0 < b \ll a$, deduce that

$$\det(bI + aA_q) \approx ba^{q-1} \prod 4\sin^2(\pi k/q).$$

Looking directly at this determinant, check that, in fact

$$\det(bI + aA_q) \approx q^2 \cdot ba^{q-1}.$$

Conversely,, if a and b are fixed and q is large, we can approximate the determinant by

$$\det(bI + aA_q) = e^{\sum_{k=0}^{q-1} \log(b + 4a\sin^2(\pi k/q))} \approx e^{\frac{q}{\pi} \int_0^{\pi} \log(b + 4a\sin^2(\theta))d\theta}.$$

Use your favorite table of integrals to evaluate this integral, finding

$$\det(bI + aA_q) \approx \left(\tfrac{1}{2}(b + 2a + \sqrt{b^2 + 4ab})\right)^q \quad \text{if } q \gg 0.$$

5. Computational Experiments on Segmenting One-Dimensional Signals

The purpose of these problems is to segment various one-dimensional signals by implementing the many models introduced in this chapter. We give general guidelines here, but note that it is always more instructive to play around, especially looking for the causes of *failures*—segment boundaries that are not found or extra boundaries that ought not to be there.

Parsing Beran's oboe cadenza

Find the individual notes in the simple piece of music that started this chapter. The music is on our website in the file cadenza.au, in a compact format readable by MATLAB via auread. Alternatively, it is also there as an ascii file of 40,000 numbers. The numbers are samples of the sound wave taken at 8,000 Hertz for an interval of 5 seconds (see Figure 2.13). In the course of this interval, the oboe plays 22 notes (some are double—the same note with a short rest in between). The goal is to find the note boundaries and the frequencies of the notes themselves by implementing the piecewise Gaussian model, where discrete time is divided into segments $\{0 = n_0 < n_1 < \cdots < n_K = 40000\}$ and the kth note is played in the interval $[n_{k-1}, n_k]$. Let the approximate discrete *period* of the kth note be the integer p_k, and model the note in this interval by the Gaussian distribution

$$p(s|_{[n_{k-1}, n_k]}) = \frac{1}{Z} e^{-\sum_{n_{k-1}}^{n_k - p_k} (s(n+p_k) - s(n))^2}.$$

2.7. Exercises

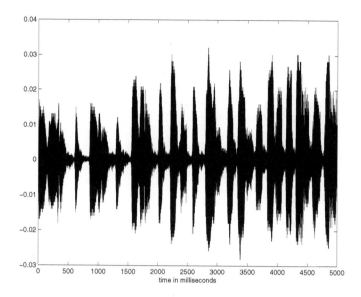

Figure 2.13. The 5-second interval of oboe music from Exercise 5.

Here you may assume that any period will be in the range $5 \leq p \leq 40$ (which corresponds to frequencies of $8000/p$, or the three-octave range $[200, 1600]$ Hertz). In addition, you should put a simple exponential prior on the number K of notes by $\mathbb{P}(K = k) = \frac{1}{Z} e^{-ak}$. To find the most probable sequence of notes in this model, we suggest precalculating the table by using

$$C(n, p) = \sum_{k=1}^{n} s(k) s(k+p).$$

Then we can approximate the exponent by

$$\sum_{k=1}^{K} \sum_{n_{k-1}}^{n_k - p_k} (s(n+p_k) - s(n))^2 \approx 2 \sum_{n=1}^{40000} s(n)^2 - 2 \sum_{k=1}^{K} (C(n_k, p_k) - C(n_{k-1}, p_k)).$$

Your algorithm should proceed by dynamic programming, calculating by induction on the time step n the best segmentation of the music from the beginning through sample n. To do this, you must loop only over the immediately preceding note boundary (including 0) *and* possible p's in the last time segment. If you run out of memory, a good way to simplify the calculation is to restrict the time steps to multiples of some small number (e.g., 10). Find the optimal segmentation of the whole sequence with various values of the parameter a in the prior. Values of a that ar too large lead to segmentations with too few notes and values of a that are too small may lead to too many notes. Remember, there should be 22 notes. Check your results by running `specgram(data,[],8000)` or simply `plot(data), axis([a b -.03 .03])` to see what is happening.

Speech

Speech signals are, in general, not at all easy to segment. Figure 2.12 was a misleadingly simple example. However, it is easy to play with models to find the easy parts in speech: the silence preceding each stop consonant, the huge difference between unvoiced fricatives (such as sh) and voiced sounds such as vowels. Vowels, for example, are similar to music with the vocal chords contributing a fundamental with a frequency of 90–150 Hertz for males, 150–250 for females, and the rest of the mouth emphasizing or suppressing its harmonics. Unvoiced fricatives, in contrast, are colored noise with broad bands of power. For making these distinctions, the models we have introduced are sufficient. For subtle sounds (e.g., differences of stop consonants, semi-vowels, etc.) the standard HMM may be needed. But that is not a small project.

Streaks in baseball

A better way to experiment with EM is to take a much simpler example than speech: streaks. Basic baseball lore says that batters have streaks in which they get runs for many games in a row (with perhaps a few "off" days), followed by fallow periods without any runs. Fortunately, there are quite a lot of boxscores online these days, and you can find many at http://www.baseball-boxscores.com/. The idea, then, is to fit a hidden Markov model to an individual batter, with observables x_i being either the number of hits in the ith game, or more simply, 1 if the batter got a hit and 0 if not. The hidden variable y_i is 1 if the player is on a streak, 0 if not. You can initialize the probabilities by assuming a 0.6 chance of continuing in a streak/nonstreak and taking, say, 2 times the player's batting average for the probability of a hit on a streak, and 0.5 times the batting average for nonstreaks. But all these constants are the λ's that are to be fit with EM. Find the iterative update rule for them given by the EM algorithm and see what you get!

6. Period Doubling: Abstract Analysis

An interesting phenomenon should have occurred if you analyzed the oboe cadenza: instead of the perceptually correct notes, you sometimes come up with a period that is 2, 3, or 4 times larger. In fact, writing the true periods of the notes as $8000/f$, the true frequencies are $f \approx 447 \cdot 2^{k/12}$ and $-2 \leq k \leq 12$. If you calculate these, you will see that some of the resulting periods are not close to integers in our sampling. This causes a problem in our approach.

1. Consider sampling a pure cosine function as follows:

$$s(n) = \cos\left(2\pi \frac{n}{K + 0.5}\right),$$

where K is an integer and $K + 0.5$ is the true period of s. Window the cosine so it has a large but finite support. Calculate the autocorrelation function of s and find its max among all positive integer shifts p.

2. The above example suggests that selecting the first largish peak rather than the absolute biggest peak of the autocorrelation may be a better strategy with discrete data. Here's another problem. Consider

$$s(n) = \cos\left(2\pi \frac{2n}{K}\right) + a \cos\left(2\pi \frac{n}{K}\right),$$

2.7. Exercises

made up of a note of frequency $2/K$ plus a small contribution of frequency $1/K$. This will sound like frequency f, *but* again calculate the autocorrelation function and see where its peak is.

7. Shannon-Nyquist Sampling Theorem

The Shannon-Nyquist sampling theorem gives conditions to exactly recover a function from its samples at discrete points on a regular grid. More precisely, the result is given by Theorem 2.17.

Theorem 2.17. *Let $f \in L^2(\mathbb{R})$ be a band-limited function, which means that there exists $W > 0$ such that its Fourier transform $\widehat{f}$ is 0 outside $[-W, W]$:*

$$\text{Support}(\widehat{f}) \subset [-W, W].$$

Then, f can be recovered from its samples at points $n/2W$, $n \in \mathbb{Z}$, via an interpolation with the sine cardinal function:

$$\forall t \in \mathbb{R}, \quad f(t) = \sum_{n \in \mathbb{Z}} f\left(\frac{n}{2W}\right) \frac{\sin \pi(2Wt - n)}{\pi(2Wt - n)}.$$

1. Prove the theorem. First, recall that the Fourier transform $\widehat{f}$ of f is defined by

$$\widehat{f}(\omega) = \int_{\mathbb{R}} f(t) e^{-2i\pi\omega t} dt.$$

Let $x \to F(x)$ be the function defined on $[-\frac{1}{2}, \frac{1}{2}]$ by $F(x) = \widehat{f}(2Wx)$, and then defined on all $\mathbb{R}$ by 1-periodicity. Using the Fourier series for periodic functions, we have

$$F(x) = \sum_{n \in \mathbb{Z}} c_n e^{-2i\pi nx}, \text{ where } c_n = \int_{-1/2}^{1/2} F(x) e^{2i\pi nx} dx.$$

Compute the coefficients c_n and prove that

$$c_n = \frac{1}{2W} f\left(\frac{n}{2W}\right).$$

2. Use again the inverse Fourier transform, and exchange sum/integral to get the result: for all $t \in \mathbb{R}$,

$$f(t) = \sum_{n \in \mathbb{Z}} f\left(\frac{n}{2W}\right) \frac{\sin \pi(2Wt - n)}{\pi(2Wt - n)}.$$

3. What makes things work in the Shannon-Nyquist sampling theorem is that Support($\widehat{f}$) and Support($\widehat{f}$) + $2W$ don't overlap. When f is sampled at points of the form $n/2W'$, where W' is such that Support($\widehat{f}$) and Support($\widehat{f}$) + $2W'$ overlap, then the sampling is too sparse, and f cannot be recovered from its samples; this is called *aliasing*. In this case, some strange geometric distortions can appear. This can be easily observed with MATLAB, for instance, in the following way: plot as an image the function $f(x, y) = \cos(\alpha x + \beta y)$, with, say, $\alpha = .2$ and $\beta = .4$,

at points x=[0:1:N] and y=[0:1:N] with $N = 200$. Call A this image. Then subsample A by taking B=A(1:k:N,1:k:N). Try this for different values of k. The geometric distortions that appear for some values of k are typical of the artifacts often observed with digital images.

8: Segmentation of One-Dimensional Slices of Images

Image slices

Segmenting a full image, which we consider in Chapter 4, requires serious computing power. However, the one-dimensional signals obtained by restricting to a slice of an image (i.e., $I(i_0, *)$ or $I(*, j_0)$) are easy to handle and experiment with. Figure 2.11 showed three such slices. The horizontal or vertical white lines are the slices. The grey values of the pixels along these lines are shown. The first one is a range image, so its values are not light intensities but ranges, actually the logs of the distance from the camera to the nearest surface. Anyone can find an abundance of images on a computer. We have also posted several range images on our website: range images have very clear discontinuities at the edges of objects and are well adapted for the weak string or $u + v$ model described in Section 2.6.1.

1. Make paper plots of your range or grey-value slice functions $f(x)$. Examine both the images and the plots, identifying which parts of the plots corresponding to which parts of the image crossed by the line. Then mark what you think is the natural segmentation of the line—mark the points on the image where the line crosses some boundary and the corresponding point on the plot (as closely as you can find it). Note that some segments may be distinguished by having different means, some by having different smooth fits and some by having different variances.

2. What ideally would be the best stochastic model for each segment of $f(x)$ that you have found? Consider constants, linear functions, and smoothly/slowly varying functions with or without additive or salt/pepper noise and various possible noise models: (i) white Gaussian, (ii) colored Gaussian, (iii) other, with large or small variances. In some places, you should find noticeable departures from the Markov assumption, that is, noncontiguous segments of the plots that are related to each other. Describe the hidden variables you need to flesh out the stochastic model and set up the "energy" functionals (i.e., log probabilities) to make the model precise. Do you want to make the breakpoints into a Poisson process or do you see other structure that might be modeled?

3. Following the techniques introduced in this chapter, define the simplest piecewise constant model, or, if ambitious, more complex models. This model should put a Poisson prior on the breakpoints (segment boundaries), a uniform prior on the mean values in each segment and assume the observation in each segment comes from a Gaussian model, this constant plus white noise. You can find its minimum by dynamic programming, running from left to right across the signal. At each point k, assume there is a breakpoint between k and $k + 1$ and consider only the possible positions for the one preceding breakpoint at some position between ℓ and $\ell + 1$. Then E breaks up into the best energy E from 1 to ℓ (already calculated) plus a term in the variance of the signal from $\ell + 1$ to k. If you are careful, the vector

2.7. Exercises

of these variances, as ℓ varies, can be calculated in one line of MATLAB code. (*Hint*: precalculate cumsum of the square of the data.)

Apply this model to the one-dimensional sampled image data $f(k)$ of the slices, found on our web-site and experiment with the choice of various constants in E, the density of the Poisson process, the range for the means, and the variance of the data term. Solve for the resulting optimal segmentation. When is the segmentation closest to the one you chose above as the "correct" segmentation?

4. An interesting variant of this model is allowing the variance to be a hidden variable and letting it vary from segment to segment. The minimization takes place as before, except that you have to be careful with very small variances; sometimes it works better if you add a small amount of noise to the data.

5. A completely different method for solving for a piecewise constant model is to use the *Perona-Malik* differential equation, which is very popular in image processing. The idea behind this equation is to regard the values of $f(x)$ to stand for *heat* and imagine the heat is diffusing along the x-axis. In that case, we would solve for the heat $f(x,t)$ at time t by solving the usual heat equation $\frac{\partial f}{\partial t} = \frac{\partial^2 f}{\partial x^2}$. What Perona and Malik [177] proposed was to decrease the conductance wherever there was a large heat gradient. This has no physical meaning, but it allows compartments to form across whose endpoints the heat jumps and there is no conductance. The equation is

$$\frac{\partial f}{\partial t} = \frac{\partial}{\partial x} \left(\frac{\frac{\partial f}{\partial x}}{1 + c\left(\frac{\partial f}{\partial x}\right)^2} \right).$$

Unfortunately, this equation is ill-posed, and there is no satisfactory theory for it on the continuum. But if we make x discrete and replace the x-derivatives with finite differences, we get a fine system of ordinary differential equations (ODEs). Solving these, you will find that the sampled function f creates piecewise constant functions with sharp jumps, which then gradually erode away. This defines a segmentation of the domain of the function depending on the constant c chosen and the time at which the run is terminated. This is not difficult to implement in MATLAB by using the Runge-Kutta ODE solver ode45, for example. Try this on the same image slices used earlier in this exercise. Note that $1/c$ is roughly the value of the gradient squared at which the equation switches over from smoothing to creating jumps. Compare the results with the energy minimization methods from parts 3 and 4 of this exercise. In fact, Perona-Malik can be interpreted as gradient descent for a variant of these energies, so the methods are not so different.

Figure 3.1. Three license plates, which are trivial for a human to read but difficult for a computer due to the shadows, dirt, and graininess (courtesy of S. Geman [114]).

- 3 -
Character Recognition and Syntactic Grouping

In Chapters 1 and 2, we were interested in one-dimensional signals (English strings and music). One pattern that occurred in both cases was that the signals could be naturally decomposed into parts, separated by breaks across which there was less mutual information and the statistics of the signal changed; these were the words in the strings and the notes in the music. Now we look at two-dimensional signals given by images of the world. The domain of this signal is called the image plane, either an open set $R \subset \mathbb{R}^2$ or a discrete sampled version of this in the lattice $\mathbb{Z}^2$. The most obvious pattern we see in most such signals is that the image contains various *objects*. The outline of such an object is a one-dimensional contour in the image domain Ω, and thus studying these contours and the shapes they enclose is a transition from the one-dimensional world we have been looking at and the fully two-dimensional world considered in Chapter 4. We call these *salient* or *boundary* contours. We wish here to take a specific simple example to motivate the analysis we will do and the stochastic models we will construct. For this we use images of text, either handwritten or printed, made up of isolated characters. The problem is to read the characters, for example, to decide whether a character is a *G* or a 6 (see Figure 3.1).

The simplest description of any planar shape is by its boundary curve: the basic pattern of a boundary is a discontinuity (or a steep gradient) in brightness across this boundary. In Section 3.1, we see how to use hypothesis testing to identify these salient curves in images. In Section 3.2, we will stochastic models for plane curves using Brownian motion and "snakes," enabling us to use Bayes' theorem for finding boundary curves. Note that these types of contours combine what we called *value patterns* in Chapter 0 with *geometric patterns*: they are recognized by their distinctive high-gradient gray-levels, but their shapes are mostly smooth curves—a distinctive geometric structure. Modeling the boundary of a shape is not enough, however; we will also need to look at the interior of the shape. To do this, in Section 3.3 we study the medial axis description of shapes—a higher-order one-dimensional geometric pattern found in images.

In all of this, hierarchical geometric patterns, especially the grouping principles investigated by the Gestalt school of psychology and outlined in Section 3.4, play an important part. Contours are often ambiguous because they are noisy, blurry, incomplete, or confused by clutter; hence, the full contour has to be inferred by linking contour fragments. This is the first, most basic grouping principle. The medial axis is, by definition, a grouping of opposite sides of a shape in a symmetrical fashion. Grammars are the quintessential hierarchical formalism, and when applied to complex compound objects in images are a formal embodiment of more of the grouping principles. Finally, Section 3.5 shows how grammars may be used for assembling shapes out of parts, leading us finally to models for recognizing alphanumeric characters. Thus, every ingredient of pattern theory plays a role.

To make all of this more tangible, let's start by giving a simple mathematical formulation of the problem of character recognition. The huge literature on character recognition has been used as a test-bed for many major theories in statistical learning theory (e.g., Yann LeCun's work [134]), which we will not attempt to summarize here. It is also an excellent example in which we can incorporate all the ideas outlined in the previous paragraph. The problem may be described in the following way. Let I be an image of a character, and let $R \subset \mathbb{R}^2$ denote its domain, so I is a function, $I : R \to \mathbb{R}$. The hidden random variables we have to deal with are the following:

- two average gray levels, a and b, one for the background and the other for the character,

- a stroke width r, lengths L_i of the strokes, and the strokes themselves given by parametric curves:
$$\gamma_i : [0, L_i] \longrightarrow R,$$

- a parse tree grouping the above strokes into composite "objects."

Let D be the unit disc; then define the shape S as a union of suitably thickened strokes:
$$S = \bigcup_i (\gamma_i([0, L_i]) + rD).$$

If this shape fits the image, we expect the following functional to be small:
$$E = \int_S (I(x) - a)^2 dx + \int_{R-S} (I(x) - b)^2 dx.$$

To find a character in the image I, we can minimize E over a, b, and S—that is, over a, b, r, L_i, and γ_i. We must also add to this energy a term in $\dot{\gamma}_i$ and $\ddot{\gamma}_i$ expressing the fact that we want the strokes and the parameterization γ_i to be smooth. We also need to express by an energy how the strokes are supposed to

3.1. Finding Salient Contours in Images

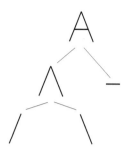

Figure 3.2. Example of the decomposition of a character into a parse tree.

fit together. This is done by a parse tree, the sort of thing seen in grammars for languages, which represents the decomposition of a character into its elementary parts. Figure 3.2 is a simple illustration with the character A.

The parse tree is a labeled tree such that the leaves are the strokes γ_i and the root is labeled by the name of the character. Some constraints must be satisfied in order to reconstruct reasonable-looking characters. For example, there is the constraint that the horizontal middle stroke of the character A must be between the two oblique strokes. These are expressed by using an analog of the binding energy from Chapter 1 to identify word boundaries. Putting all this together, the energy we have to minimize is

$$E = \int_S (I(x) - a)^2 dx + \int_{R-S} (I(x) - b)^2 dx$$
$$+ \sum_i \int_0^{L_i} \left(\alpha^2 \| \dot{\gamma}_i \|^2 + \beta^2 \| \ddot{\gamma}_i \|^2 \right) ds$$
$$+ \sum \text{ parse tree binding-energy terms.}$$

If the image and the strokes are sampled with a finite set of pixels, we can easily form a probability model with probability proportional to e^{-E}. Figure 3.23 at the end of the chapter shows some samples from such a model.

3.1 Finding Salient Contours in Images

Images are full of contours that separate regions with significantly different gray levels. They occur in all images as the boundaries of objects present (where foreground ends and background begins). They also occur as a consequence of illumination (e.g., at the edge curves of a polyhedral object where the tangent plane to the surface has a discontinuity), as well as at shadow edges. And they occur where the surface reflectance—its albedo—suddenly changes (e.g., due to paint or discoloration or moisture). Such contours appear in the license plate images of Figure 3.1 as the edges of the alphanumeric characters on the plates and the edges of the plates. This section and Section 3.2 consider the detection of such contours.

The local evidence for contours is given by the discrete image gradient of the image. Discrete derivatives can be calculated in several ways:

$$\nabla_{\text{centered}} I(i,j) = \left(\frac{I(i+1,j) - I(i-1,j)}{2}, \frac{I(i,j+1) - I(i,j-1)}{2} \right), \text{ or}$$

$$\nabla_{2 \times 2} I(i,j) = \tfrac{1}{2}(I(i+1,j) + I(i+1,j+1) - I(i,j) - I(i,j+1),$$
$$I(i,j+1) + I(i+1,j+1) - I(i,j) - I(i+1,j)). \quad (3.1)$$

These amount to convolving the image I with the filters with values

$$\tfrac{1}{2} \cdot \begin{pmatrix} 0 & 0 & 0 \\ -1 & 0 & 1 \\ 0 & 0 & 0 \end{pmatrix}, \tfrac{1}{2} \cdot \begin{pmatrix} 0 & 1 & 0 \\ 0 & 0 & 0 \\ 0 & -1 & 0 \end{pmatrix}, \tfrac{1}{2} \cdot \begin{pmatrix} -1 & 1 \\ -1 & 1 \end{pmatrix}, \tfrac{1}{2} \cdot \begin{pmatrix} 1 & 1 \\ -1 & -1 \end{pmatrix}$$

Using either variant, the discrete image gradient can be expected to be large at the pixels on such contours:

$$\|\nabla I(i,j)\| > \text{some threshold.}$$

Such a test gives you a set of points in the image containing large parts of most important contours that you would pick out, but it also gives lots of small fragments of contours, broad bands about the true contours, and gaps in these; see Section 3.1.1, Figure 3.4. Calculating the gradient is only the first most basic step in finding edges.

Here we take up a new method, different from those in previous chapters, of using statistics to locate salient patterns. This is the method of hypothesis testing, in which we do not model the distribution of the patterns but rather use a model for the signals in the *absence* of any pattern. We show two such methods for detecting salient contours. Then Section 3.2 seeks stochastic models for the contours themselves.

3.1.1 Hypothesis Testing: Straight Lines as Nonaccidental Groupings

Hypothesis testing, also known as *a contrario* algorithms, is based on a probability model not for the patterns to be recognized, but for random samples without

3.1. Finding Salient Contours in Images

the pattern: the so-called *null hypothesis*. Instead of seeking patterns in our sample that are highly likely in the model, we seek patterns in the sample that are highly *unlikely* under the null hypothesis. See also the book of A. Desolneux, L. Moisan, and J.-M. Morel [59] for a detailed presentation of a large number of *a contrario* methods for image analysis.

Straight contours are the simplest kind of contours we may find in an image. But these straight contours are often cut into pieces due to, for example, foreground objects occluding parts of the line, noise, or effects of illumination. We have to complete them by the binding of shorter segments according to some grouping principles. We describe here a method [56] based on hypothesis testing that is used, at the same time, for straight contour detection and for straight contour completion. This method takes as its input an orientation assigned to each pixel (e.g., the orientation can be the direction orthogonal to the gradient of the image), and its output is a set of straight line segments that are believed to be nonaccidental groupings of these points.

Let us first describe more precisely the framework of this method. We consider a discrete image I with size $N \times N$ pixels. At each pixel x, we can compute as in Equation (3.1) the discrete gradient $\nabla_{2 \times 2} I(x)$, and then, at the points where this gradient is not zero, we define an orientation $\theta(x) \in [0, 2\pi)$ as the orientation of the gradient rotated by $\pi/2$. This orientation is equal to the orientation of the level line $\{y | I(y) = I(x)\}$ of the image I passing through x (in other words, if the pixel x is on the boundary of an object, then the gradient $\nabla I(x)$ is orthogonal to the boundary of the object and the orientation $\theta(x)$ is the orientation of the tangent to the boundary at this point).

A discrete straight segment S in the image is determined by its beginning and ending pixels (x_b, x_e). Along its length, we can define the sequence of pixels $S = (x_b = x_1, x_2, \ldots, x_{l(S)} = x_e)$ closest to the straight line and at distance 2 along the line (so that the gradients are computed on disjoint 2×2 neighborhoods). Let $l(S)$ denote the length of this sequence and θ_S denote the orientation of the segment (see Figure 3.3). A "straight contour" in the image may be defined as a straight segment along which pixels have their orientation aligned with that of the segment according to a given precision $\alpha\pi$. Since images do not have infinite resolution and are often blurry and noisy, and since human vision also cannot see orientations with infinite accuracy, we will not take α too small. Generally, in the experiments, we take $\alpha = 1/16$. On each straight segment $S = (x_1, x_2, \ldots, x_{l(S)})$, we can count the number $k(S)$ of aligned orientations it contains:

$$k(S) = \sum_{i=1}^{l(S)} \mathbb{I}_{\{|\theta_S - \theta(x_i)| \leq \alpha\pi\}},$$

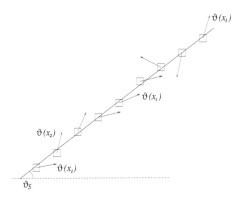

Figure 3.3. A discrete straight segment S is a sequence of pixels $S = (x_1, x_2, \ldots, x_l)$ at distance 2 along the line. Each pixel has an orientation $\theta(x_i)$. We count the number $k(S)$ of pixels that have their orientation aligned with that of the segment, according to a given precision $\alpha\pi$.

where $\mathbb{I}_{\{\cdot\}}$ denotes the indicator or characteristic function of an event: it is 1 if the condition is satisfied and 0 otherwise.

In previous chapters we have used stochastic models for the signal and likelihood ratios to determine whether a hypothetical structure is present or not. Here we introduce the *a contrario* or null-hypothesis method, the standard method of frequentist statistics. We must introduce a model for *the absence of the structure*, the so-called *null hypothesis*. In the case of aligned pixels, the null hypothesis is the following:

All image orientations are independent uniform random samples from $[0, 2\pi)$.

The method of hypothesis testing is based on calculating the probability p of the data under the null hypothesis. If this probability is quite small, we say it is quite likely that the null hypothesis is false, and hence that the structure is really present and not an accident. Conventionally, values such as $p < .05$ or $p < .01$ are considered strong evidence for rejecting the null hypothesis.

In our case, the event that is unusual under the null hypothesis is that $k(S)$ is large, specifically $k(S) \geq k_0$. The probability of $k(S) \geq k_0$ under the null hypothesis is:

$$\mathcal{B}(l(S), k_0, \alpha) = \sum_{k=k_0}^{l(S)} \binom{l(S)}{k} \alpha^k (1-\alpha)^{l(S)-k}.$$

Thus, $\mathcal{B}(l, k, \alpha)$ denotes the tail of the binomial distribution with parameters l and α. We could then declare a straight segment S to be *significant*, that is, due

3.1. Finding Salient Contours in Images

to something real in the world and not to an accidental alignment in the image, if $\mathcal{B}(l(S), k(S), 1/16) \leq .01$.

However, this is not right: we would be falling into a common error in hypothesis testing. We are testing not one event, the significance of one line segment, but a whole collection of N^4 events (indeed, we have N^2 choices for the starting point and N^2 for the endpoint). So under the null hypothesis, some of these N^4 events will have large values of $k(S)$ purely by accident. The remedy is to make the Bonferroni correction: if some trial has probability p of being positive and we have m trials, then an upper bound for the probability of one of these m trials being positive is mp. (See, for a reference, [106].) This upper bound comes from the fact that for any set of events $A_1, A_2, \ldots, A_m$, then

$$\mathbb{P}(\cup_{i=1}^m A_i) \leq \sum_{i=1}^n \mathbb{P}(A_i).$$

Notice that the Bonferroni correction is valid for any set of trials; they do not need to be independent. This is what happens here for straight segments since they can overlap or intersect.

Therefore, we associate with each segment S the expected number of line segments under the null hypothesis that are at least as unlikely as S—or the "number of false alarms" (NFA)— defined by

$$\text{NFA}(S) = N^4 \times \mathcal{B}(l(S), k(S), \alpha).$$

Then, if $\text{NFA}(S) < \varepsilon$, we say that the segment S is an *ε-significant straight contour* in the image. This is an estimate for the number of false alarms: defining a significant straight segment as a segment S such that $\text{NFA}(S) < \varepsilon$ ensures that in an image of noise (where the orientations are iid uniformly on $[0, 2\pi)$) with the same size $N \times N$, the expected number of significant segments will be $< \varepsilon$. Significant segments are observed events with p-values less than ε/N^4. Two examples are shown in Figure 3.4 to illustrate (a) how long straight edges even with low contrast can be detected and (b) how straight edges that are partially occluded can be found and completed behind the foreground object.

If S is a line segment that we perceive in the image, not only is S likely to be ε-significant for small ε, but all line segments that are slightly longer or shorter than S are likely to be significant too. So the most useful concept is that of a *maximally significant* line segment: one whose NFA is less than that of both longer and shorter line segments in the same line. Note that maximally significant line segments can often jump gaps where the image orientation is not aligned with the segment. This contour completion is a very important feature in edge detection.

118 3. Character Recognition and Syntactic Grouping

Figure 3.4. Image of an aqueduct (top row), and scan of a painting by Uccello (bottom row; "Presentazione della Vergine al tempio," Paolo Uccello, c. 1435, Duomo di Prato, Italy). On the left is the original image; in the middle, the gradient thresholded at what seemed to us the most informative value; and on the right, the maximally significant straight lines (for the aqueduct, these are in white on top of the original image to clarify their location). Note how the thresholded gradient misses highly salient lines such as the top of the aqueduct and has gaps such as the stairs in the Uccello, while picking up random-seeming bits on the hills behind the aqueduct and the wall to the left in the Uccello. The aqueduct illustrates how long straight contours are significant even if their gradients are small. The Uccello illustrates how occluded contours can be completed by the method: here black points are points that belong to a significant segment where the image orientation is aligned with that of the segment, and gray points are points that belong to a significant segment but do not have their orientation aligned. The stairs are obtained as significant segments despite occlusion caused by the child.

3.1.2 Significant Level Lines

Most contours in images, however, are not straight. Can we extend the null-hypothesis method to curved lines? A difficulty is that there are an astronomical number of curved lines. A better approach is to seek salient contours given by the *level lines* of the image. The level lines of a gray-level image I provide a complete description of the image. Their set is called the "topographic map," by analogy with geography, where the gray level is seen as an elevation. For a gray-level image I, we first define the upper and lower closed level sets by

$$\forall \lambda \in \mathbb{R}, \quad \chi^\lambda = \{x \text{ such that } I(x) \geq \lambda\} \quad \text{and} \quad \chi_\lambda = \{x \text{ such that } I(x) \leq \lambda\}.$$

The level lines are then defined as the boundaries of the connected components of these level sets. The level lines, indexed by the associated gray level, entirely determine the image I. Moreover, level lines are simple closed curves (which may get closed by including part of the boundary of the image), and if we denote by D and D' the domains, respectively, enclosed by two level lines, we either have

3.1. Finding Salient Contours in Images

$D \subset D'$ (or $D' \subset D$), or $D \cap D' = \emptyset$. Thus, linking level lines to embedded level lines puts a tree structure on the set of all level lines that enables us to recover I up to a monotone contrast change. The level lines contain all of the geometric information of the image, but not all of them are relevant to our perception of objects. Level lines, moreover, often follow perceptually salient edges for a while, but then, when both sides of this edge become darker or lighter, the level line meanders off into regions with no strong edges. This is clear in the middle image in Figure 3.5(c) showing the level lines of the letter F from the license plate. Thus, segments of level lines need to be studied. The ones that are important for the perception and recognition processes are the ones that are highly contrasted (the difference between the gray-level values of both sides of the curve is large). Another argument in favor of segments of level lines is that they are the curves that locally maximize contrast. The reason for this comes from the following simple calculation. Let $t \to \gamma(t)$ be a curve parameterized by arc length. The contrast of the image I across the curve γ at t can then be defined as $|I(\gamma(t) + \varepsilon \vec{n}(t)) - I(\gamma(t) - \varepsilon \vec{n}(t))|$, where $\vec{n}$ denotes the unit normal to the curve and ε is small. Using a first-order approximation, we have

$$|I(\gamma(t) + \varepsilon \vec{n}(t)) - I(\gamma(t) - \varepsilon \vec{n}(t))| \simeq 2\varepsilon |<\nabla I(\gamma(t)), \vec{n}(t)>|.$$

Thus, the contrast is maximal when $\nabla I(\gamma(t))$ and $\vec{n}(t)$ are collinear, and this turns out to be the definition of a level line. Indeed, we have

$$\forall t, I(\gamma(t)) = \text{cnst} \iff \forall t, <\nabla I(\gamma(t)), \gamma'(t)> = 0$$
$$\iff \forall t, \nabla I(\gamma(t)) \text{ and } \vec{n}(t) \text{ collinear}.$$

Now, how can these significant curves be automatically extracted from an image? Let us look at criteria for the significance of level lines depending on contrast [57]. Given a gray-level image I made of $N \times N$ pixels, let H denote the cumulative distribution function or repartition function of the norm of the gradient in the image:

$$\forall t, \quad H(t) = \frac{1}{N^2} \#\{\text{pixels } x \text{ such that } \|\nabla I(x)\| \geq t\}.$$

Given a segment S of a level line C with length $l(S)$ counted in "independent" pixels (i.e., S is a sequence of pixels $x_1, x_2, \ldots x_{l(S)}$ taken at distance 2, in order to compute the gradients on disjoint neighborhoods), let μ be its minimal contrast defined by $\mu = \min_{1 \leq i \leq l(S)} \|\nabla I(x_i)\|$. The null hypothesis for the sequence of pixels $x_1, x_2, \ldots x_{l(S)}$ is that they sample the distribution of contrast randomly.[1]

---

[1] This null hypothesis does not seem to come from a probability distribution on images. Indeed, it is a probability distribution on all possible functions from the set of pixels to the reals, which we apply to the map $x \mapsto \|\nabla I(x)\|$.

In images, $H(\mu)^{l(S)}$ is the probability under the null hypothesis of their minimum contrast being at least μ.

Let N_{LL} denote the number of segments of level lines in the image. If ℓ_i is the length of the ith connected component of a level line C_i, then

$$N_{LL} = \sum_i \frac{\ell_i(\ell_i - 1)}{2}.$$

Then we can apply the Bonferroni correction, as in Section 3.1.1, to bound the likelihood, under the null hypothesis, of any level curve being contrasted at level μ by $N_{LL} \cdot H(\mu)^{l(S)}$. Therefore, we say the level line segment S is said to be an ε-*significant edge according to contrast* if

$$N_{LL} \cdot H(\mu)^{l(S)} < \varepsilon. \tag{3.2}$$

As in the case of significant straight contours, this definition means that we are interested in events that have a p-value less than ε/N_{LL}. Since N_{LL} can be seen as a number of tests, the above definition implies that the number of false positives in a noise image (with gradient distribution given by H and the same number of curves N_{LL}) will be less than ε. Moreover, we restrict our attention to *maximally significant edges*, segments of level lines that neither contain nor are contained in segments of the same level line that are more significant. Typically, if a segment of one level line is maximally significant, there will also be segments of adjacent level lines that are maximally significant. This situation can be eliminated by ignoring maximally significant edges if they are wholly included in some small neighborhood of the union of all the more significant maximally significant edges.

Figure 3.5 shows these edges for several images (for the license plate image, a Gaussian smoothing has been performed first). There are two basic problems with the definition of Equation (3.2). One is that in highly textured regions, it picks up many short level lines, none of which are especially salient. One approach is to ask for level lines that are also improbably smooth [41, 42]. The second problem is that even a very salient region may have some places where its boundary has low contrast. The letter F in Figure 3.5 is a good example. The F blurs into the plate along the top, and since there are no level lines that follow the top as well as the sides, this part of the edge is not boosted up in significance by being attached to any of the more significant edges.

The hypothesis-testing approach is extremely elegant and gives simple straightforward tests for many kinds of patterns. But it has its limitations: the best contour detector would not be limited to level lines. Segments of level lines are a good starting point, but they need then to be allowed to get grouped together according to other Gestalt principles such as "good continuation" (continuity) or closure.

3.1. Finding Salient Contours in Images

(a)

(b)

(c)

Figure 3.5. Maximally significant edges in three images: (a) a church, (b) a license plate, and (c) the F in the license plate. The left column shows the original images, the right column contains the edges. For (c), we also show in the middle all the level lines that touch the boundary of the letter F. Note that some parts of the level lines do complete the gap at the top of the letter, but they are not long enough to overcome their weak gradient. Other Gestalt principles are needed to find the top edge.

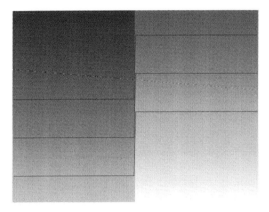

Figure 3.6. Example of an edge with varying illumination, together with three level lines of the image. Every part of the edge is now given by a level line but no single level line tracks the full length of the edge.

Consider, for example, an image of a surface that is a bit darker on one side of some edge and lighter on the other. The edge is given by any level line at an intensity between that of the two sides. Now suppose the surface is illuminated by a light that is brighter at one end of the edge and darker at the other, so that at the dark end of the edge, its bright side reflects less light than the dark side does at the brightly lit end (see Figure 3.6). Every part of the edge is now given by a level line but no single level line tracks the full length of the edge. Section 3.2 takes up a method that can find such edges.

3.2 Stochastic Models of Contours

In Section 3.1, we detected salient contours in images as exceptions from a noise-like probability model. But what are salient contours really like? Can we model them stochastically? In Chapter 2, we studied discretely sampled random processes $\{f(k\Delta t)\}$, which were given by defining probability densities on the point $(f(0), f(\Delta t), \cdots, f(n\Delta t)) \in \mathbb{R}^{n+1}$. Now we want to pass to the continuum limit and study *random functions* f given by probability measures on some space $\mathcal{F}$ of functions: these are called stochastic processes. A random curve can then be described by a pair $(\mathcal{X}(t), \mathcal{Y}(t))$ of random real-valued functions or by random orientation as a function of arc length. The simplest and most basic of all stochastic processes is Brownian motion, which we introduce next. Although the theory of Brownian motion is mathematically more complex than the concepts we have used so far, the basic ideas behind it are not complicated.

3.2. Stochastic Models of Contours

3.2.1 Basics VII: Brownian Motion

Definition 3.1 (Brownian Motion). A one-dimensional standard Brownian motion is a random continuous function $t \mapsto \mathcal{B}(t)$ defined for $t \geq 0$ such that $\mathcal{B}(0) = 0$ (that is, a probability measure[2] on the Banach space $\mathcal{F}$ of continuous functions on $\mathbb{R}_+$ with $f(0) = 0$) and

- $\mathcal{B}$ has independent increments: for all $0 = t_0 < t_1 < \ldots < t_n$, then $\mathcal{B}(t_1) - \mathcal{B}(t_0), \ldots, \mathcal{B}(t_n) - \mathcal{B}(t_{n-1})$ are independent.
- For all $s, t \geq 0$, the increment $\mathcal{B}(s+t) - \mathcal{B}(s)$ has normal distribution with mean 0 and variance t (n.b. NOT t^2!).

Is there a simple reason why the variance of an increment $\mathcal{B}(s+t) - \mathcal{B}(s)$ has to be set equal to t and not to t^2? Setting it equal to t^2 would have seemed at first more natural since it would have implied that the increment is of the order t. Instead, taking the variance equal to t implies that the increment is of the order of $\sqrt{t}$. The reason we must do this is the following: consider an increment $\mathcal{B}(s+t+t') - \mathcal{B}(s)$ and write it as the sum $\mathcal{B}(s+t+t') - \mathcal{B}(s+t) + \mathcal{B}(s+t) - \mathcal{B}(s)$. The variance of a sum of two independent random variables is the sum of the variances, which implies that we have to set the variance of $\mathcal{B}(s+t) - \mathcal{B}(s)$ equal to some fixed multiple of t for all $s, t \geq 0$.

Brownian motion is constructed as follows. As above, let $\mathcal{F}$ be the set of continuous functions $t \mapsto f(t)$ defined for $0 \leq t < \infty$ such that $f(0) = 0$. For all sequences of sample values $0 < a_1 < a_2 < \cdots < a_n$, let $\pi_{\vec{a}} : \mathcal{F} \to \mathbb{R}^n$ be the linear map

$$\pi_{\vec{a}}(f) = (f(a_1), f(a_2), \cdots, f(a_n)).$$

To define a probability measure P on $\mathcal{F}$, we first define its marginals $\pi_{\vec{a}*}P$, which are probability measures on $\mathbb{R}^n$. That is, we will require

$$P(\pi_{\vec{a}}^{-1}(U)) = \pi_{\vec{a}*}P(U)$$

for all open sets U in $\mathbb{R}^n$. From the two properties in the definition, we see that the measure $\pi_{\vec{a}*}P$ must be given by the density

$$d\pi_{\vec{a}*}P(x) = \frac{1}{Z} e^{-\left(\frac{x_1^2}{2a_1} + \frac{(x_2 - x_1)^2}{2(a_2 - a_1)} + \cdots + \frac{(x_n - x_{n-1})^2}{2(a_n - a_{n-1})}\right)} dx_1 \cdots dx_n.$$

This means that at time $t = a_1$, $f(t)$ has drifted a distance $x_1 = f(a_1)$, which is normally distributed with mean 0 and variance a_1; at time $t = a_2$, $f(t)$ has drifted

---

[2] A major technicality here is that one must specify the measurable sets in this function space, that is, the σ-algebra on which the probability measure is defined. We refer the reader to [23] for this. All the sets that arise in our discussion will be measurable.

further by a distance $x_2 - x_1 = f(t_2) - f(t_1)$, which is normally distributed with mean x_1 and variance $a_2 - a_1$; and so on. We get in this way a Gaussian measure on each space of values $(f(a_1), \cdots, f(a_n))$. As these sample points are augmented, it is not difficult to check that these measures are mutually compatible. Then *Kolmogorov's Existence Theorem* (see, for example, Chapter 7 of [23]) says that they all come from a unique measure on some space of functions. That, in fact, they are induced from a probability measure on the space of continuous functions $\mathcal{F}$ is a bit more difficult to prove. We omit this proof, but note that it is the converse to Theorem 2.13 in Chapter 2, which dealt with what happened when we have a random variable $\mathcal{X} = f(t+s) - f(s)$ that could be divided up as a sum of n independent $\mathcal{Y}_i$. For Brownian motion, we have

$$\begin{aligned} \mathcal{X} &= \mathcal{B}(t+s) - \mathcal{B}(s) \\ &= \left(\mathcal{B}(s+\frac{t}{n}) - \mathcal{B}(s)\right) + \left(\mathcal{B}(s+\frac{2t}{n}) - \mathcal{B}(s+\frac{t}{n})\right) \\ &\quad + \cdots + \left(\mathcal{B}(s+t) - \mathcal{B}(s+\frac{(n-1)t}{n})\right) \\ &= \mathcal{Y}_1 + \mathcal{Y}_2 + \cdots + \mathcal{Y}_n, \end{aligned}$$

so $\mathcal{X}$ can be divided into the sum of n independent increments. In Theorem 2.13, we showed that if the kurtosis of $\mathcal{X}$ was greater than 3, then there existed a and b independent of n such that $\mathbb{P}(\max_i |\mathcal{Y}_i| > a) > b$. This would clearly preclude the limiting function $\mathcal{B}(t)$ from being continuous with probability 1. But Gaussian variables have kurtosis equal to 3 and then the opposite happens: for any a and b, when n is large enough, $\mathbb{P}(\max_i |\mathcal{Y}_i| > a) < b$, which is the key point in checking that the limiting function $\mathcal{B}(t)$ is almost surely continuous.

The above may be called a construction of P by its marginals. Note that because of the uniqueness in Kolmogorov's Theorem, Brownian motion itself is unique. We can rewrite the basic formula for the marginals in a very suggestive form:

$$d\pi_{\vec{a}*}P(x) = \frac{1}{Z}e^{-\frac{1}{2}E(x_1,\cdots,x_n)}dx_1 \cdots dx_n,$$

where

$$E(x_1, \cdots, x_n) = \sum_{k=1}^{n}\left(\frac{x_k - x_{k-1}}{a_k - a_{k-1}}\right)^2 \cdot (a_k - a_{k-1}),$$

with $a_0 = x_0 = 0$. Note that E is just the Riemann sum for the integral $\int_0^{a_n} \mathcal{B}'(t)^2 dt$ with the intervals $0 < a_1 < \cdots < a_n$. So, in the limit, it is tempting

3.2. Stochastic Models of Contours

to write

$$dP(B\,|_{[0,a]}) = \frac{1}{Z}e^{-\frac{1}{2}E(B)} \prod_{t=0}^{t=a} dB(t),$$

$$\text{where} \quad E(B) = \int_0^a B'(t)^2 dt.$$

Indeed, physicists like to do this, but, strictly speaking, this is nonsense! As you subdivide further and further, Z goes to zero and the Riemann sum in the exponent goes to infinity because the limit $\mathcal{B}(t)$ turns out to be almost surely nowhere differentiable. This nondifferentiability is immediately clear by inspecting the sample Brownian path in (a) of Figure 3.7. But the formula does serve well as a shorthand to understand Brownian motion intuitively.

3.2.2 Using Brownian Motion to Synthesize Contours and Snakes

Brownian motion can be used as the basis for many stochastic models for contours in images. In order to synthesize plane curves, the simplest thing we can do is to let the curve be traced out by two Brownian motions,

$$\mathcal{X} = \mathcal{B}_1 \quad \text{and} \quad \mathcal{Y} = \mathcal{B}_2,$$

where $\mathcal{B}_1$ and $\mathcal{B}_2$ are two independent one-dimensional Brownian motions. The curves this defines look like tangled balls of wool, however (see graph (b) of Figure 3.7).

A better model is obtained by letting the direction Θ of the curve follow a Brownian motion:

$$\frac{d\mathcal{X}}{dt} = \cos\Theta, \quad \frac{d\mathcal{Y}}{dt} = \sin\Theta, \quad \text{and} \quad \Theta = \mathcal{B},$$

where $\mathcal{B}$ is a one-dimensional Brownian motion. This is called the *elastica model*, and it is shown in graph (c) of Figure 3.7. If a curve is sufficiently differentiable, the derivative $\Theta'(t)$ is its curvature $\kappa(t)$. Thus, by using the heuristic formula for the probability measure defining Brownian motion, the elastica model can be described by the probability density:

$$dP = \frac{1}{Z}e^{-\frac{1}{2}\int \Theta'(t)^2 dt} \prod d\Theta(t) = \frac{1}{Z}e^{-\frac{1}{2}\int \kappa(t)^2 dt} \prod d\Theta(t).$$

This is usually described as saying "$\kappa(t)$ is white noise." Although the formula is meaningless (because it equals 0/0), it does allow us to define *modes* of this

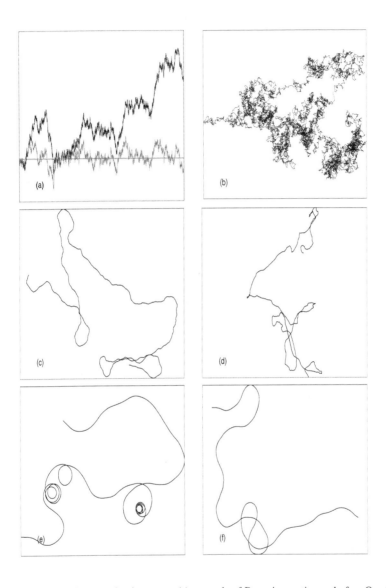

Figure 3.7. Synthesizing stochastic curves: (a) a sample of Brownian motion and of an Ornstein-Uhlenbeck driven by the same sample (the Ornstein-Uhlenbeck mirrors the Brownian but is nearly always closer to zero); (b) a curve such that the two coordinates $\mathcal{X}$ and $\mathcal{Y}$ are independent Brownian motions; (c) a curve such that the direction Θ is a Brownian motion; (d) a curve whose x- and y-velocities are independent Ornstein-Uhlenbeck processes; (e) a curve such that the curvature κ is Brownian motion; and (f) the curvature is an Ornstein-Uhlenbeck process.

3.2. Stochastic Models of Contours

distribution. These are the curves that minimize (or are critical points of) the curve energy:

$$E(\Gamma) = \int_\Gamma \kappa(t)^2 dt$$

(e.g., with fixed endpoints and fixed initial and final direction). This variational problem and its solutions were first studied by Euler in 1744 [68]. He called these curves *elastica*, hence the name given to this model. They were introduced into computer vision by Horn [108] and, later, Mumford [167]. A discrete approximation (which was used to produce Figure 37) to samples from the elastica model is given by polygons $P_k = (x_k, y_k)$, where

$$x_{k+1} = x_k + \cos(\theta_k) \cdot \Delta t,$$
$$y_{k+1} = y_k + \sin(\theta_k) \cdot \Delta t,$$
$$\theta_{k+1} = \theta_k + n_k \cdot \sqrt{\Delta t},$$

and where $\{n_k\}$ are independent Gaussian variables of mean 0 and variance 1 and the square root is needed on the last line because that is how much Brownian motion moves in time Δt.

Letting direction be Brownian makes nice curves, so let's allow curvature to be Brownian motion, or

$$\frac{d\mathcal{X}}{dt} = \cos\Theta, \quad \frac{d\mathcal{Y}}{dt} = \sin\Theta, \quad \frac{d\Theta}{dt} = \kappa, \quad \text{and} \quad \kappa = \mathcal{B}.$$

Unfortunately, this is not so interesting; it produces huge coils of rope because Brownian motion gets large (both positively and negatively) and stays large in absolute value for longer and longer periods as time goes on. If we make curvature do this, the curve will spiral around in longer and longer, tighter and tighter spirals, both clockwise and counterclockwise (see graph (d) of Figure 3.7).

The way to fix this is to put a restoring force on Brownian motion, dragging it back to the origin. In the discrete case, we can achieve this by the difference equations:

$$g_{k+1} = x_k + \cos(\theta_k) \cdot \Delta t$$
$$y_{k+1} = y_k + \sin(\theta_k) \cdot \Delta t$$
$$\theta_{k+1} = \theta_k + \kappa_k \cdot \Delta t$$
$$\kappa_{k+1} = \kappa_k - \kappa_k \cdot \Delta t + n_k \sqrt{\Delta t}$$

The new term $-\kappa_k \cdot \Delta t$ has the effect that it tries to shrink κ exponentially to zero while it is being stochastically perturbed by the white noise. Such equations are called *stochastic difference equations*. There is a highly developed theory of *stochastic differential equations*, which says that we may pass to the limit as $\Delta t \to 0$ in such equations and get solutions to equations written as

$$\frac{dx}{dt} = \cos(\theta), \quad \frac{dy}{dt} = \sin(\theta), \quad \frac{d\theta}{dt} = \kappa,$$
$$d\kappa = -\kappa \cdot dt + d\mathcal{B}.$$

The equations in the first line are standard ODEs produced by passage to the limit in the first three difference equations. The equation in the second line, however, is a new sort of limiting equation due to the square root in the last difference equation. This specific equation for κ generates what is called the Ornstein-Uhlenbeck process. It has nothing specifically to do with curvature κ; rather it is a general stochastic process like Brownian motion except that its values are pulled back to 0. Two excellent textbooks about this are Øksendal [170] or Volume 3 of Hoel, Port and Stone [107]. More concretely, this process—let's call it $\mathcal{X}(t)$ now and assume it satisfies the stochastic differential equation $d\mathcal{X} = -a\mathcal{X} \cdot dt + d\mathcal{B}$—may also be defined by its marginals, the same way Brownian motion was defined, but with new marginals:

$$dP(\mathcal{X}(t+s) = x \mid \mathcal{X}(s) = x_0) = \frac{1}{\sqrt{2\pi t_a}} e^{-(x - e^{-at}x_0)^2 / 2t_a} dx,$$
$$\text{where} \quad t_a = \frac{1 - e^{-2at}}{2a}.$$

Note that we are (a) shrinking the expected value of $\mathcal{X}(t+s)$ from $\mathcal{X}(s)$ to $e^{-at}\mathcal{X}(s)$, and (b) shrinking the variance of the random change from t to t_a because during time t, the random changes at first get discounted by the shrinkage. [We can check that this is right by checking that the variance of $\mathcal{X}(t_1 + t_2 + s) - e^{-a(t_1+t_2)}\mathcal{X}(s)$ is the sum of that of $e^{-at_1}(\mathcal{X}(t_2 + s) - e^{-at_2}\mathcal{X}(s))$ and $\mathcal{X}(t_1+t_2+s) - e^{-at_1}\mathcal{X}(t_2+s)$.] Samples of this process are illustrated in graph (a) of Figure 3.7. Here we have plotted the Brownian motion which drives this sample of the Ornstein-Uhlenbeck process as well as the sample process itself. We can see that they share the same small features, but where the Brownian motion makes large excursions from zero, the Ornstein-Uhlenbeck is held back and never wanders far from zero. If we let the x- and y-velocities of a path follow independent Ornstein-Uhlenbeck processes, we get samples such as graph (d). They are roughly similar to letting the direction Θ be Brownian, as in graph (c), but differ in that the Ornstein-Uhlenbeck pair can both be small, and hence the curve slows down and often makes a corner, or they can both be large, and hence the curve has extended parts with similar direction as though it was following a

3.2. Stochastic Models of Contours

highway. Finally, we can let the curvature evolve by an Ornstein-Uhlenbeck process, and we get samples that are quite nice smooth random-looking curves; see graph (f) of Figure 3.7.

3.2.3 Snakes and Their Energy

The above stochastic models are actually not the most common and simplest stochastic models for contours, namely the *snake models*. We introduce these in a different way and then make the connection with the elastica model of which it is a simple variant. This approach uses parameterized curves $\gamma(t)$, for $t \in [0, L]$, t not necessarily arc length. Let $P_i = \gamma(i\Delta t)$, where $\Delta t = L/N$, be uniformly placed samples so that $\Gamma = \{P_0, \ldots, P_N\} \in \mathbb{R}^{2N+2}$ is a polygonal approximation to the continuous curve. Notice that modulo translations of Γ are in $\mathbb{R}^{2N}$. In the discrete case, we put a Gaussian model on the set of Γ's modulo translations:

$$p(\Gamma) = \frac{1}{Z} e^{-\frac{1}{2}E(\Gamma)}$$

$$E(\Gamma) = \alpha^2 \sum_{k=1}^{N} \left\| \frac{P_k - P_{k-1}}{\Delta t} \right\|^2 \Delta t + \beta^2 \sum_{k=2}^{N} \left\| \frac{P_k - 2P_{k-1} + P_{k-2}}{\Delta t^2} \right\|^2 \Delta t.$$

This snake energy E is the discrete version of the continuous energy $\tilde{E}$ defined by

$$\tilde{E}(\gamma) = \alpha^2 \int_0^L \| \dot{\gamma} \|^2 \, dt + \beta^2 \int_0^L \| \ddot{\gamma} \|^2 \, dt.$$

Note that if the curve is parameterized by arc length, then $\| \dot{\gamma}(t) \| \equiv 1$ and $\| \ddot{\gamma}(t) \| = \kappa(t)$, the curvature. Then, if L is fixed, the snake energy reduces to the elastica energy up to a constant.

To describe the continuous stochastic model given by snakes, we need to reduce it in some way to Brownian motion. The simplest method is to diagonalize the discrete energy. First, let us write the velocity of the curve as $(u_k, v_k) = \frac{P_k - P_{k-1}}{\Delta t}$ so that

$$E(\Gamma) = \alpha^2 \sum_{k=1}^{k=N} (u_k^2 \Delta t + v_k^2 \Delta t) + \beta^2 \sum_{k=2}^{k=N} \frac{(u_k - u_{k-1})^2}{\Delta t} + \frac{(v_k - v_{k-1})^2}{\Delta t}.$$

Start with the identity

$$\sum_{k \in \mathbb{Z}} (au_k + bu_{k-1})^2 = (a^2 + b^2) \sum_{k \in \mathbb{Z}} u_k^2 + 2ab \sum_{k \in \mathbb{Z}} u_k u_{k-1}$$

$$= (a+b)^2 \sum_{k \in \mathbb{Z}} u_k^2 - ab \sum_{k \in \mathbb{Z}} (u_k - u_{k+1})^2.$$

To make the right-hand side coincide with E (except for endpoint terms), we want $(a+b)^2 = \alpha^2 \Delta t$ and $-ab = \frac{\beta^2}{\Delta t}$, which is solved by setting

$$a = \tfrac{1}{2}\left(\alpha\sqrt{\Delta t} - \sqrt{4\beta^2/\Delta t + \alpha^2 \Delta t}\right),$$
$$b = \tfrac{1}{2}\left(\alpha\sqrt{\Delta t} + \sqrt{4\beta^2/\Delta t + \alpha^2 \Delta t}\right).$$

Then, we calculate the endpoint terms and find

$$E(\Gamma) = \sum_{k=2}^{k=N} (au_{k-1} + bu_k)^2 + (ab+b^2)u_1^2 + (ab+a^2)u_N^2$$
$$+ (av_{k-1} + bv_k)^2 + (ab+b^2)v_1^2 + (ab+a^2)v_N^2.$$

Ignoring the endpoint terms again, the stochastic model is now a diagonal Gaussian model:

$$p(\Gamma) = \frac{1}{Z} e^{-\frac{1}{2} \sum_{k=2}^n (au_{k-1}+bu_k)^2 + (av_{k-1}+bv_k)^2}.$$

That is, the terms $au_{k-1} + bu_k$ and those for v are independent standard Gaussian variables n_k (n for "noise"). Therefore, setting $au_{k-1} + bu_k = n_k$, we get

$$u_k - u_{k-1} = (-\frac{a}{b} - 1)u_{k-1} + \frac{1}{b}n_k$$
$$\approx -\frac{\alpha}{\beta}u_{k-1}\Delta t + \frac{\sqrt{\Delta t}}{\beta}n_k \quad \text{if } \Delta t \ll \beta/\alpha,$$

which is now in the form of the Ornstein-Uhlenbeck stochastic difference equation. Passing to the limit as $\Delta t \to 0$, this becomes the Ornstein-Uhlenbeck stochastic differential equation for the random velocities $u(t), v(t)$:

$$du = -\frac{\alpha}{\beta}u \cdot dt + \frac{1}{\beta}d\mathcal{B}_1$$
$$dv = -\frac{\alpha}{\beta}v \cdot dt + \frac{1}{\beta}d\mathcal{B}_2,$$

where $\mathcal{B}_1$ and $\mathcal{B}_2$ are independent Brownian motions. We have now shown that the Gaussian snake model on polygons with which we started has a continuous limit given by letting the velocities follow Ornstein-Uhlenbeck processes (integrating to get the random curve). A sample is shown in graph (d) of Figure 3.7.

3.2.4 Contour Detection

We now have several simple stochastic models of contours. We can apply these models to detect contours Γ in images I by the use of Bayes' rule (see Equation 1, Section 0.3):

$$P(\Gamma|I) \propto P(I|\Gamma) \cdot P(\Gamma).$$

3.2. Stochastic Models of Contours

 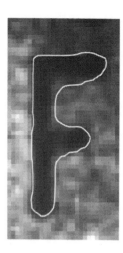

Figure 3.8. Two examples of snakes. The windows in the church shown in Figure 3.5 (left) and the letter F from the same figure. Both are local minima for geometric snake energy with $\beta = 0$ and $h =$ identity, but the snake around the F is not very robust—small perturbations take us out of its basin of attraction. In all cases, α must be chosen carefully to balance the shrinking force with the attraction of high gradients.

The first factor should express the fact that images are usually contrasted across contours (i.e., their gradients should be large). The second factor can be any of the contour models discussed so far. This method is known as "active contour models," especially when it is applied by starting with some specific contour and evolving the contour to gradually increase $P(\Gamma|I)$ (see, for instance, [122], [76], or [45]). If we take the negative logs, we get an energy version of each method. For a given I, associate a geometric energy $E_g(\gamma)$ to an unparameterized contour γ, and a parametric energy $E_p(\gamma)$ to a parameterized contour γ :

$$E_g(\gamma) = \int_0^{\ell(\gamma)} (\alpha^2 + \beta^2 \kappa(s)^2) \cdot ds + \int_0^{\ell(\gamma)} h(|\nabla I(\gamma(s))|) \cdot ds,$$

where s is arc length (elastica model)

$$E_p(\gamma) = \int_0^L (\alpha^2 \|\dot{\gamma}\|^2 + \beta^2 \|\ddot{\gamma}\|^2) \cdot ds + \int_0^L h(|\nabla I(\gamma(s))|) \cdot ds,$$

where s is arbitrary (snake model)

and $\kappa(s)$ denotes the curvature of γ. We seek contours that minimize these energies locally. Examples are shown in Figure 3.8. Note that, in general, the obtained curves do not coincide with level lines of the image I, but for some choices for the constants and the function h, snakes and significant level lines become very close [58].

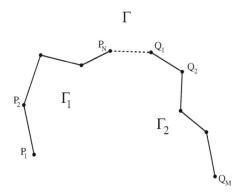

Figure 3.9. Example of binding of two snakes Γ_1 and Γ_2.

3.2.5 Likelihood Ratio for Grouping Snakes

Snakes are, by definition, connected. We have seen, however, that salient contours in images almost always have breaks caused by noise, low contrast, occlusion, or misleading illumination. To complete the contour, we must jump across these breaks. Just as for straight contours, to complete arbitrary contours, we need a way to group snakes. We describe here one way to decide whether to group two snakes together by using a likelihood ratio, just as we did in Chapter 1 to group letters within words. Let $\Gamma_1 = \{P_1, P_2, \ldots, P_N\}$ and $\Gamma_2 = \{Q_1, Q_2, \ldots, Q_M\}$ be two snakes. We want to compute the probability of the composite structure $\Gamma = \{P_1, P_2, \ldots, P_N, Q_1, Q_2, \ldots, Q_M\}$ made by concatenation of the two curves Γ_1 and Γ_2 (Figure 3.9). More precisely, we want to compute the likelihood ratio $P(\Gamma)/P(\Gamma_1)P(\Gamma_2)$ to know when Γ_1 and Γ_2 should be grouped in order to have a gain (e.g., of code length). Although this calculation is rather special, it demonstrates how the idea of likelihood ratios can be used to quantify grouping principles; hence let's take the time to work this out.

Following Section 3.2.3, ignoring the image, the prior probability of the snake Γ (modulo translations) is given by

$$P(\Gamma) = \frac{1}{Z_{N+M}} e^{-\alpha^2 \sum_{k=2}^{N} \|P_k - P_{k-1}\|^2 - \beta^2 \sum_{k=3}^{N} \|P_k - 2P_{k-1} + P_{k-2}\|^2}$$

$$\times\, e^{-\alpha^2 \sum_{k=2}^{M} \|Q_k - Q_{k-1}\|^2 - \beta^2 \sum_{k=3}^{M} \|Q_k - 2Q_{k-1} + Q_{k-2}\|^2}$$

$$\times\, e^{-\alpha^2 \|P_N - Q_1\|^2 - \beta^2 \|(P_N - P_{N-1}) - (Q_1 - P_N)\|^2 - \beta^2 \|(Q_1 - P_N) - (Q_2 - Q_1)\|^2}.$$

3.2. Stochastic Models of Contours

Thus, the likelihood ratio for comparing grouped with ungrouped snakes is

$$\frac{p(\Gamma)}{p(\Gamma_1)p(\Gamma_2)} =$$
$$\frac{Z_N Z_M}{Z_{N+M}} e^{-\alpha^2 \|P_N - Q_1\|^2 - \beta^2 \|(P_N - P_{N-1}) - (Q_1 - P_N)\|^2 - \beta^2 \|(Q_1 - P_N) - (Q_2 - Q_1)\|^2}.$$

We have to be a bit careful here: the snake model is defined on polygons mod translations and when we glue two smaller snakes to form one big snake, we must fix the *relative* translation. The simplest approach is to beef up the snake models to probability models on snakes with position by assuming one of its points lies in some area D and follows a uniform probability distribution in D. This simply adds a factor area(D) to the normalizing constant Z. To get a likelihood ratio, let's assume the final endpoint P_N of Γ_1 lies in D, as does the initial endpoint Q_1 of Γ_2. For the big curve Γ, we can assume either that P_N or Q_1 lies in D: if D is sufficiently large compared to $1/\alpha$ *i.e., how close the vertices of almost all snakes are) this makes no appreciable difference. Now we have probability models both of Γ and of the pair (Γ_1, Γ_2) in $\mathbb{R}^{2N+2M}$, each with one extra factor area$(D)^{-1}$, and we can take the ratio of their densities.

Notice that this likelihood ratio is only a function of the endpoints (P_{N-1}, P_N, Q_1, and Q_2) and of the lengths N and M of the two snakes Γ_1 and Γ_2. The next step is to compute the normalizing constants Z_N, Z_M, and Z_{N+M}. Since the probability model is Gaussian and uniform, these constants are related to the determinant of quadratic forms. More precisely, we have $Z_N = \text{area}(D) \cdot \pi^{N-1} \cdot \det(Q_N)^{-1}$, where Q_N is the matrix of the quadratic form

$$(x_1, \ldots x_{N-1}) \mapsto \alpha^2 \sum_{k=1}^{N-1} x_k^2 + \beta^2 \sum_{k=2}^{N-1} (x_k - x_{k-1})^2.$$

Let us now denote $r = \alpha^2/\beta^2$ and assume that $N \geq 3$; we then have

$$Q_N = \beta^2 \begin{pmatrix} r+1 & -1 & 0 & 0 & 0 & \cdots \\ -1 & r+2 & -2 & 0 & 0 & \cdots \\ 0 & -1 & r+2 & -1 & 0 & \cdots \\ 0 & \cdots & \cdots & \cdots & \cdots & 0 \\ \cdots & 0 & 0 & -1 & r+2 & -1 \\ \cdots & 0 & 0 & 0 & -1 & r+1 \end{pmatrix}.$$

The determinant of Q_N is not especially nice, but miraculously, the ratio $\det(Q_{N+M})/(\det(Q_N) \cdot \det(Q_M))$ can be given by a relatively simple explicit formula. This is done in Exercise 2 at the end of the chapter. The result is expressed in terms of the number

$$\lambda = (r + 2 + \sqrt{r^2 + 4r})/2 \in (r+1, r+2)$$

whose inverse governs how far influence spreads along the snake (e.g., the correlation between tangent vectors will decrease like λ^{-k} for k intervening vertices):

$$\frac{\det(Q_{N+M})}{\det(Q_N) \cdot \det(Q_M)} = \alpha^2 \cdot \frac{\lambda^2 - 1}{r}$$
$$\cdot \left(1 + \lambda^{-2N} + \lambda^{-2M} + \text{higher powers of } \lambda^{-1}\right).$$

An exact formula is derived in Exercise 2. For N and M not too small, this estimates the log-likelihood ratio:

$$\log\left(\frac{p(\Gamma)}{p(\Gamma_1)p(\Gamma_2)}\right) \approx \left(\log\left(\frac{\text{area}(D).\alpha^2}{\pi}\right) - \alpha^2 \parallel P_N - Q_1 \parallel^2\right)$$
$$+ \left(\log\left(f(\frac{\beta^2}{\alpha^2})\right) - \beta^2 \parallel P_{N-1} - 2P_N + Q_1 \parallel^2 - \beta^2 \parallel P_N - 2Q_1 + Q_2 \parallel^2\right),$$

where $f(t) = (t + \frac{1}{2})\sqrt{4t + 1} + 2t + \frac{1}{2}$. The first line contains the α terms and simply measures the proximity of the endpoints; the second line, which vanishes if $\beta = 0$, contains the second derivative terms. Given numerical values for α and β, the above formula provides a threshold to decide when two snakes Γ_1 and Γ_2 should be grouped together.

3.3 The Medial Axis for Planar Shapes

Up to now, we have focused on salient contours in images. But images are functions on a two-dimensional domain Ω, and most of their patterns are associated with regions $S \subset \Omega$. In our motivating example, S will be the black area of the character. The boundary of such regions is a salient contour γ and is easily detected (in the absence of noise) by any of the methods studied so far. But this contour is not always a great descriptor for S; for instance, we do not know when two points $\gamma(s_1)$ and $\gamma(s_2)$ are nearby, as happens, for example, on opposite sides of a "neck" or on opposite sides of an elongated stroke in the shape. A much more effective way to describe what a shape S is like is to use a different curve, the axis of the shape.

3.3.1 Definitions

We give here the definition and main properties of the medial axis, or skeleton, of a shape. The medial axis was first introduced by the mathematical biologist Harry Blum [25] and has been used since then in many areas of applied mathematics [199]. It is a good shape descriptor and is frequently used in shape categorization. It is very effective, in particular, for our application to character recognition. In

3.3. The Medial Axis for Planar Shapes

the following, a *shape* S will be defined as a bounded connected open subset of $\mathbb{R}^2$ such that its boundary ∂S is the finite union of smooth curves, each one having a finite number of local maxima of curvature. These are the generic shapes from a deformation point of view. We also denote the open unit disk of the plane $\mathbb{R}^2$ by Δ.

Definition 3.2 (Medial Axis). Let S be a shape. We say that an open disk is a maximal disk if it is contained in the shape S and if there exists no other disk inside the shape that properly contains it. The medial axis $Ax(S)$ of the shape S is then defined as the set of the centers of all maximal disks:

$$Ax(S) = \{P \in \mathbb{R}^2 \mid \exists r \geq 0 \text{ such that } (P + r\Delta) \text{ is a maximal disk}\}.$$

The same definition works, of course, for any open subset of the plane. It has been investigated in this case in [49], but these generalizations, for which the medial axis can get quite complex, are not relevant here.

Notice that given the medial axis $Ax(S)$ and given for each $P \in Ax(S)$ the radius $r(P)$ of the maximal disk centered in P, then S may be recovered by

$$S = \bigcup_{P \in Ax(S)} (P + r(P)\Delta).$$

From the point of view of differential geometry, $Ax(S)$ is a subset of the "symmetry set" of S, which is defined as the locus of the centers of all bitangent circles (not only the ones contained in the shape).

There are various equivalent definitions of the medial axis of a shape, two of which follow.

- For each point P in the shape S, consider its distance from the boundary

$$d(P, \partial S) = \min_{Q \in \partial S} \| P - Q \|.$$

Then the medial axis $Ax(S)$ is defined as the set of points where the function $P \mapsto d(P, \partial S)$ is not differentiable.

- Another possible definition of the medial axis is based on curve evolution. Let the curve ∂S shrink by the following evolution:

$$\forall P \in \partial S, \ f_t(P) = P + t \vec{n}_{P,\partial S},$$

where $\vec{n}_{P,\partial S}$ denotes the inward unit normal to ∂S at point P. The medial axis is then the set of the "shocks" of this propagation (i.e., the first points

136   3. Character Recognition and Syntactic Grouping

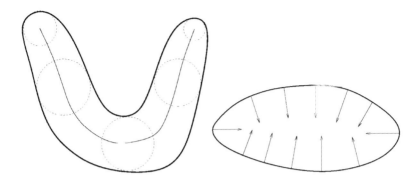

Figure 3.10. An example of the medial axis and of some maximal disks of a shape (left); the idea of "prairie fire" defining the medial axis as the locus of quench points (right).

$f_t(P)$ that are equal to another $f_t(P')$). This equation of evolution is sometimes called "prairie fire," and $Ax(S)$ is the set of quench points (see the right-hand diagram in Figure 3.10). If we modify $f_t(P)$ to stop as soon as $f_t(P) = f_t(P')$, we get a deformation retraction of the whole region onto its medial axis. In particular, *the medial axis is connected, is a tree if the region is simply connected, and otherwise has the same number of loops as the region has holes.*

Yet another way to define the medial axis is by lifting the curve to the 2-sphere via stereographic projection and taking its convex hull—see Exercise 3 at the end of the chapter. This construction has become a key tool in the bridge between complex analysis and three-dimensional topology pioneered by Thurston (see [148], page 144). Several interesting mathematical properties of the medial axis can be investigated. We will not give the proofs of these results here; instead we refer, for example, to [49] and to the book by J. Serra *Image Analysis and Mathematical Morphology* [192].

3.3.2 Properties of the Medial Axis

The medial axis is a graph with smooth branches and two types of singular points: the end points and the bifurcation points (see examples in Figure 3.11).

- An endpoint of $Ax(S)$ corresponds to the center of an osculating circle to ∂S at a point where the curvature is positive and locally maximal (i.e., a "vertex" of the shape). Recall that we have assumed there are only a finite number of these.

3.3. The Medial Axis for Planar Shapes

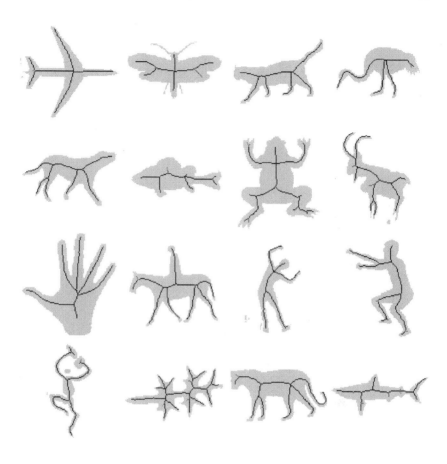

Figure 3.11. A set of outlines of natural shapes and their medial axes, slightly improved following [240] (courtesy of Song-Chun Zhu [235, Figure 21]).

- If a circle is tangent at k points, where $k > 2$, the medial axis has k branches meeting at the center of the circle (see Figure 3.12). Note that k is also finite because the axis has only a finite number of endpoints.

We can make explicit the reconstruction of the parts of the shape S corresponding to the nonsingular parts of $Ax(S)$. The nonsingular parts of $Ax(S)$ can be described by a curve $\gamma(t)$, where t is the arc length, and by a function $r(t)$ corresponding to the radius of the maximal disk centered in $\gamma(t)$ (see also Figure 3.13). The two sides of the boundary curve ∂S can be described explicitly in terms of

3. Character Recognition and Syntactic Grouping

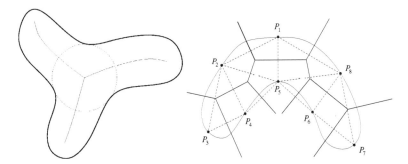

Figure 3.12. Example of singularities of the medial axis: three endpoints and a triple point (left); the Voronoi decomposition and Delaunay triangulation for eight points on the boundary of a shape (right). The obtained result V is an approximation of the medial axis and other pieces of curves called "hairs." The Delaunay triangulation of this set of points is plotted with dashed lines: it is the "dual" of the Voronoi decomposition. The centers of the circumscribed circles to the triangles are the vertices of the medial axis.

these functions by the parameterization:

$$P_\pm(t) = \gamma(t) + r(t)\left(-r'(t)\gamma'(t) \pm \sqrt{1-r'(t)^2}\gamma'(t)^\perp\right).$$

Differentiating we find that arc length $s_\pm$ on them (increasing in the same direction as on the axis) is given by

$$ds_\pm = \left(\frac{1-(r^2/2)''}{\sqrt{1-r'^2}} \pm r\kappa\right) \cdot dt,$$

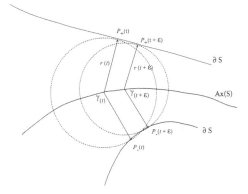

Figure 3.13. Description of a shape using the medial axis (curve $t \mapsto \gamma(t)$) and a function $r(t)$ corresponding to the radius of the maximal disk centered in $\gamma(t)$.

3.3. The Medial Axis for Planar Shapes

where κ is the curvature of the axis γ. It follows that γ and r have to satisfy the following three conditions:

$$\begin{cases} |r'| < 1 \\ (r^2/2)'' < 1 \\ |\kappa| r\sqrt{1 - r'^2} < 1 - (r^2/2)'' \end{cases} \quad (3.3)$$

Proof: Let x be a contact point corresponding to the point $(\gamma(t), r(t))$ of the medial axis, and let f be the function defined for $|\epsilon|$ small by

$$f(\epsilon) = |x - \gamma(t + \epsilon)|^2 - r(t + \epsilon)^2.$$

Then, by definition of x, $f(0) = 0$ and, moreover, f has a local minimum for $\epsilon = 0$ (because for all $|\epsilon|$ small, x has to be outside the disk of center $\gamma(t + \epsilon)$ and radius $r(t + \epsilon)$). Thus $f'(0) = 0$ and $f''(0) \geq 0$. We compute the derivatives of f, and the previous conditions become

$$\langle x - \gamma(t), \gamma'(t) \rangle = -r'(t) r(t) \quad \text{and}$$

$$|\gamma'(t)|^2 - \langle x - \gamma(t), \gamma''(t) \rangle - r'(t)^2 - r(t) r''(t) \geq 0.$$

Then, using the fact that $|x - \gamma(t)| = r(t)$ and $|\gamma'(t)| = 1$, we get

$$|r'(t)| \leq 1 \text{ and } x = \gamma(t) + r(t) \left(-r'(t) \gamma'(t) \pm \sqrt{1 - r'(t)^2} \gamma'(t)^\perp \right).$$

Computing dx/dt and using $\gamma''(t) = \kappa(t) \gamma'(t)^\perp$, we obtain the formula for $ds_\pm$. □

Having these conditions, it is not difficult to generate random ribbon-like shapes; see Exercise 4 at the end of the chapter.

3.3.3 The Voronoi Decomposition and the Delaunay Triangulation

Some fast algorithms exist for computing the medial axis of a shape by using the Voronoi decomposition and the Delaunay triangulation. These dual constructions have been thoroughly studied in computational geometry; see [172, 180].

Let $\mathbb{S} = \{P_i\}$ be a finite set of points in $\mathbb{R}^2$. Define $d(P, \mathbb{S}) = \min_i d(P, P_i)$. The Voronoi decomposition of the plane associated with $\mathbb{S}$ is the set of disjoint subsets defined by:

$$C_i = \{ P \in \mathbb{R}^2 \mid \| P - P_i \| = d(P, \mathbb{S}) < \min_{j \neq i} \| P - P_j \| \},$$

$$V(\mathbb{S}) = \{ P \in \mathbb{R}^2 \mid \exists i \neq j, \| P - P_i \| = \| P - P_j \| = d(P, \mathbb{S}) \}.$$

The result is a decomposition of the plane into "cells": each cell C_i is a "locus of proximity" of some P_i, that is, it is the set of points that are closer to P_i than to any other points of $\mathbb{S}$. $V(\mathbb{S})$ is the locus of centers of maximal open disks in the complement of $\mathbb{S}$. The Delaunay triangulation is the dual of the Voronoi decomposition; namely, it is the union, for all $P \in V(\mathbb{S})$, of the convex hull of $\{P_i \mid \parallel P_i - P \parallel = d(P, \mathbb{S})\}$. For each edge in $V(\mathbb{S})$, there will be a dual edge in the Delaunay triangulation, and for each vertex in $V(\mathbb{S})$, there will be a polygon in the Delaunay triangulation (k-sided if the vertex has degree k).

The link between these constructions and the medial axis is the following. Given a shape S, we can choose a large finite number of points $\mathbb{S}$ on the boundary ∂S, and then compute the Voronoi decomposition to this set of points. The maximal circles in $\mathbb{R}^2 - \mathbb{S}$ will be of two types: if they meet two nonadjacent points of $\mathbb{S}$, this will approximate the maximal circles in $\mathbb{R}^2 - \partial S$ but if they meet only two adjacent points P_i, P_{i+1}, their centers will lie on the perpendicular bisector of the line segment $\overline{P_i, P_{i+1}}$; see Figure 3.12 for an example with eight points $\mathbb{S}$. Thus, the obtained result V is a union of (i) an approximation of the medial axis, and (ii) other pieces of curves, which we call "hairs." The basic result is that the medial axis $Ax(S)$ is then recovered by

$$Ax(S) = \lim_{\mathbb{S} \to \partial S} [V(\mathbb{S}) - h(\mathbb{S})],$$

where the set of hairs $h(\mathbb{S})$ is the set of the points $P \in V(\mathbb{S})$ that are equidistant from adjacent points P_k, P_{k+1}. This way of computing the medial axis is very efficient since very fast algorithms to compute the Voronoi decomposition (algorithms in $n \log n$, where n is the number of points) have been found in computational geometry. In Exercise 5 at the end of the chapter, we give a MATLAB code for computing the medial axis of a shape (which is in fact a polygon when discretized) from its Delaunay triangulation.

The Delaunay triangulation also passes to the limit as the set $\mathcal{P}$ gets denser and denser and gives a way to decompose the closed shape $\overline{S}$ into a continuous family of disjoint polygons such as

$$\overline{S} = \bigcup_{P \in Ax(S)} \left(\text{convex hull of the finite set } (P + r(P)\overline{\Delta}) \cap \partial S \right).$$

Figure 3.14 shows three examples of this. Notice that for $P \in Ax(S)$, the intersection $(P + r(P)\overline{\Delta}) \cap \partial S$ is generally made of two points, and thus their convex hull is just a segment. But for triple points, there are three contact points, and their convex hull is then a triangle.

3.3. The Medial Axis for Planar Shapes

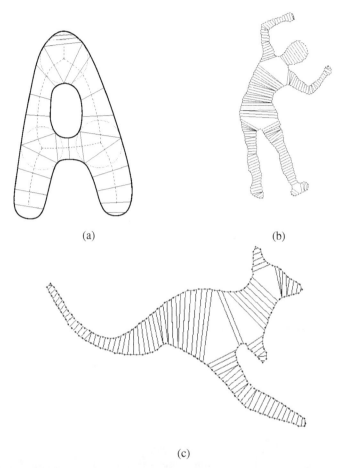

(a) (b)

(c)

Figure 3.14. Three examples of the dual of the medial axis: decomposing a shape via its Delaunay decomposition. (a) A small set of simplices. (b) and (c) Denser selections in natural shapes. ((b) and (c) courtesy of Song-Chun Zhu [240, Figure 4].)

3.3.4 Medial Axis and MDL

The broader significance of $A(S)$ is that if we seek to describe S with as few bits as possible, then the axis replaces the two-dimensional shape with a one-dimensional structure that is often very informative (see Figure 3.11) and can be described compactly. This is the minimum description length (MDL) approach. One problem, however, is that the medial axis is very sensitive to small irregularities of ∂S and this can make it a complex tree. For example, if the shape S

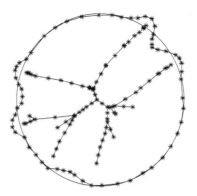

Figure 3.15. The medial axis is very sensitive to small irregularities of the boundary of the shape since they create many branches. (This figure was obtained by using the MATLAB code given in Exercise 5 at the end of the chapter).

is a perfect disk, then its medial axis is reduced to a single point (the center of the disk), but if S is more or less a disk with many small irregularities, then the medial axis has many branches; see Figure 3.15. To get the *essential* part of the skeleton, the medial axis $A(S)$ has to be "pruned." For example, if S is more or less a blob, we first find one circular approximation to S and then describe the deviation from the circle. In that case, the pruned $A(S)$ should be just one point (the center of the circle). The same principle holds for ribbons: if $|r'|$ is very small, we first approximate the complex ribbon S by one with constant radius and then give the description of the small deviations.

One possible approach to "pruning" $A(S)$ is to use the geometric heat equation on ∂S to smooth it (see Section 5.3.4). If the axis is suitably pruned, it can be used for many other applications, such as matching two shapes by matching their medial axes. These ideas have been extensively developed by Ben Kimia [191, 198]. In what seems to be a very promising direction, the mathematical analysis of optimal shape encoding via the medial axis has been analyzed by using ϵ-entropy by Kathryn Leonard [141].

3.4 Gestalt Laws and Grouping Principles

Let's summarize what we have found so far. First, given an image, we should search for salient contours. Such contours are salient because there is a strong

3.4. Gestalt Laws and Grouping Principles

contrast between the black/white intensity on its sides at many points on the curve, and the curve is reasonably smooth. A key grouping principle is that two such contours may be joined together if they align well. Second, contours often surround a shape that is the two-dimensional representation of a three-dimensional object— or it may just be a two-dimensional object such as a character. In either case, the contour is closed and the axis of the object groups pieces of the bounding contour in pairs. Such bounding contours may be parallel, if the object is ribbon-like, or mirror symmetric, if the axis is straight. How many such grouping principles are there?

At the beginning of last century, in Germany and later in Italy, a group of psychophysicists called the "Gestalt school" (notably Metzger [156], Wertheimer [227], and Kanizsa [119–121]) tried to understand the way human vision proceeds and to give laws for the constitution of visual objects. According to Gestalt theory, grouping is the main law of visual perception: whenever points (or previously formed objects) have a characteristic in common, they get grouped and form a new, larger visual object. The main criteria for grouping are:

- in general, *proximity*;

- for edges, *alignment, parallelism, closedness, pointing to a common point*;

- for regions, *similar color or texture, convexity, constant width and symmetry*.

This grouping principle is illustrated by Figure 3.16. In Figure 3.16 (a), we perceive a black rectangle, even though it is strongly occluded by several white

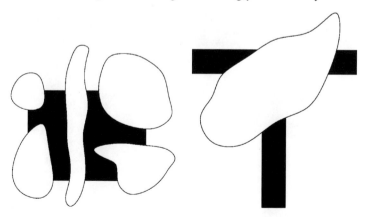

Figure 3.16. (a) A black rectangle is perceived due to grouping principles (same color, alignment, parallelism, T-junctions). (b) The character T is recognized despite the occlusion. (Inspired by Gaetano Kanizsa [121].)

shapes, its angles are not visible, and the rectangle is cut in two pieces. This can be explained in terms of grouping principles: we first group all black points together, then we group aligned segments across the white occluders, and finally we group the edges in pairs by the parallelism of the opposite sides of the occluded rectangle.In Figure 3.16 (b), we can perceive the letter T despite the occlusion for the same reasons.

Note that what we perceive are parts of the surfaces of objects in our field of vision. The task of visual perception is to separate the surfaces that belong to distinct objects and to combine the surfaces that are parts of the same object. In both examples in Figure 3.16, we first combine the black regions into parts of the surface of the one connected black object, and then we reconstruct the missing parts of this object. A key step is to decide, for each edge, whether it "belongs" to the surface on one side or the other; that is to decide which surface is nearer in three-dimensional space and, from the our viewpoint, comes to an end at the edge.

The grouping point of view is closely related to the MDL principle. The aim of MDL is to choose among different possible descriptions of an image (according to given classes of models) the one that leads to the smallest expected code length (see also the discussion in Chapter 0, where the MDL principle is related to the problem of the *maximum a posteriori* estimate). We give an example here of this link between grouping principles and MDL. Assume that we want to encode the image shown in Figure 3.17. This is an image of some size $N \times N$ that contains six black points. If we just code the positions of the six points as if they were independent and uniformly distributed on the image, we would get (using

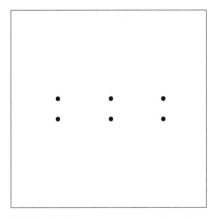

Figure 3.17. An image depicting six dots; we may encode it by six coordinate pairs or by using grouping principles, such as symmetry. The latter needs far fewer bits.

3.4. Gestalt Laws and Grouping Principles

Shannon's optimal coding theorem) the following expected code length:

$$\text{Length}(\text{code}_1) = -6\log_2(1/N^2) = 12\log_2(N).$$

But if we encode this same figure using its symmetries, we can significantly decrease the expected code length. We can, for example, do it in the following way: first, give two perpendicular axis of symmetry (we have N^3 choices—N^2 for the position of the intersection of the two lines and N for the orientation), then just give the coordinates of one corner point (N^2 choices), and finally code the fact that we use symmetry to get the four corners and add midpoints of the long edges. Thus, we can expect

$$\text{Length}(\text{code}_2) = \log_2(N^3) + \log_2(N^2) + \log_2(M) = 5\log_2(N) + \log_2(M),$$

where M is the number of grouping rules we might have used. Since this number of grouping rules is quite small, the previous formula shows that the gain of code length may be very large!

Some of the most startling demonstrations found by the Gestalt theorists involve perceiving "invisible" objects—objects for which the image gives no direct evidence. In Gestalt theory, one distinguishes two different types of contours. First, "modal contours" are edges that are present in our field of view, so they should show up in the image; however, by accidents of illumination and reflectance, they may be very difficult to see. Second, "amodal contours" are edges that are occluded by a nearer object and hence are not visible in the image; however, these can be part of our mental reconstruction of the three-dimensional geometry of the scene. This is a very specific fact about vision: the world is three-dimensional and the image is only a two-dimensional signal created by the visible surfaces. Very often, parts of some of the visible objects are hidden behind closer objects. Then the occluded contours of these partly hidden objects form amodal contours.

This is illustrated by the examples in Figure 3.18. Figure 3.18 (a) shows the famous Kanizsa triangle, in which connecting the straight edges of the "pac men" creates a triangle. But by accident, the color of the triangle and of the background are identical so the edges of the triangle are modal contours, which happen to have left no trace in the image gray levels. The triangle is then assumed to occlude full disks at each vertex, and the occluded parts of the boundaries of the disks are amodal contours. In Figure 3.18 (b), first the endpoints of the visible earlike arcs are grouped because they are close and form a smooth contour. Then these contours are connected, forming the full outline of a pear, a modal contour that also has left no trace in the image. Finally, assuming the pear is there, the earlike arcs are closed up into sets of concentric circles, half visible, half behind

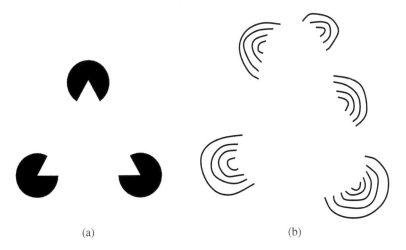

Figure 3.18. (a) The famous Kanizsa phantom triangle, an example of "amodal completion;" (b) another Kanizsa demonstration, a phantom pear occluding six ring-like structures

the pear and amodal. In natural images, visual contours are often cut into several pieces due to occlusions. But just as in Kanizsa's demonstrations, we reconstruct in our minds the full contour even though the amodal part is totally absent in the actual image and the modal part may be indistinct. This is a great example of our use of Bayesian priors: without some prior knowledge of what object boundaries usually are like and of how objects occlude each other, we would be unable to guess what the hidden parts of the object were or even that there were hidden parts. We need to combine this prior knowledge with the evidence given by the image to reconstruct the hidden parts of the objects.

Grouping principles are also present in the problem of character recognition. We begin with a grouping of the edges of a character into extended contours, then we group together the contours on opposite sides of each stroke and model this group with the axis. Each alpha-numeric character is then made up of a small number of elementary parts: straight lines and smooth nearly circular arcs. Then these must be grouped by using relations between these elementary parts: which strokes connect to which strokes and with what proportions and angles. In addition, some general Gestalt rules are used. For capital letters, in some cases edges are supposed to be (roughly) parallel: horizontal edges in E, F, and Z; vertical edges in H, M, N, and U. In some cases, there is reflectional symmetry about a vertical axis: A, H, M, O, T, U, V, X, and Y. In Figure 3.1, it is clear that where shadows and dirt obscure the characters; completing the pattern to form the appropriate parallel and symmetric strokes will often find the full character.

3.5. Grammatical Formalisms

Figure 3.19. Graphic artists enjoy playing with human perception and testing its limits. Above are six numerals from a student calendar. (Design by George Tscherny, Inc.; published by School of Visual Arts.)

But, just so we do not get lulled into a sense of how straightforward things are, Figure 3.19 shows playful versions of the first six numerals. Note the principles that are employed: the 1 is an example of how edges can be compound instead of being a simple large gradient from a white to a black region; the 2 shows how many parallel edges form a texture, which is then treated as a single region; the 3 plays on our memory of quill pens and medieval manuscripts, where lines in one direction are thin and in another are thick—here the medial axis will still work; the 4 shows how a sloppy gray filling is mentally extended to the outer contour and how the dots are grouped into the axis; the 5 incorporates modal and amodal contours, mostly but not all parallel to visible edges—it requires knowledge of extreme shadow effects to be properly perceived; the loops of the spiral in the 6 need to be grouped, collapsed, or maybe cropped to find the true 6—it suggests a flourish of the pen. What is striking is how easily we can recognize all of these numerals.

The Gestaltists created a nearly purely qualitative theory: they described multiple rules but not how to decide when one rule is applied and not another and how they interact. Their rules apply recursively and create what we call a hierarchical geometric pattern. To make some mathematics out of it, we must first ask how to describe hierarchical structures, and this leads us to grammars.

3.5 Grammatical Formalisms

The basic idea of grammar is that, given a signal $s : \Omega \to V$ defined on a domain Ω with values in some fixed set V, then the signal, *restricted to parts of the domain*, has patterns that occur, with some variations, in many other signals of the same class. Thus, these parts of the signal are *reusable* parts. Some of these subsets are small and represent "atoms" of the signal; others are found by grouping the atoms, or grouping some groups. Thus, we get a *hierarchy of subsets*, each being a reusable part of the signal, that is, a part that may be plugged in with some variations in another signal. These special subsets may be organized in a *parse tree*, with the whole domain being the root, and vertical arrows put between any two subsets S, T if $S \subsetneq T$ for which there is no intermediate subset R with $S \subsetneq R \subsetneq T$. Many books on grammars exist; we cite [52] for a cross-linguistic analysis of language grammars, [197] for an introduction to problems of context sensitivity, and [234] for grammars of images.

We saw in the introduction to this chapter that a character could be represented as a parse tree. The most familiar examples, however, are the parse trees of language. To set the stage, we first give an example from English grammar, which will illustrate most of the key ideas in this section. The example is the sentence:

Students, who love a book, often buy it.

(This is a belief dear to the heart of all textbook writers such as ourselves.) The result is a labeled tree with labels such as S = sentence, NP = noun phrase,

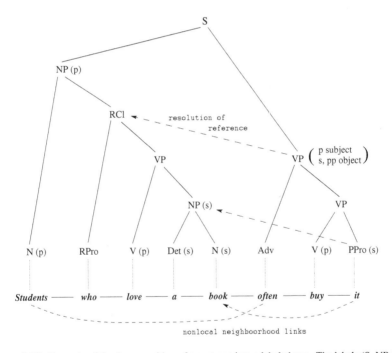

Figure 3.20. Example of the decomposition of a sentence into a labeled tree. The labels (S, NP, VP, N, RCl, Adv, RPro, V, Det, PPro) stand for sentence, noun phrase, verb phrase, noun, relative clause, adverb, relative pronoun, verb, determiner, and personal pronoun, respectively. Number is indicated in parentheses as s or p, and the dashed line indicates the resolution of the reference of the personal pronoun it. Note also that the verb phrase has the specific attributes of having a plural subject and a singular personal pronoun object, given in parentheses, which are needed to ensure agreement with the nouns in the relative clause.

3.5. Grammatical Formalisms

VP = verb phrase, Det = determiner, Adv = Aaverb, amd PPro = personal pronoun, shown in Figure 3.20. Note that the children of each node of the tree is a subset of the sentence that makes sense on its own. Thus, the tree is simply a two-dimensional visualization of a set of nested intervals in the string of words of the sentence.

Note that the sentence incorporates some grammatical rules to prevent wrong groupings, especially the agreement of number between "students" and "love" (both plural), and "book" and "it" (both singular). This is the first lesson: we cannot combine pieces arbitrarily. We picked this sentence because it also shows two more complex effects: (a) the subject of the sentence, "students," is separated from the verb "buy" by a clause; and (b) the reference of the pronoun "it" is the object "book" embedded in this clause, thus linking the object in the whole sentence with the object in the clause. Complexities such as these have analogs in vision, which are discussed in Section 3.5.2.

We first will develop several formalisms on trees and grammars. Trees are very useful structures generated by several different models. We give some examples of these here. The first example is a random branching process modeling population growth. The second example is a probabilistic context-free grammar (PCFG). Then we consider context sensitivity, and finally we go back to the problem of character recognition.

3.5.1 Basics VIII: Branching Processes

Let's forget about parse trees for a moment and develop the mathematics of the simplest stochastic labeled trees. References for this topic are Chapter 8 in [212] and in much more depth in [10, 101].

Example 1 (Population Growth). We can model the growth of a population in the following way. We ignore sex and assume the population is all female. We assume further that a random woman has k children with probability p_k, where $k = 0, 1, 2, \cdots$. Let $\mathcal{X}$ be a random variable with $\mathbb{P}(\mathcal{X} = k) = p_k$, for all $k \geq 0$. Then, every woman generates a tree of her offspring. We may measure the size of the tree by letting $\mathcal{N}_n$ denote the size of the population at the nth generation. Then $\mathcal{N}_0 = 1$ and

$$\mathcal{N}_{n+1} = \mathcal{X}_1^{(n)} + \mathcal{X}_2^{(n)} + \ldots + \mathcal{X}_{\mathcal{N}_n}^{(n)},$$

where the $\mathcal{X}_i^{(n)}$ are independent copies of the random variable $\mathcal{X}$ representing the number of children of the i person in the nth generation. By convention, the sum above has value zero if $\mathcal{N}_n = 0$. This model defines a probability distribution on the set of all rooted trees, which is called a *random branching process*.

We let μ denote the mean number of children of any person; $\mu = \sum_{k \geq 0} k p_k$, and σ^2 denote the variance of the number of children, $\sigma^2 = \sum_{k \geq 0} (k - \mu)^2 p_k$. In all the following, we assume that μ and σ^2 are finite. Then we have $\mathbb{E}(\mathcal{N}_{n+1}) = \mu \mathbb{E}(\mathcal{N}_n)$ and $\text{Var}(\mathcal{N}_{n+1}) = \mu^2 \text{Var}(\mathcal{N}_n) + \sigma^2 \mathbb{E}(\mathcal{N}_n)$. (It is a useful exercise to verify these facts.) By induction, this leads to

$$\mathbb{E}(\mathcal{N}_n) = \mu^n \text{ and } \text{Var}(\mathcal{N}_n) = \sigma^2 \mu^{n-1} \left(1 + \mu + \mu^2 + \ldots + \mu^{n-1}\right).$$

One key issue in this model is the study of the extinction of the population: what is the probability that the tree is finite? The answer is given by the following theorem.

Theorem 3.3. *Let $H_n = \mathbb{P}(\mathcal{N}_n = 0)$ denote the probability of extinction at time n. Then the sequence $\{H_n\}_{n \geq 0}$ is nondecreasing, and its limit η represents the limit probability of extinction. Moreover, we have:*

- *If $\mu < 1$ or $\mu = 1$ and also $\mathbb{P}(\mathcal{X} = 1) < 1$, then $\eta = 1$, which means that we have almost surely extinction.*

- *If $\mu > 1$ or $\mu = 1$ and also $\mathbb{P}(\mathcal{X} = 1) = 1$, then $\eta < 1$, which means that we can have infinite trees with positive probability.*

Proof: We use the generating function of the process defined for $s \geq 0$ by $G(s) = \mathbb{E}(s^{\mathcal{X}}) = \sum_{k \geq 0} p_k s^k$. The main properties of this function are: G is nondecreasing, convex, $G(0) = p_0$, $G(1) = 1$, and $G'(1) = \mu$. Let G_n denote the generating function of $\mathcal{N}_n$; then

$$G_{n+1}(s) = \mathbb{E}(s^{\mathcal{N}_{n+1}}) = \sum_{k \geq 0} \mathbb{E}(s^{\mathcal{X}_1 + \mathcal{X}_2 + \ldots \mathcal{X}_k}) \mathbb{P}(\mathcal{N}_n = k)$$

$$= \sum_{k \geq 0} G(s)^k \mathbb{P}(\mathcal{N}_n = k) = G_n(G(s)).$$

Since the generating function of $\mathcal{N}_1$ is G, by induction we get that the generating function of $\mathcal{N}_n$ is $G(G(G(..)))$ (n times). If $H_n = \mathbb{P}(\mathcal{N}_n = 0)$ denotes the probability of extinction at time n, then $H_n = G_n(0)$ and thus $H_{n+1} = G(H_n)$ and $H_1 = G(0)$. Let α denote the smallest positive solution of the equation $G(s) = s$; then for all s in $[0, \alpha]$, we have $G(s) \in [s, \alpha]$. This shows that the sequence $\{H_n\}_{n \geq 0}$ is nondecreasing, and its limit is the fixed point α. Using the convexity of G, we see that $\alpha < 1$ is equivalent to $\mu = G'(1) > 1$ or $G(s) = s$ for all $s \in [0, 1]$ (i.e., $\mathbb{P}(\mathcal{X} = 1) = 1$). See Figure 3.21 for two examples of generating functions. □

3.5. Grammatical Formalisms

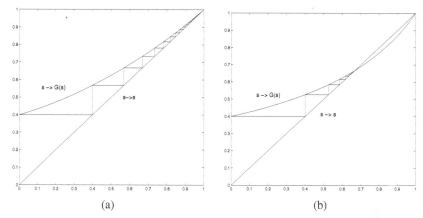

Figure 3.21. Two examples of random branching processes. On the same figure, we plot the generating function G, and the sequence $(H_n)_{n \geq 0}$ of probabilities of extinction at time n. (a) Example with $G(s) = 0.4 + 0.3s + 0.3s^2$; in this case, $\mu = 0.9$ and we thus almost surely have extinction. (b) Example with a geometric distribution $p_k = (1-p)p^k$ with $p = 0.6$. In this case, $\mu = p/(1-p) > 1$ and the generating function is $G(s) = (1-p)/(1-ps)$. The limit probability of extinction is $\eta = \min(1, 1/p - 1)$, which here becomes $\eta = 2/3$.

Example 2 (Probabilistic Context-Free Grammars). Assume now that our tree is labeled; the labels ℓ are of two types, $\ell \in N$ are called *nonterminals*, and $\ell \in T$ are called *terminals*. One of the labels, $S \in N$, is called the *start state*. For each nonterminal label ℓ, we assume we are given a nonempty finite set of *production rules*:

$$R_{\ell,k} : \ell \longrightarrow (m_1, \cdots, m_{n(k)}), \quad 1 \leq n(k) < \infty, m_i \in N \cup T.$$

Moreover, for each $\ell \in N$, we assume we are given a probability distribution on the set of production rules $P_\ell(R_{\ell,k})$, where

$$\sum_k P_\ell(R_{\ell,k}) = 1, \quad \forall \ell.$$

The terminal labels form the leaves of the tree and have no associated productions.

Then we may generate trees just as above. Start at a root node with label S. Choose randomly one of its production rules and generate the first level of the tree—a set of new nodes whose labels are given by the right-hand side of the production rule. For each node with a nonterminal label (if any) on this level, choose randomly one of its production rules and form level two in the same way. Continue indefinitely. Any node with a terminal label becomes a leaf. The resulting trees may, of course, be finite or infinite. This algorithm produces a random tree and thus a probability distribution of the set of all $N \cup T$-labeled trees[3] with root S. For grammar applications, we want the trees to be almost surely finite; there is a simple criterion for this in terms of the probability distributions P_ℓ, which is worked out in Exercise 6 at the end of this chapter. When this holds, we call the set $(N, T, S, \{R_\ell\}, \{P_\ell\})$ a *probabilistic context-free grammar*.

Suppose we have a finite tree β, with nodes $v \in V$ that have labels $L(v)$ and are expanded by the rules $R(v)$. Then the probability of the whole tree is

$$P_{\text{tree}}(\beta | \text{root } S) = \prod_{v \in V} P_{L(v)}(R(v)).$$

We can equally well start at any other label ℓ_0 instead of S and produce random trees with a root labeled ℓ_0. Let $P_{\text{tree}}(\beta | \text{root } \ell_0)$ be the probabilities we get. (We will leave out the word "root.") Then the recursive rule is as follows.

If the root expands by $\ell_0 \xrightarrow{R} m_1, \cdots, m_n$, then

$$P_{\text{tree}}(\beta | \ell_0) = P_{\ell_0}(R) \cdot \prod_{i=1}^{n} P_{\text{tree}}(\alpha_i | m_i),$$

where α_i = subtree with root labeled m_i.

Alternatively, we can assume we are given a probability distribution Q on $N \cup T$ and then generate an *extended* probability distribution on finite trees *with arbitrary roots* by

$$P_{\text{ext}}(\beta) = Q(L(\text{root})) \cdot \prod_{v \in V} P_{L(v)}(R(v)).$$

In the applications to grammars of languages, the terminals will be the words of the language. The nonterminals in Example 2 were the categories (S, NP, VP, N, RCl, Adv, RPro, V, Det, PPro). In applications to images, the terminals could be single pixels or they could be small parts of the image containing the simplest structures, the smallest pieces of edges, bars, blobs, and so forth. These applications be discussed further in Chapter 4. Nonterminals are the sort of things we have been discussing: extended lines and curves; elongated ribbon-like objects with a simple axis; more complex shapes, possibly classified as the letter A, for example, or even more complex objects such as images of faces.

---

[3]If the trees can be infinite, we are dealing with an uncountable set and we need to say on what set the probability measure is defined. We take a basis of our σ-algebra of measurable sets to be those defined by fixing some initial part of the tree.

3.5.2 Context Sensitivity

PCFGs can and are being used to model a large part of natural language grammar. But note that, as we have presented grammars, there is no way to insure that "a book" occurs more frequently than the ungrammatical "a books" or "many book." If we simply have rules, NP $\to$ Det, N, Det $\to$ a, Det $\to$ some, N $\to$ book, and N $\to$ books, then

$$\frac{P(\text{a books})}{P(\text{a book})} = \frac{P(\text{books})}{P(\text{book})},$$

whereas we would like the left-hand side to be much less than the right, and, in fact, 0 for grammatical text. Agreement violates context freeness in its simplest version. What linguists do is to endow nonterminals with a list of *attributes*, such as number and gender (in many languages). Then they replace the nonterminal NP (noun phrase) with a family of nonterminals labeled by NP plus attributes. Moreover, they replace the single production rule NP $\to$ (Det,N) with a family of production rules:

$$R_\alpha : \text{NP} + (\text{attr}.\alpha) \to (\text{Det} + (\text{attr}.\alpha), \text{N} + (\text{attr}.\alpha)), \quad \forall \alpha,$$

which forces the determiner and the noun to have the same attributes. Notice, though, that the dashed arrow in Figure 3.20 is still not covered: we need to determine the NP to which the pronoun "it" refers and then make their numbers equal. It gets worse: pronouns can link distinct sentences and can link to proper names, in which case their genders in all these sentences must agree. It is not clear to us whether computational linguists have any good way of handling this issue. We have proposed in Figure 3.20 two ad hoc solutions: (1) the top-level VP carry as an attribute the fact that its object is a pronoun, hence in need of finding its reference; (2) a link be established to the relative clause where the reference lies. We are proposing that attributes can make explicit certain long-distance links as well as the standard links of words to their immediate neighbors.

In the grammar of images, there is major increase in the complexity of this problem. Speech and text are one-dimensional signals, and each atom—be it a phoneme or word—has an immediate predecessor and an immediate successor. A large proportion of the agreement requirements are between immediate neighbors[4] and this order relationship is always implicit in the tree as long as each production rule is considered to deliver a *linearly ordered* sequence of children.

But an image is two-dimensional and there is no linear ordering to subsets of its domain. Instead, each subset has a larger set of neighbors. For example, if $\Omega = \coprod S_k$ is a decomposition of the domain of an image into disjoint subsets,

---

[4]But this is not true in the so-called free word order languages, such as literary Latin and some Australian languages.

then one can form a *region adjacency graph* (RAG), which is a graph with one node for each subset S_k and with a horizontal undirected edge between all pairs of adjacent regions, that is, regions sharing part of their boundaries. This RAG replaces the linearly ordered strings found in language, which leads to several changes in the formalism of grammar:

1. The entire parse tree should have horizontal undirected edges expressing adjacency as well as its vertical directed edges, making it into a *parse graph*.

2. Production rules need to carry more structure:

   $$R : S_0 \longrightarrow < V_R, E_R >, V_R \text{ a set of child nodes } S_k,$$
   $$E_R \text{ a set of edges connecting some pairs } (S_k, S_l).$$

   Furthermore, S_0 will be connected to neighbors, and such edges must descend to neighbor edges between their children; for this, the child graph $< V, E >$ should be given open edges and rules specified so that if (S_0, S_0') is an edge at the higher level, then this edge spawns one or more edges connecting some open edges of $< V_R, E_R >$ with open edges of $< V_R', E_R' >$.

3. Nodes v of a parse graph should carry attributes $\alpha(v)$ as well as labels $L(v)$ and rules $R(v)$. For images, these attributes typically include *position, orientation, scale, and color*, but in some cases much more (e.g., texture for larger regions; see Chapter 4). For every production rule

   $$R : S_0 \longrightarrow < V_R, E_R >,$$

   the attributes $\alpha(v_0)$ of the parent node v_0 should be given by a function $\alpha(v_0) = f_R(\alpha(v_1), \cdots, \alpha(v_n))$ of the attributes of its child nodes $v_i \in V_R$.

4. We now see the proper setting for the grouping attributes of Gestalt theory: when qualities such as similar color, orientation, or symmetry cause parts of an image to be grouped, they should create additional horizontal edges in the parse graph. For example, where two shorter image contours align and are to be linked across a gap, there should be a parse graph edge joining them, and when there are pairs of parallel or symmetric contours, they should also be joined by a parse graph edge.

This is not meant to be a precise mathematical definition, but rather a framework for context-sensitive image grammars. Theories of such parse graphs are still in their infancy, although ideas of this type have been studied for quite a

3.5. Grammatical Formalisms

while. Both faces and bodies have a hierarchical set of parts and have been used as test-beds for grammatical style models. These go back to Fischler and Elschlager [73] and Marr [152] and have been developed more recently by Ullman, Huttenlocher, and their associates [2, 29, 71, 190]. An early attempt to combine the n-gram models of Chapter 1 with probabilistic context-free grammars produced the context-sensitive grammars in [149]. The major work on this problem has been done by Geman and his team [22, 84, 114], who have used character recognition as a test-bed, and by Zhu and his team [48, 100, 179, 234]. Both of these groups use some form of the above definition, but there are multiple ways to set up the full formalism. A key difficulty involves the many structures present in images, in which geometry, illumination, and semantic knowledge all combine to produce contextual constraints.

The *parse graphs* in our example below consist of:

1. a tree β generated by the PCFG;

2. a set of horizontal edges E, including those produced by the enhanced production rules but possibly more;

3. values of the attributes $\alpha_v \in A_{\ell(v)}$ for each vertex, which are computed by the f's from the attributes of the leaves under v.

The probability model requires knowing additional binding factors B_e for each horizontal edge e, which indicate to what degree the two pieces, given their attributes "fit" in the context of the parse. Then the probability that we use, up to a normalizing constant, is of the form

$$P(\omega) \propto \prod_{v \in V} P_{\ell(v)}(R(v)) \cdot \prod_{e=<v,w>\in E} B_e(\alpha(v), \alpha(w), \alpha(\text{parents of } v, w))$$

Thus, altogether, our probabilistic context-sensitive grammar consists of

$$\{N, T, S, \{R_\ell, \text{ each with } V_R, E_R, f_R, B_e\text{'s}\}, \{A_\ell\}, \{P_\ell\}\}$$

Example (Snakes). The binding of two shorter snakes into one long one is a simple example of the theory. Let T consist of one label pt with attributes $P = (x, y)$, its position. Let $N = \{sn_n | n \geq 2\}$ be labels for snakes of length n and let the attributes of a snake be its first and last points, P_1 and P_N and the velocity vectors $\vec{t_1} = P_2 - P_1$, and $\vec{t_n} = P_n - P_{n-1}$ at the beginning and end.

The production rules are

$$R_1 : sn_2 \longrightarrow (pt \leftrightarrow pt), \quad \text{prob} = 1$$

and for $n \geq 3$:

$$R_1 : sn_n \longrightarrow (pt \leftrightarrow sn_{n-1}), \quad \text{prob} = 1/(n-1)$$
$$R_m : sn_n \longrightarrow (sn_m \leftrightarrow sn_{n-m}), \quad \text{prob} = 1/(n-1)$$
$$R_{n-1} : sn_n \longrightarrow (sn_{n-1} \leftrightarrow pt), \quad \text{prob} = 1/(n-1)$$

Given two snakes, $\Gamma_1 = \{P_1, P_2, \ldots, P_N\}$ and $\Gamma_2 = \{Q_1, Q_2, \ldots, Q_M\}$, we want the binding factor B_e of the edge in the composite structure

$$\Gamma = (\Gamma_1 \stackrel{e}{\leftrightarrow} \Gamma_2) = \{P_1, P_2, \ldots, P_N, Q_1, Q_2, \ldots, Q_M\}.$$

The attributes of Γ_1 are just $\alpha(\Gamma_1) = (P_1, P_N, \vec{s}_1 = P_2 - P_1, \vec{s}_N = P_N - P_{N-1})$ and those of Γ_2 are $\alpha(\Gamma_2) = (Q_1, Q_M, \vec{t}_1 = Q_2 - Q_1, \vec{t}_M = Q_M - Q_{M-1})$. B_e is just the likelihood ratio defined in Section 3.2.5, namely:

$$\frac{p(\Gamma)}{p(\Gamma_1)p(\Gamma_2)} = \frac{Z_N Z_M}{Z_{N+M}} e^{-\alpha^2 \|P_N - Q_1\|^2 - \beta^2 \|(\vec{s}_N - (Q_1 - P_N)\|^2 - \beta^2 \|(Q_1 - P_N) - \vec{t}_1\|^2}.$$

Notice that the binding function B_e is indeed a function of the attributes of Γ_1 and Γ_2.

3.5.3 Factoring B and "Compositional" Grammars

Here we present a somewhat simplified version of another way of analyzing the binding factors in context-sensitive grammar, due to Stuart Geman and his collaborators [84]. First, assume we fix, as they do, a probability distribution $Q(\ell)$ on $N \cup T$, so we may define an extended probability model that assigns probabilities to all trees with any label ℓ on its root:

$$P_{\text{ext}}(\omega) =$$
$$\frac{1}{Z(\ell)} \cdot Q(\ell) \cdot \prod_{v \in V} P_{L(v)}(R(v)) \cdot \prod_{e=<v,w>\in E} B_e(\alpha(v), \alpha(w), \alpha(\text{parents of } v, w))$$

$$Z(\ell) = \sum_{\omega \text{ with root label } \ell} \prod_{v \in V} P_{L(v)}(R(v)) \cdot \prod_{e \in E} B_e(\alpha(v), \alpha(w), \alpha(\text{parents of } v, w)).$$

Now we want to assume that all horizontal edges $e = <u, v>$ in our parse graph connect two child nodes in a production; in other words, there are no long-distance edges connecting nodes lying in trees that don't get connected until you

3.5. Grammatical Formalisms

go up more than one step. This holds in many situations. Now, if we write r for the root of some parse graph ω, $\ell = L(r)$, $s_k, 1 \leq k \leq n$ for the children of the root, and γ_k for the entire subtree with root s_k and label $m_i = L(s_i)$, we can factor the probability:

$$P_{\text{ext}}(\omega) = \frac{Q(\ell)/Z(\ell)}{\prod_k Q(m_k)/Z(m_k)} \cdot P_{L(r)}(R(r)) \cdot \prod_{e=<u,v>\in E_R(r)} B_e(\alpha(u), \alpha(v), \alpha(r))$$
$$\cdot \prod_k P_{\text{ext}}(\gamma_k).$$

Therefore, the likelihood ratio satisfies

$$\frac{P_{\text{ext}}(\omega)}{P_{\text{ext}}(\gamma_1) \cdots P_{\text{ext}}(\gamma_n)} = F(R(r), \alpha(w_1), \cdots, \alpha(w_n)).$$

(Note that $\alpha(r)$ is a function of the attributes $\{\alpha(w_i)\}$, so we can omit it here.)

What does the binding factor F mean? To analyze F, define the subsets of the set of all parse trees as

$$\Omega_\ell = \{\omega | \ell = L(\text{root } \omega)\}$$
$$\Omega_R = \{\omega | \ell = L(\text{root } \omega), R = \text{production at root}\}.$$

Note that $\Omega_R \cong \prod_i \Omega_{m_i}$ because every n-tuple of trees γ_i with root labels m_i can be assembled—no matter now implausibly—into a tree with root ℓ and children γ_i. Thus we can look at the ratio of measures,

$$\frac{P_{\text{ext}}|_{\Omega_{R(r)}}}{\prod_i P_{\text{ext}}|_{\Omega_{m_i}}},$$

and it is a function that depends only on $\alpha(w_1), \cdots, \alpha(w_n)$. Let π_ℓ and π_R be the maps

$$\pi_\ell : \Omega_\ell \longrightarrow A_\ell, \qquad \pi_R : \Omega_R \longrightarrow A_{m_1} \times \cdots \times A_{m_n},$$
$$\pi_\ell(\omega) = \alpha(\text{root}), \qquad \pi_R(\omega) = (\alpha(w_1), \cdots, \alpha(w_n)).$$

What we saw means that the two measures are constant multiples of each other on each fiber of the map π_R, so their ratio can be calculated as the ratio of the direct image measures or of the conditional measures:

$$\frac{P_{\text{ext}}(\omega)}{P_{ext}(\gamma_1) \cdots P_{ext}(\gamma_n)} = \frac{\pi_{R,*}(P_{\text{ext}}|_{\Omega_R})}{\prod_i \pi_{m_i,*}(P_{\text{ext}}|_{\Omega_{m_i}})}\Big(\alpha(w_1), \cdots, \alpha(w_n)\Big)$$
$$= \frac{Q(\ell) \cdot P_\ell(R) \cdot \pi_{R,*}(P_{ext}(\cdot|\Omega_R))}{\prod_i Q(m_i) \cdot \pi_{m_i,*}(P_{ext}(\cdot|\Omega_{m_i}))}\Big(\alpha(w_1), \cdots, \alpha(w_n)\Big),$$

that is,

$$F(R, \alpha(w_1), \cdots, \alpha(w_n)) = \frac{Q(\ell) \cdot P_\ell(R) \cdot \pi_{R,*}(P_{\text{ext}}(\cdot|\Omega_R))(\alpha(w_1), \cdots, \alpha(w_n))}{\prod_i Q(m_i) \cdot \pi_{m_i,*}(P_{\text{ext}}(\cdot|\Omega_{m_i}))(\alpha(w_i))}.$$

This has a simple intuitive meaning. On the one hand, $\pi_{R,*}(P_{\text{ext}}(\cdot|\Omega_R))$ is the distribution of the n-tuples of attributes of the components of the object represented by a random ω whose head production is R. This is something we can hope to estimate empirically. Take many examples of some structure ω from a big database and check how closely their parts align or are symmetric. We expect that the attributes will be very constrained when the structure ω is present. On the other hand, $\prod_i \pi_{m_i,*} P_{\text{ext}}(\cdot|\Omega_{m_i}))$ represents independent distributions of each α_i coming from all examples of the ith component γ_i. The classic example of this was given by Laplace in 1812 in his *Essay on Probability* [133]. Take ω to be the word CONSTANTINOPLE, and let its constituents be the 14 letters that compose it; the word CONSTANTINOPLE has a much larger probability in text than the probability that this word appears by randomly choosing 14 independent letters. The result gives the binding factor F a simple intuitive meaning. If the fraction is greater than 1, the probability of the components coming from the larger structure ω is greater than these coming together by chance.

Using Shannon's optimal coding, we can consider the $-\log_2$ of these probabilities and give an interpretation of the likelihood ratio in terms of code length. First, we simplify the notation a bit by writing α_i for the attributes $\alpha(w_i)$, simply P for P_{ext} and Q_R for the probability distribution of n *bound* attributes $\pi_{R,*}(P_{\text{ext}}(\cdot|\Omega_r))$. Then

$$-\log_2(Q(\ell)) = \text{bits to encode the label } \ell$$
$$-\log_2(P_\ell(R)) = \text{bits to encode the choice of production rule } R \text{ given } \ell,$$
$$-\log_2(Q_R(\alpha_1, \cdots, \alpha_n)) = \text{bits to encode the } n\text{-tuple of attributes,}$$
$$\text{given that they are part of some } \omega$$
$$-\log_2(P_{\text{ext}}(\gamma_i)) = \text{bits to encode } \gamma_i \text{ as an object}$$
$$-\log_2\left(\frac{P(\gamma_i)}{Q(m_i)}\right) = \text{bits to encode } \gamma_i \text{ given its label } m_i$$
$$-\log_2\left(\frac{P(\gamma_i)}{Q(m_i).\pi_{m_i,*}(P(\cdot|m_i))(\alpha_i)}\right) = \text{bits to encode } \gamma_i \text{ given its label } m_i$$
$$\text{and attributes } \alpha_i.$$

3.5. Grammatical Formalisms

Putting it all together, we thus have the identity

$$\begin{aligned}
\text{bits to encode } \omega &= -\log_2(P(\omega)) \\
&= \text{bits to encode } \ell + \text{bits to encode } R \\
&\quad + \text{bits to encode the bound } n\text{-tuple } (\alpha_1, \cdots, \alpha_n) \\
&\quad + \sum_i \text{bits to encode } \gamma_i, \text{ given } m_i, \alpha_i.
\end{aligned}$$

3.5.4 The Example of Character Recognition

We now want to return to the identification of characters and pull together almost every model considered so far. Here we find for the first time in this book, a strongly hierarchical structure with four quite distinct levels—an example of a hierarchical geometric structure as described in Chapter 0. We will choose the letter A as in the introduction to this chapter. It exemplifies the way constraints interact: when the three strokes of the A are put together into the letter A, the proximity constraints, orientation constraints, and symmetry constraints are redundant. The parse graph is shown in Figure 3.22. We will work our way up from the bottom, describing the binding factors that make up the probability model. Rather than writing everything as exponentials, we deal with energies $E_e = -\log(B_e)$. The whole model has quite a few constants in it, which we denote by $a, b, c, \cdots$.

The nodes of the parse tree are (i) the whole letter, (ii) its three strokes and a "bounding box"(a rectangular area S containing the letter), (iii) a set of maximal disks covering each stroke, and (iv) the pixels themselves. Note that we must include the background rectangular area: the letter is instantiated not merely by certain black areas but by being surrounded by a white (or contrasting) background. We can, of course, still recognize the letter if the background is messy or partially occluded, as in the license plates at the beginning of the chapter or in Kanizsa's demonstration in Figure 3.16. But as the probability decreases, recognition can be uncertain, and at least some homogeneous and contrasting background is essential to recognize the letter. Note also that we are avoiding building the axes of the strokes by assembling linelets. This is a good alternative but it is also possible to assemble the stroke directly from the maximal disks that it contains. In that case, all the snake energy terms are thrown together when the disks are bound into a stroke. This also connects the parse directly to the theory of the medial axis.

Starting at the bottom, factors B relate the intensity of adjacent pairs of pixels p_1, p_2 with the color co (color outside) of the background rectangle—which is a parent of every pixel—and, if the pixel is inside the letter, with the color ci (color inside) of the parent disk above it. Letting $I(p)$ be the image intensity at pixel p,

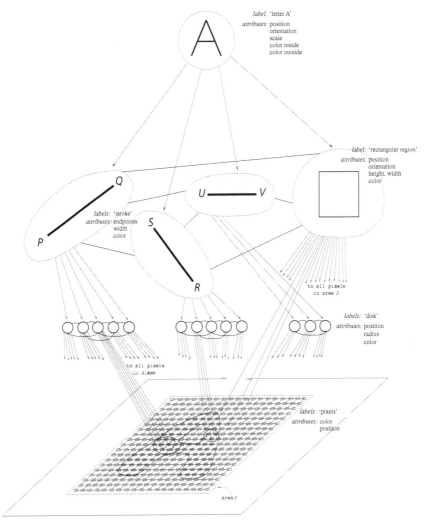

Figure 3.22. Parsing the letter A. There are four levels: the whole letter, its component strokes and background rectangle, the maximal disks of the strokes, and the pixels. Labels, attributes, and horizontal edges connecting adjacent structures are indicated.

we define

$$E_{p_1,p_2}(I(p_1),I(p_2),ci,co) = \begin{cases} a^2((I(p_1)-co)^2+(I(p_2)-co)^2)+b^2(I(p_1) \\ \quad -I(p_2))^2 \\ \quad \text{if } p_1 \text{ and } p_2 \text{ are both outside the letter;} \\ a^2((I(p_1)-co)^2+(I(p_2)-ci)^2) \\ \quad \text{if } p_1 \text{ is outside and } p_2 \text{ is inside the letter;} \\ a^2((I(p_1)-ci)^2+(I(p_2)-ci)^2)+b^2(I(p_1) \\ \quad -I(p_2))^2 \\ \quad \text{if } p_1 \text{ and } p_2 \text{ are both inside the letter.} \end{cases}$$

3.5. Grammatical Formalisms

Thus, all pixels inside the characters try to match their intensity with the color of the stroke of which they are part; and those outside try to match the background color. Also, all adjacent pairs of pixels, both of which are either inside or outside, try to have similar colors. We study such models, which go back to the physicist Ising, in Chapter 4. For now, we can readily guess that such energies will produce some mottling in the colors inside and outside around two different gray levels. Of course, this sort of binding can be used for many situations in which there is a background on which some objects are superimposed. Note that the adjacent pixel pairs that cross the boundary of the character violate the simplifying assumption made in our compositional model treatment: this horizontal edge joins nodes whose first common ancestor is the full letter itself.

Next, we look at the disks. Their attributes are computed from those of the pixels beneath them: their intensity can be taken to be the mean of the intensities of these pixels, their position the mean of the pixels' positions, and their radius the maximum of the distances of the pixels from the disk's position (i.e., its center). We link not merely adjacent disks but disks with one disk between them because the curvature constraint on the axis, treated as a snake, binds these disks. Let $\{D_1, \cdots, D_n\}$ be the disks in a stroke, and let P_k, r_k, c_k be their positions, radii, and colors, respectively. The energy is

$$E(\{D_1, \cdots, D_n\}) = \sum_{k=1}^{n-1} c^2 ||P_k - P_{k+1}||^2 + d^2(r_k - r_{k+1})^2 + e^2(c_k - c_{k+1})^2$$
$$+ \sum_{k=2}^{n-1} f^2 ||P_{k-1} - 2P_k + P_{k+2}||^2.$$

Note that this energy is not actually a sum of terms associated with each edge because the last term, the curvature term, involves three nodes all connected to each other, a so-called *clique* in the graph of horizontal edges. Such clique energies are seen again in Chapter 4. There are some alternatives here, one of which is that we could require $||P_k - P_{k+1}||$ to be constant and leave out the c-term. This is an elastica model instead of snake, however.

Next, let's look at the three strokes. Their endpoints obviously come from the attributes of the first and last disks that compose them. Their radii and colors can be taken as the mean of those of their disks. These radii and colors may vary over the stroke. How much depends on the constants c and e in the previous energy. There should be a set $\{R_n\}$ of production rules for strokes, where n is the number of disks that cover it. Each n is essentially the length of the stroke. For each n, the rule produces a string of n disk nodes, linked linearly, plus extra edges joining nodes with one node between them. Assigning probabilities to these rules penalizes long strokes and has the same sort of effect as the term $c^2||P_k - P_{k+1}||^2$ in the energy.

The position, orientation, height, and width attributes of the bounding box clearly are read off from the positions of the pixels to which it links. Its intensity attribute can be taken, for instance, as the *median* of the pixel intensities in its rectangle (assuming there are more background pixels than letter pixels).

The meat of what makes an A is contained in the energies joining the three strokes, as well as those joining the strokes to the bounding box. Let the left stroke begin at P at the bottom left and end at Q at the top; let the right stroke begin at R at the bottom right and end at S at the top; and let the middle stroke go from U near the middle of $\vec{PQ}$ to V near the middle of $\vec{RS}$. Then we certainly want energy terms to bring the endpoints of the strokes to the right places:

$$E_1 = g^2(||Q - S||^2 + ||U - (P + Q)/2||^2 + ||V - (R + S)/2||^2),$$

and we want the angle at the top to be about 45°:

$$E_2 = h^2(\text{angle}(P - Q, R - S) - \pi/4)^2.$$

Now let $\vec{a}$ be the orientation of the rectangle. Then we also want symmetry terms such as

$$E_3 = j^2(||P - Q|| - ||R - S||)^2 + k^2((U - V) \cdot \vec{a})^2 + ((P - R) \cdot \vec{a})^2),$$

which seek two sides of equal length and a middle stroke and base horizontal with respect to the axis. Finally, we need constraints that ensure that P, Q, R, S, U, and V are all in the rectangle and that the rectangle is, say, 10 percent larger than the smallest rectangle containing these six points.

The exact form of all these terms is quite arbitrary. For instance, we could also take a model letter A and simply take P, Q, R, S, U, and V to be these points on the model plus a Gaussian displacement. The ideal approach is to use a big database of letters and measure how precisely they are made. For numerals, the National Institute of Standards and Technology (NIST) database of handwritten zip codes is an amazing resource for poorly formed numerals. Given a large database, one can actually estimate the marginal statistics on the 6-tuple P, Q, R, S, U, V and make a better model by combining the ideas presented in Section 3.5.3 with those of exponential models to be discussed in Section 4.6.1. A similar approach has been implemented successfully to read license plates [114] and in learning (and synthesizing!) the structure of clock faces [179]. Figure 3.23 shows three samples from the above model using the A from the Calibri font as a model and not allowing the radius of the disks to vary along each stroke. We also

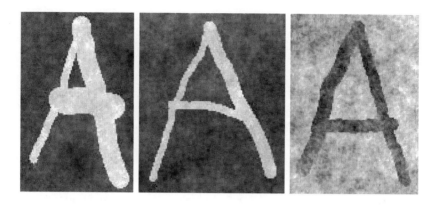

Figure 3.23. Three samples from the final stochastic model of images of the letter A.

used a Mumford-Shah (MS) model for the gray levels (see Section 4.2.3). The first two samples can be thought of as letters with a bit of fancy penmanship, and the last one as a muddy license plate.

The model combines (i) the modeling of contours or axes by shapes, (ii) the modeling of regions by sweeping out circles (i.e., by medial axis), and (iii) the use of context-sensitive grammar rules to finally assemble the letter. Doing inference with a model this complex is not easy. There are no longer any tricks like dynamic programming. The most promising approach is to use a combination of top-down and bottom-up approaches; see, for example, [29] and Chapter 8 in [234]. These algorithms are *not* guaranteed to find the minimum of the energy, but they seek a good guess by refining hypotheses over multiple top-down and bottom-up passes. This means that, starting from evidence at the image level, we seek higher-level hypotheses that integrate this evidence—in our example, first medial points, then strokes, then full letters. Given some support for a high-level hypothesis, we then ask for weaker evidence in the image that confirms the hypothesis. Thus, one line may be clear but another indistinct (as in the letter F in the bottom of Figure 3.5) and, once having proposed that a certain letter may be present, we check for further evidence. These algorithms are still experimental and will not be pursued here.

3.6 Exercises

1. Edge Detectors

We have presented a level-lines algorithm and snakes as methods for tracing contours in images. We chose these because they exemplified basic statistical techniques and stochastic models. A multitude of other approaches also exist. In this exercise, we sketch three other approaches and propose that the reader take his or her favorite image(s) and compare them. Only by looking very closely at specific contours found in specific images can one come to appreciate the difficulty in tracing the "correct" contours.

Canny's edge detector

As in Section 3.1, we start by Gaussian filtering of the image followed by centered x- and y-derivatives. At this point, we have estimates of ∇I and hence the magnitude of the gradient $|\nabla I(i,j)|$ and the orientation angle $\theta(i,j)$ at each pixel (i,j). But instead of merely thresholding, Canny introduced two key ideas [40]. The first is nonmaximum suppression which proceeds as follows: At each pixel (i,j), we draw a square around the pixel bounded by the adjacent grid lines $x = i \pm 1$ or $y = j \pm 1$. We then take the line through (i,j) along the gradient (i.e., perpendicular to the level line) and find the two opposite points P_1, P_2 where the line hits this square. We interpolate the magnitude of the gradient $|\nabla I|$ from the corners of the square to these points P_i, P_2 on the edges, and declare the gradient of I at the pixel (i,j) to be *locally maximal* if $|\nabla I(i,j)| \geq \max(|\nabla I(P_1)|, |\nabla I(P_2)|)$. This takes the typical fuzzy edge and finds the pixels along that edge where the magnitude of the gradient is bigger than at all the pixels to the sides.

The second idea is hysteresis, which uses both a low t_{lo} and a high t_{hi} threshold. Once and for all we throw out any pixels where the magnitude of the gradient is less than t_{lo}. Then we start at all pixels where $|\nabla I| \geq t_{hi}$, and we follow contours as long as the $|\nabla I| \geq t_{lo}$. Quite simply, we grow a set of pixels by starting with those where $|\nabla I| \geq t_{hi}$ and, at each stage, add any of its 8-neighbors where $|\nabla I| \geq t_{lo}$. Canny proposed to set the thresholds adaptively, using statistics of the magnitude of the image gradient: take the whole image and compute the mean of $|\nabla I|$ and multiply this by some preassigned constants such as 0.5 and 1.25.

Two further ideas due to Canny are usually not implemented, being computationally much slower, but they are quite interesting. One is to use mean gradient statistics in *local* neighborhoods of each pixel to assign different thresholds in different parts of the image. This will suppress small wiggly contours in textured areas where the adaptive thresholds are set high but retain weak contours in uniform areas where the adaptive thresholds are set low. A second refinement that Canny proposed is to check the consistency of contours by using filters that are not isotropic, but extended along ellipses to integrate evidence of contrasting black/whiteintensities between the two sides of a putative extended smooth contour. There are various such filters, such as Gabor filters, steerable filters, and curvelets, which we discuss briefly in Chapters 4 and 6. The simplest is just the derivative of an eccentric Gaussian:

$$F(x,y) = x \cdot e^{-((\rho x)^2 + y^2)/2\sigma^2}$$

with an eccentricity $\rho > 1$. This filter would be used to check the strength of a vertical contour. A rotated version checks other orientations.

Marr's zero-crossing of the Laplacian method

A major drawback to Canny's method is that it delivers many incomplete contours that end fairly arbitrarily. One of the pluses for the level-line approach discussed in Section 3.1.2 is that level lines do not end abruptly; except for a few that hit saddle points, they are all closed curves or curves that end at the boundary of the image. Marr's idea [151] was to use

3.6. Exercises

the zero-value level lines of the Laplacian of the Gaussian filtered image, which are also usually closed curves:

Marr's edges = connected components of the locus $\triangle G_\sigma * I = 0$.

Note that $\triangle(G_\sigma * I) = (\triangle G_\sigma) * I$ and that

$$\triangle G_\sigma = \frac{x^2 + y^2 - 2\sigma^2}{2\pi\sigma^6} e^{-\frac{x^2+y^2}{2\sigma^2}},$$

and if we choose a discrete version of the function $\triangle G_\sigma$, we can form the Laplacian of the Gaussian in the discrete setting.

Why the zero-crossings of the Laplacian? This is a version of nonmaximum suppression: at most nontextured pixels P, an image is approximately a function of one variable in the direction of its gradient, $I(x, y) \approx f(ax + by)$, where $a = \nabla_x I(P), b = \nabla_y I(P)$. Nonmaximum suppression means we seek pixels P where $|f'|$ is a local maximum, so that $\triangle I(P) = f''(P) = 0$.

The second idea Marr's approach is to consider these contours for all σ, or all amounts of blurring. MArr and Hildreth [151] proposed that the most significant contours would be those that persisted for a large range of scales but that they would then be best spatially localized by following the contour down to the smallest σ.

Nitzberg-Shiota's method

In [168], Nitberg, Mumford, and Shiota introduce a version of Canny's edge detection with one important change. This is based on the idea that we are interested in the tangent to the contour line but not in an orientation of the contour. That is, we are interested in tracing the contour by following the line perpendicular to the gradient but not in assigning a forward or backward direction on it. This direction is the principal direction of the 2 × 2 rank 1 matrix

$$(\nabla I)^{\otimes 2} = \begin{pmatrix} (\nabla_x I)^2 & (\nabla_x I) \cdot (\nabla_y I) \\ (\nabla_x I) \cdot (\nabla_y I) & (\nabla_y I)^2 \end{pmatrix}.$$

To merge data from nearby points, we can Gaussian filter these matrices, getting a 2 × 2 positive semidefinite matrix $G * (\nabla I)^{\otimes 2}$. The eigenvector for the largest eigenvalue of this matrix is the strongest local gradient direction and its perpendicular is the candidate for the direction of a contour through this pixel. But we can now assess the evidence for this being a strong salient contour by whether this eigenvalue is bigger and, better, much bigger than the smaller one. If both are roughly equal, this must be caused by noise or texture, not a contour.

This method then used nonmaximum suppression and tracing with hysteresis, as did Canny. The main point of this method is to find the T-junctions in an image where one contour ends, because, in three dimensions, it is an edge going behind another object. This is not easy but the two eigenvalues of the above matrix assist in finding corners and T-junctions; see Section 3.2 of [168].

Exercise 1, then, is to implement these methods in MATLAB using the built-in `CannyStanford` in the image toolbox if you have it, and to experiment with it, comparing it with the algorithms given in the text. Looking closely at what it does right and what it does wrong is the only way to appreciate the subtleties of contour-finding in images. Code for our significant level-line finder can be found on our website.

2. Some Useful Determinants

Some nearly diagonal matrices play a big role with snakes. It is not difficult to work out all the determinants that come up.

1. In Chapter 2, we analyzed the characteristic polynomial of the matrix

$$A_q = \begin{pmatrix} 2 & -1 & 0 & \cdots & 0 & 0 & -1 \\ -1 & 2 & -1 & \cdots & 0 & 0 & 0 \\ & \cdots & & \cdots & & \cdots & \\ 0 & 0 & 0 & \cdots & -1 & 2 & -1 \\ -1 & 0 & 0 & \cdots & 0 & -1 & 2 \end{pmatrix}.$$

In this chapter, we need a slightly modified matrix:

$$B_q = \begin{pmatrix} 1 & -1 & 0 & \cdots & 0 & 0 & 0 \\ -1 & 2 & -1 & \cdots & 0 & 0 & 0 \\ & \cdots & & \cdots & & \cdots & \\ 0 & 0 & 0 & \cdots & -1 & 2 & -1 \\ 0 & 0 & 0 & \cdots & 0 & -1 & 1 \end{pmatrix} =$$

$$A_q + \begin{pmatrix} -1 & 0 & 0 & \cdots & 0 & 0 & 1 \\ 0 & 0 & 0 & \cdots & 0 & 0 & 0 \\ & \cdots & & \cdots & & \cdots & \\ 0 & 0 & 0 & \cdots & 0 & 0 & 0 \\ 1 & 0 & 0 & \cdots & 0 & 0 & -1 \end{pmatrix}.$$

This is a perturbation of A_q by a rank 1 matrix. The characteristic polynomial of such a matrix can be computed by using Lemma 3.4:

Lemma 3.4. : *If A is an invertible matrix and $a \otimes b$ is a rank 1 matrix, then*

$$\det(A + a \otimes b) = \det(A) \cdot (1 + <A^{-1}a, b>).$$

Hence, if D is a diagonal matrix with entries λ_k and E is the rank 1 matrix $E_{ij} = a_i b_j$, then

$$\det(D + E) = \prod_i \lambda_i \cdot \left(1 + \sum_k \frac{a_k b_k}{\lambda_k}\right).$$

Prove this. Then apply this to B_q, using the fact that A_q has eigenvalues $\lambda_k = -(\zeta^k - 2 + \zeta^{-k}) = 4\sin^2(\pi k/q)$ and eigenvectors e_k with ith-entry ζ^{ik}. First, rewrite B_q in the basis $\{e_k\}$. Show that

$$B_q(e_k) = \lambda_k \cdot e_k + \frac{1 - \zeta^{-k}}{q} \sum_l (-1 + \zeta^l) e_l.$$

Hence, using the lemma, show

$$\det(bI + aB_q) = \prod_i (b + a\lambda_i) \cdot \left(1 - \frac{1}{q} \sum_k \frac{a\lambda_k}{b + a\lambda_k}\right).$$

3.6. Exercises

For large q and $b/a > 0$, the second term is quite simple to approximate. Substitute in the value of λ_k and note that for large q,

$$\frac{1}{q} \cdot \sum_{k=0}^{q-1} \frac{4\sin^2(\pi k/q)}{b/a + 4\sin^2(\pi k/q)} \approx \frac{1}{\pi} \int_0^\pi \frac{4\sin^2(\theta)}{b/a + 4\sin^2(\theta)} d\theta.$$

Evaluate the integral, using a table of integrals if necessary, and show that for large $q, b/a > 0$,

$$\det(bI + aB_q) \approx \det(bI + aA_q) \cdot \left(1 + \frac{4a}{b}\right)^{-1/2}.$$

2. In the text, however, we need the ratio

$$\frac{\det(rI + B_{N+M-1})}{\det(rI + B_{N-1}) \cdot \det(rI + B_{M-1})}.$$

Combine the calculation in Chapter 2 with question 1 to find that for large N, M, $r > 0$, this approximates

$$\frac{r + 2 + \sqrt{r^2 + 4r}}{2} \cdot \sqrt{1 + \frac{4}{r}}.$$

One can get finer results by applying the idea of Lemma 3.4—namely, writing $\det(D + E) = \det(D) \cdot \det(I + D^{-1} \cdot E)$—to the matrices

$$D = \begin{pmatrix} rI_{N-1} + B_{N-1} & 0 & 0 \\ 0 & r & 0 \\ 0 & 0 & rI_{M-1} + B_{M-1} \end{pmatrix},$$

E = zero except 3×3 block $\begin{pmatrix} 1 & -1 & 0 \\ -1 & 2 & -1 \\ 0 & -1 & 1 \end{pmatrix}$

at rows/columns $N-1, N, N+1$

$E + D = rI_{N+M-1} + B_{N+M-1}.$

Here E is rank 2, not rank 1, but the method works and will give the needed ratio, but requires the vector

$$u = B_q^{-1} \cdot \begin{pmatrix} 0 \\ \vdots \\ 0 \\ 1 \end{pmatrix}.$$

This sort of question is solved by looking at vectors that are geometric series. Applying B_q to a vector $v_k = \lambda^k$, all but the first and last components are multiples of $-1 + (r+2)\lambda - \lambda^2$, and this factor is zero if

$\lambda = \lambda_+ \stackrel{\text{def}}{=} \frac{1}{2} \cdot (r + 2 + \sqrt{r^2 + 4r})$, or its inverse, $\lambda_- \stackrel{\text{def}}{=} \frac{1}{2} \cdot (r + 2 - \sqrt{r^2 + 4r})$.

168   3. Character Recognition and Syntactic Grouping

Use this to solve for u as a combination of ascending and descending powers of λ, then compute $D^{-1} \cdot E$ and its determinant. Prove:

$$\frac{\det(rI + B_{N+M-1})}{\det(rI + B_{N-1}) \cdot \det(rI + B_{M-1})} = \frac{1}{r}\left((r+1+\beta_N).(r+1+\beta_M) - 1\right),$$

$$\text{where } \beta_N = (\lambda_+ - r - 1)\left\{\frac{1 + \lambda_+^{-2N+1}}{1 - \lambda_+^{-2N}}\right\}.$$

Hence, we get a correction for the determinant ratio for finite N and M, $r > 0$:

$$\frac{\det(rI + B_{N+M-1})}{\det(rI + B_{N-1}) \cdot \det(rI + B_{M-1})} \approx$$

$$\frac{r + 2 + \sqrt{r^2 + 4r}}{2} \cdot \sqrt{1 + \frac{4}{r}} \cdot \left(1 + \lambda_+^{-2N} + \lambda_+^{-2M}\right).$$

3. The Medial Axis as a Convex Hull

Stereographic projection is the map from the unit sphere to the plane as shown on the left in Figure 3.24. The unit sphere maps to the plane tangent to it at the South Pole, taking every point P of the sphere to the intersection of the plane with the line joining P to the North Pole. If (x, y, z) are three-dimensional coordinates with the origin at the South Pole (so $x^2 + y^2 + (z-1)^2 = 1$) and (u, v) are two-dimensional coordinates in the tangent plane there, then the projection is given by

$$u = \frac{2x}{2-z}, \quad v = \frac{2y}{2-z}$$

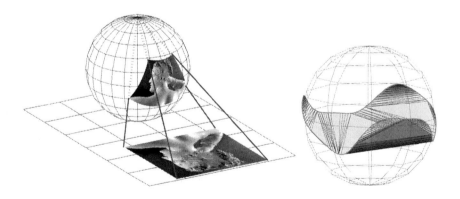

Figure 3.24. Stereographic projection from the North Pole illustrated by lifting an image of a marble bust from the plane to the sphere (left). The lower boundary of the convex hull of a curve on the sphere (right). It is a developable surface with the triangular faces corresponding to each of the tritangent circles inside the curve.

3.6. Exercises

or inverting

$$x = \frac{4u}{u^2 + v^2 + 4}, \quad y = \frac{4v}{u^2 + v^2 + 4}, \quad z = \frac{2(u^2 + v^2)}{u^2 + v^2 + 4}$$

If we let C be a simple closed curve in the plane and C' be its lift to the 2-sphere by stereographic projection, the convex hull of C' is a lens-shaped object with an upper and lower boundary, each a topological disk with boundary C'. This is shown on the right in Figure 3.24.

Prove that the Delaunay triangulation of the interior of C is essentially the same as the lower convex hull boundary. For each bitangent circle inside C, you have an edge of the Delaunay triangulation and an edge of the lower convex hull boundary; for each tritangent circle inside C, you have a triangle of the Delaunay triangulation and a triangular face of the hull, and so on. Each of these corresponds under stereographic projection.

4. Random Ribbons via the Medial Axis

A simple approach to generating random shapes is to create a shape by starting with a random curve $\Gamma = \text{locus } \{\vec{\gamma}(s)\}$, which we take as the axis, creating a random radius function $r(s)$, and defining the shape as the union of the disks centered on Γ with radius r. There are many choices. One way to do this is to use *two* Ornstein-Uhlenbeck processes, $\mathcal{X}_k(t), k = 1, 2$, and take their exponentials $e^{\mathcal{X}_k(t)}$. We then use them as the relative speeds $ds_\pm/dt$ of movement of the contact points $P_\pm$ with respect to arc length on the medial axis (see Section 3.3.2). Using the formulas of that section, we can add and subtract and get the two ODEs:

$$\frac{1 - r(t).r''(t) - r'(t)^2}{\sqrt{1 - r'(t)^2}} = \frac{e^{\mathcal{X}_1(t)} + e^{\mathcal{X}_2(t)}}{2},$$

$$r(t)\kappa(t) = \frac{e^{\mathcal{X}_1(t)} - e^{\mathcal{X}_2(t)}}{2}.$$

1. Sample Ornstein-Uhlenbeck processes in MATLAB. There is a simple way to do this with a*cumsum(fac.*randn(1:n))./fac, fac = exp(b*1:n)). But be careful; this and everything that follows is very sensitive to the choice of constants a,b.

2. Use one of its ODE integrators to solve the first equation for $r(t)$ until you get $|r'(t)| = 1$, which marks an endpoint of the shape.

3. Solve the second equation for κ and integrate to find the axis up to a translation and rotation.

4. Plot the sample ribbon shapes that you obtain.

5. Computing the Medial Axis with MATLAB

We assume that N points are given on the boundary ∂S of a shape S. These points can also be selected manually, by using the MATLAB function ginput. The following code is a Matlab program that computes the medial axis of the polygon made by the N points.

The main steps of this code are:

1. Write the N points as a vector z of complex numbers and plot the polygon.
2. Use the function delaunay to get the Delaunay triangles, and compute the centers and radii of the circumscribed circles.
3. Prune the exterior triangles. To do this, use the fact that if the points of the polygon are counterclockwise ordered, then a Delaunay triangle with ordered vertices u, v, w is interior if and only if the angle between $u - w$ and $v - w$ is positive.
4. The circumcenters of the remaining triangles are now the vertices of the medial axis. Finally, find the edges of the medial axis.

```
% This code assumes a polygon is given by a vector z of
% complex numbers
plot([z z(1)],'-*'), hold on, axis equal
xmax = max(real(z)); xmin = min(real(z));
ymax = max(imag(z)); ymin = min(imag(z));
maxlen = max((xmax-xmin), (ymax-ymin))/20;

% Get Delaunay triangles and compute their circumcenters
tri = sort(delaunay(real(z),imag(z))')';
u = z(tri(:,1)); v = z(tri(:,2)); w = z(tri(:,3));
dot = (u-w).*conj(v-w);
m = (u+v+i*(u-v).*real(dot)./imag(dot))/2;
r = abs(u-m);

% Prune the exterior triangles. The m's ares now the vertices
% of the medial axis
inside = imag(dot) < 0;
% if the polygon is clockwise, then put:
%inside = imag(dot) > 0;
triin = tri(inside,:);
m = m(inside); r = r(inside);
nt = size(triin,1);

% Find edges of the medial axis
B = sparse([1:nt 1:nt 1:nt], triin(:), ones(1,3*nt));
[a,b,c] = find(B*B'>1);
ind = a>b; a = a(ind); b = b(ind);
plot([m(a); m(b)], '-')

% Subdivide long segments in the medial axis
numsub = ceil(abs(m(a)-m(b))/maxlen);
ind = numsub > 1; delta = 1./numsub(ind);
newm = m(a(ind))*(delta:delta:1-delta) + m(b(ind))
   *((1-delta):-delta:delta);
newr = abs(newm - z(dsearch(real(z), imag(z), tri, real(newm),
   imag(newm))));
m = [m newm]; r = [r newr];
nm = size(m,2);
plot(m,'*')
```

3.6. Exercises

The medial axis can be a good description of a shape S, but one problem is that it is very sensitive to small irregularities of ∂S. For example, if the shape S is a disk, then its medial axis is reduced to a single point (the center of the disk); but if S is more or less a disk with many small irregularities, then the medial axis has many branches. Check this by drawing manually (by using the MATLAB function ginput) an irregular circle.

A good test-bed for medial axes as shape descriptors is the case of *leaves*. We suggest the following project. Gather a bunch of maple, oak, elm, and other leaves with interesting shapes. Put them on a scanner backed with contrasting paper, scan to get the leaf image, threshold to get the leaf shape, and sample the boundary to get an approximating polygon. The code above will give you the medial axis. You will probably find it is very complex, although there are "main branches" too that reflect the overall shape. You can "melt" the shape a bit to get at the main branches by using the geometric heat equation described in Section 5.3.4. To implement this on your leaf polygons, compute the orientation $(\theta_C)_k$ for each edge and the curvature $(\kappa_C)_k$ at each vertex as the first difference of θ divided by the appropriate arc length. Then you can solve numerically a finite-difference form of the geometric heat equation. To make this stable, you just need to be sure to take your time steps small enough compared to the edge lengths.

6. Finiteness of Branching Processes

First consider unlabeled branching:

1. Prove the lemma: if $\mathcal{X}$ is a nonnegative random variable with mean μ and variance σ^2, then

$$\mathbb{P}(\mathcal{X} > 0) \geq \frac{\mu^2}{\sigma^2 + \mu^2}.$$

2. If $\mu > 1$, apply the above inequality to $\mathcal{N}_k$ to give an upper bound to the probability of the tree being finite in terms of μ and $\mathrm{Var}(\mathcal{N}_1)$.

Next, we consider labeled branching. As in Section 2, let $P_\ell(R_{\ell,k})$ be the probability of choosing production rule k from a nonterminal with label ℓ. Let $N_{\ell,k}(m)$ be the number of nodes (terminal or nonterminal) with label m produced by rule $R_{\ell,k}$. Define the matrix

$$M_{\ell,m} = \begin{cases} \sum_{\text{rule } k} N_{\ell,k}(m) P_\ell(R_{\ell,k}) & \text{if } \ell \text{ is nonterminal,} \\ 0 & \text{if } \ell \text{ is terminal,} \end{cases}$$

representing the expected number of label m nodes from a label ℓ node.

1. Show that if all the eigenvalues of M have absolute value less than 1, the expected size of a parse tree is finite and equal to

$$e_S \cdot (I - M)^{-1} \cdot (1\,1\,\cdots\,1)^t, \quad \text{where } e_S = \text{vector with 1 at the start state}.$$

Conclude that it is almost surely finite.

2. Show, moreover, that if any eigenvalue of M has an absolute value greater than 1, then there is a positive probability of generating infinite parse trees.

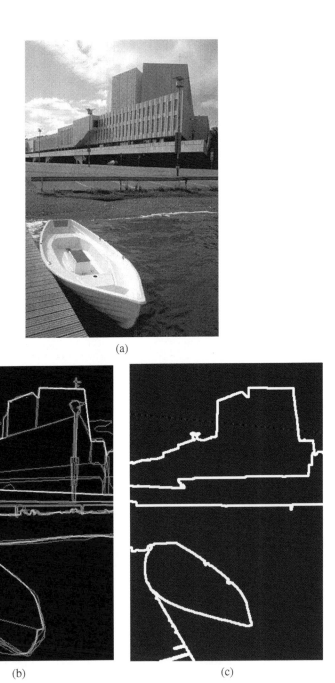

Figure 4.1. (a) A gray-level image of an urban scene of a waterfront. (b) Segmentations of the image by human subjects (the thickness of contours indicates how often contours were chosen). (c) A machine segmentation of the scene. Note that the subtle changes between the beach and the water are *not* recorded as a boundary by the machine. If the algorithm were set to detect more boundaries, it would likely pick up the clouds first. The machine also has no models of three-dimensional objects; thus, it fails to complete the boat and places the boundary between the building and the parking lot quite strangely. (Courtesy of J. Malik's lab.)

- 4 -
Image Texture, Segmentation and Gibbs Models

A FUNDAMENTAL ontological fact is that the world is made of objects, and thus the domain of an image can be partitioned into regions imaging the various visible objects. But what is an object? Although this is basically a philosophical question, thinking about it will help focus our minds on what this segmentation should look like.

The ideal exemplar of the concept of an object is a *compact, rigid, movable thing*. For a human baby, the original archetypal "object" may be its rattle hanging above its head or clasped in its fist. When we look closely, however, we find a whole set of things that are objectlike but lose some of the characteristics of ideal objects:

1. objects with multiple parts that move partially independently, such as an animal with limbs;

2. objects that are totally flexible, such as clothes;

3. objects (referred to by what linguists call *mass nouns*) such as water, sand, as in "a pile of sand," or grapes, as in "a bunch of grapes";

4. things that appear as distinct objects in images but not in three-dimensional space: "surface objects" (e.g., albedo patterns, such as a flower painted on a tablecloth[1]) and illumination effects (e.g., shadows and highlights)). In a sense, we may think of these as extremely thin objects lying on top of other objects.

In an infant's development, the visual system uses three processes to recognize objects, which develop at different times in the following order:

1. motion parallax (distinguishing a moving object from a constant background—a rattle seen against the walls of the room);

---
[1] It is interesting that babies at a certain stage in their development try to pick up flowers and the like that are only patterns on the surface of a tablecloth.

2. stereoscopic vision (identifying a foreground object by triangulating to find depth);

3. two-dimensional clues in a single static image: distinct color and texture.

All of these processes allow babies—and computers—to segment a scene into different objects. Here, we study only the recognition of objects in static two-dimensional images of the world, the third of these skills. In such a situation, the various objects are usually distinguishable from each other by contrasting color and texture, and this is how the segmentation algorithms work. This fact leads to the following hypothesis:

Segmentation Hypothesis (Strong Form). *Given a single image $I(x,y)$ where $(x,y) \in R$ (R is a domain), there is a natural decomposition into regions corresponding to the visible objects,*

$$R = \coprod_i R_i,$$

such that $I|_{R_i}$ has homogeneous texture—that is, "slowly" varying intensity or "slowly" varying statistics (quasi-stationary statistics).

However, very often, the objects present in an image are quite difficult to identify without some semantic knowledge of the world—a knowledge of what typical objects, such as faces, look like. The image in Figure 4.1 (a) illustrates this problem: the shoreline causes only a small change in texture and is not found in the computer segmentation, although it is obvious to people, as Figure 4.1 (c) shows. More examples can be found in Figure 4.2. This leads us to a weakened form of the hypothesis. Thus, we consider a differerent form of decomposition that is not canonically associated with the statistics of I.

Rather a set of alternate decompositions can be computed from the statistics of I, among which the decomposition into "true" objects can be found when high-level statistics on object classes are known.

Yet another problem with the segmentation hypothesis is that many objects break up into pieces or other smaller objects sit on their surface. Then we don't know "when to stop," or what level of detail should be captured in the segmentation. This problem suggests there should a tree of finer and finer partitions of R instead of just one partition, as studied later in Chapter 6.

The goal of this chapter is to build a series of stochastic models whose random variables include the observed image $\mathcal{I}$ and a segmentation of the domain $R = \coprod_i \mathcal{R}_i$, such that $P(\{\mathcal{R}_i\} \mid \mathcal{I})$ has its mode at the most plausible decomposition. Some of these models have further hidden variables associated with each region $\mathcal{R}_i$ that describe what is homogeneous about the image on that region. They

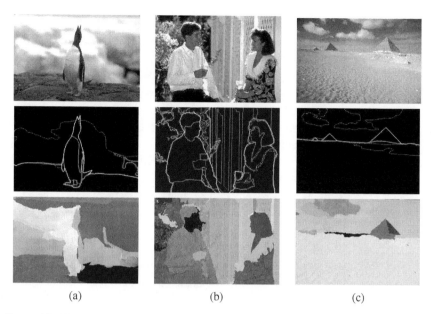

(a) (b) (c)

Figure 4.2. More examples of images, their human segmentations and their machine segmentations. Note (a) how the computer algorithm gets confused by artifacts in the jumble of ice flows; (b) how it fails to cluster the panels of the lattice fence; and (c) how it breaks the large pyramid into two segments for its two faces. The algorithm has trouble with *disconnected* regions, with texture gradients, and with discounting lighting artifacts. (See Color Plate III.) (Top row courtesy of J. Malik's lab.)

may be "cartoons," simplified versions of the image on this region, or explicit probability models for its texture. This approach to segmentation is an example of the Bayesian methods described in Chapter 0, Section 0.3 .

Since we do not believe the correct segmentation can be found with further semantic information about the world, it is not always a good idea to calculate only the mode of the model. Another way to use these models is to form small samples from this probability distribution, not committing to the mode alone. Then one can include at a later stage more sources of information, such as semantic information or global lighting information, by building on this sample to estimate the mode of the complex model with this additional information. This process is very similar to the idea of the Kalman filter, in which information about a system is coming in continuously in time and, at each point in time, we estimate both the most likely state of the system and its variance. It is also similar to what A. Blake and M. Isard [112] called "condensation," whereby the full posterior probability distribution of the state based on partial evidence is estimated by a weighted sample.

In Section 4.1, we will study a general way to create such stochastic models—Gibbs fields. Then, in Section 4.2, we study a large class of specific Gibbs fields that have been used by many groups to segment images—the $u + v$ models in which the image is separated into a fine detail component representing noise or texture and a simplified cartoon component based on its segmentation. In Sections 4.3 and 4.4, we look at ideas on how to reason with such models, finding their modes and marginals. These are all models for images and segmentations defined on a finite set of pixels. They do not, in general, have continuous limits (we study later in Chapter 6 a class of continuous limits based on scaling ideas). However, their associated energies, $E = -\log(P)$, do have continuous limits, and minimizing these gives variational problems similar to those in other applied mathematical settings. Taken on a finite set of pixels, we can sample from these models. The results are not very much like actual images of real scenes, primarily because these models have very simplified geometric pattern terms for modeling the boundary contours. The situation is very similar, however, to the machine-translation algorithm described in Section 1.6. In both cases, the priors on the hidden variables (the English text or the segmentation) are poor, and the conditional probability term (the English-to-French dictionary or the model for the gray levels of the image) is poor—but when a real image is used, the Bayesian posterior works much better. So we spend relatively little time sampling from our models and concentrate instead on using the posterior, for a fixed observed image, to find its segmentation.

In the remainder of the chapter, Sections 4.5–4.7, we study texture, textured regions being one of the main problems in finding good segmentations. What is texture, really? Modeling texture, which we look at in Section 4.5, is a pure exercise in what we called value patterns in Chapter 0. Following our pattern theory philosophy, we should be able to synthesize texture if we understand it, and exponential models, defined in Section 4.6.1, come to our rescue. We get a quite good algorithm for synthesizing texture, the basic value pattern of images. We return to the full problem of segmentation, now separating regions with distinct textures, in Section 4.7, where we describe several models of full-textured images.

4.1 Basics IX: Gibbs Fields

4.1.1 The Ising Model

The Ising model is the simplest and most intensively studied of all Gibbs fields. It has its origin in statistical mechanics—more precisely in a model for the magnetization of a lattice of iron (Fe) atoms due to Ernst Ising (pronounced "Eezing'). Applied to vision, we assume that $I(i,j)$, where $1 \leq i \leq n$ and $1 \leq j \leq m$, is a real-valued image, and that $J(i,j) \in \{0,1\}$ is a binary image supposed to

4.I. Basics IX: Gibbs Fields

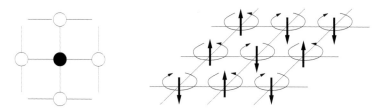

Figure 4.3. A pixel and its four nearest neighbors (left). The physical origin of the Ising model is a two-dimensional grid of iron atoms whose spins may or may not align.

model the predominantly black/white areas of I. Note that the random variable $\mathcal{I} \in \mathbf{R}^{nm}$ lies in a vector space, whereas the random variable $\mathcal{J}$ has only a finite number 2^{nm} of possible values. Let us denote the pixels simply by $\alpha = (i, j)$. We then define the energy function E by

$$E(I, J) = c \sum_{\alpha} (I(\alpha) - J(\alpha))^2 + \sum_{\alpha \sim \beta} (J(\alpha) - J(\beta))^2,$$

where $c > 0$ is a real constant and where $\alpha \sim \beta$ means that α and β are adjacent pixels (nearest neighbors) (see the left diagram in Figure 4.3).

We associate with this energy function the following probability density:

$$p(I, J) = \frac{1}{Z_T} e^{-E(I,J)/T},$$

where $T > 0$ is a constant called the temperature, and Z_T is the normalizing constant (also called the partition function) making p into a probability distribution on I and J:

$$Z_T = \sum_{J} \left(\int_I e^{-E(I,J)/T} \prod_{\alpha} dI(\alpha) \right).$$

The probability $p(I, J)$ can be decomposed in two terms:

$$p(I, J) = p(I \mid J) \times P(J),$$

with

$$\begin{aligned}
p(I \mid J) &= \left(\frac{c}{\pi T}\right)^{\frac{nm}{2}} e^{-c \sum_{\alpha}(I(\alpha)-J(\alpha))^2/T}, \quad \text{a Gaussian probability density} \\
P(J) &= \frac{1}{Z_T^0} e^{-\sum_{\alpha \sim \beta}(J(\alpha)-J(\beta))^2/T}, \quad \text{a discrete probability distribution} \\
Z_T^0 &= \sum_{J} e^{-\sum_{\alpha \sim \beta}(J(\alpha)-J(\beta))^2/T}.
\end{aligned}$$

In the physicist's version of the Ising model, $J(\alpha) \in \{-1,+1\}$, it indicates the spin of an Fe atom, and $I(\alpha)$ represents the extremal magnetic field at point α. Then E is the physical energy. Notice that $\sum_\alpha J(\alpha)^2$ is independent of J because for all α, $J(\alpha)^2 \equiv 1$, and so we have the form for E usually used in the Ising model:

$$E_{\text{phys}}(I,J) = (\text{term indep. of } J) - 2\left(c\sum_\alpha I(\alpha)J(\alpha) + \sum_{\alpha \sim \beta} J(\alpha)J(\beta)\right).$$

What sort of images does sampling from this model produce? We may sample from the pure Ising model $P(J)$ first, getting a black and white binary valued image. Then, sampling from $p(I|J)$, we add white noise independently to each pixel. Note that when T is large, $P(J)$ is nearly constant and its samples will themselves be nearly "salt-and-pepper" random black and white noise. In contrast,, for T small, the coupling between adjacent pixels of J is very strong and every boundary between black and white reduces the probability. Therefore, for T small, the samples of $P(J)$ look like fields of either all white or all black with a smattering of islands of pixels of the other color. There is a range of values of T, however, where black and white fight to make their own empires, and samples from $P(J)$ look like a mixture of black and white islands, some big, some small. These values of T are those near the so-called "critical point"; for a precise discussion, see Section 4.3.2 where we discuss specific heat. It is at this critical point that samples from $P(J)$ begin to capture the nature of real images. Examples from the slightly more complex Potts model are given in Figure 4.5. The resulting Is are noisy versions of the Js.

More interesting for the use of the Ising model to segment images is what the samples of J are *when we fix I* and consider the conditional probability

$$P_T(J \mid I) = \frac{1}{Z'_T} e^{-E(I,J)/T}.$$

Here, the normalizing constant $Z'_T = \sum_J P(I,J) \cdot Z_T$ depends on I. As before, if $T \to +\infty$, P_T approaches the uniform distribution over the finite set of all possible J. But if $T \to 0$, P_T approaches the delta distribution concentrated not on uniform Js but on whatever happens to be the minimum of $J \mapsto E(I,J)$ for this fixed I. In Figure 4.4, we give an example of an image I, the minimum of $J \mapsto E(I,J)$, and samples from $P_T(J \mid I)$ for various values of T. Note that for T large, the sampled J is indeed more or less random. As T decreases, the sample pays more and more attention to I and finally settles at the global minimum energy state for $T = 0$. How did we find such samples and guarantee that they are "average" samples from these probability distributions? This was done via Monte Carlo Markov chains, which we describe in Section 4.3.

4.1. Basics IX: Gibbs Fields

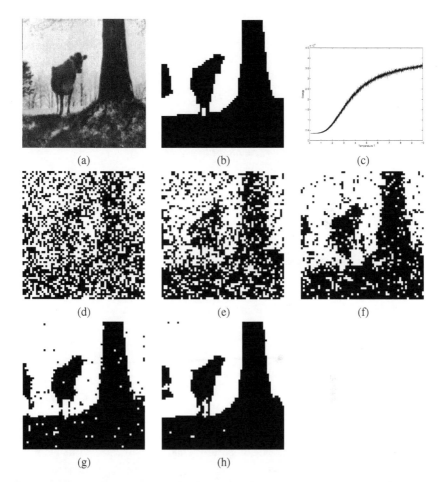

Figure 4.4. Samples from the Ising model: the gray-level image in (a) is scaled to have max/min pixel values $0, +1$ and is used as the external magnetic field I in the Ising model. The back/white image in (b) is the mode of the probability distribution, the least energy field J, when the constant $c = 4$ in the energy. The graph (c) shows values of the energy $E(I, J)$ for J sampled from $P_T(J|I)$ at various temperatures T. Images (d)–(h) are random samples J taken from the Ising model conditioned on fixed I (i.e., from $P_T(J|I)$) at temperatures $T = 10, 4, 2.5, 1.5,$ and 1, respectively. Note how the shapes emerge between (e) and (f), where the mean energy, as seen in graph (c), has a large derivative with respect to T.

At temperature $T = 0$, the Ising model is supported at the global minimum of E, the mode of all the probability distributions P_T. Surprisingly, this mode is not difficult to calculate exactly. The idea goes back to a paper by Gil Strang [206], which was rediscovered and put in a vision context by Chan and Vese [47]. In the meantime, a nearly identical method was found by Greig, Porteous, and Seheult [91]. Here's how Strang's idea works. Note that because $J(\alpha) \in \{0, 1\}$, we have

$$E(I, J) = c \sum_\alpha J(\alpha)(1 - 2I(\alpha)) + c \sum_\alpha I(\alpha)^2 + \sum_{\alpha \sim \beta} |J(\alpha) - J(\beta)|,$$

and that the middle term is independent of J. We then use the result of Theorem 4.1.

Theorem 4.1. *Let Ω be the convex set of all functions $u(\alpha)$ with values in the interval $[0, 1]$, and consider any functional*

$$F(u) = c \sum_\alpha a(\alpha) u(\alpha) + \sum_{\alpha, \beta} b(\alpha, \beta) |u(\alpha) - u(\beta)|.$$

Then its minimum is always taken on a finite union of convex sets, each spanned by vertices of Ω, that is, functions with values on $\{0, 1\}$.

Proof: Assume the minimum is taken on at a function u_0 and let $0 = y_0 < y_1 < \cdots < y_{n-1} < y_n = 1$ be the values taken on by u_0 and the values 0 and 1 (even if these are not in the range of u_0). Let $V_i = \{\alpha | u_0(\alpha) = y_i\}$. Now suppose we fix the sets V_i but let the constants $y_i, 1 \leq i \leq n-1$, vary on the set $0 = y_0 \leq y_1 \leq \cdots \leq y_{n-1} \leq y_n = 1$. On this set U of u_i, F will be linear. Since F was minimized at u_0 which is in the interior of U, it must be constant on the set U. If $n = 1$, we are done. If $n > 1$, u_0 will be a convex linear combination of vertices of U. At these vertices, u takes on only the values 0 and 1. □

The essential idea behind this theorem is that we have replaced the L^2 norm by an L^1 norm. Unit balls in L^1 norms are polyhedra whose vertices are very special points—for example, for the standard norm $\sum_i |x_i|$ on $\mathbb{R}^n$, all but one coordinate x_i vanishes at every vertex. This is part of what Donoho calls L^1-magic in the theory of compressed sensing.[2] The energies in this theorem are generalizations of the Ising energy and are known to physicists as "spin-glass" energies: nearest-neighbor interactions are not constant and can even vary in sign. An immediate corollary of this follows.

Corollary. *If $u(\alpha)$ is any function of pixels with values in the interval $[0, 1]$ at which*

$$F(u) = c \sum_\alpha a(\alpha) u(\alpha) + \sum_{\alpha, \beta} b(\alpha, \beta) |u(\alpha) - u(\beta)|$$

is minimum, then for all $\lambda \in (0, 1)$:

$$J(\alpha) = \begin{cases} 1 & \text{if } u(\alpha) \geq \lambda \\ 0 & \text{if } u(\alpha) < \lambda \end{cases}$$

also minimizes F.

---
[2] http://www.acm.caltech.edu/l1magic/

4.1. Basics IX: Gibbs Fields

Proof: Let u be a minimizer for F and let U be the set of functions defined as in the proof of Theorem 4.1. Then, for all $\lambda \in (0, 1)$, the binary function J defined here belongs to U, and since F is constant on this U, J also minimizes F. □

In case all the b are nonnegative, hence F is convex, finding such a u is a linear programming problem: introduce new variables $y_{\alpha\beta}, z_{\alpha\beta}$ for every α, β such that $b(\alpha, \beta) > 0$. Then it is easy to see that the minimum of the linear map

$$L(u, z) = c \sum_{\alpha} a(\alpha)u(\alpha) + \sum_{\alpha,\beta} b(\alpha, \beta)(y_{\alpha\beta} + z_{\alpha\beta}),$$

on the convex polyhedron
$$\begin{cases} 0 \leq u(\alpha) \leq 1, \\ y_{\alpha\beta} \geq 0 z_{\alpha\beta} \geq 0 \\ z_{\alpha\beta} = y_{\alpha\beta} + u(\alpha) - u(\beta) \end{cases}$$

is the same as that of the corollary to Theorem 4.1. Such a linear programming problem can be expected to be solved rapidly. Since $b(\alpha, \beta) = 0$ or 1 for all α, β in the Ising case, this gives a fast algorithm for finding the mode of the Ising model. The Greig-Porteous-Seheult variant of the method reduces the calculation of the mode to a "max-flow" graph problem, a special type of linear programming for which a number of fast algorithms are known. We describe this in Exercise 2 at the end of the chapter.

4.1.2 Specific Heat

In all Ising models $P_T(J|I)$ with fixed I (e.g., the cow/tree image shown in Figure 4.4), there are certain critical temperatures or ranges of temperature at which the real decisions are made. To find these, we use a number called the *specific heat*.

Lemma 4.2. *Define the mean value and the variance of the energy $E(I, J)$ at temperature T as*

$$<E>_T = \sum_J P_T(J|I)E(I, J),$$

and

$$\mathrm{Var}_T E = \sum_J (E(I, J) - <E>_T)^2 P_T(J|I).$$

Then

$$\frac{\partial}{\partial T}<E>_T = \frac{1}{T^2} \mathrm{Var}_T E,$$

and this is called the specific heat.

Proof: The proof is based on the following computation:

$$\frac{\partial}{\partial T}<E>_T = \frac{\partial}{\partial T}\left(\sum_J E(I,J)\frac{e^{-E(I,J)/T}}{Z'_T}\right)$$

$$= \sum_J E^2(I,J)\frac{1}{T^2}\frac{e^{-E(I,J)/T}}{Z'_T} - \sum_J E(I,J)\frac{e^{-E(I,J)/T}}{{Z'_T}^2}\frac{\partial Z'_T}{\partial T}.$$

Recall that Z'_T is defined by $Z'_T = \sum_J e^{-E(I,J)/T}$, and so

$$\frac{\partial Z'_T}{\partial T} = \sum_J E(I,J)\frac{1}{T^2}e^{-E(I,J)/T} = \frac{Z'_T}{T^2}<E>_T.$$

And finally

$$\frac{\partial}{\partial T}<E>_T = \sum_J E^2(I,J)\frac{1}{T^2}\frac{e^{-E(I,J)/T}}{Z'_T} - \sum_J E(I,J)\frac{<E>_T}{T^2}\frac{e^{-E(I,J)/T}}{Z'_T}$$

$$= \frac{1}{T^2}\sum_J E^2(I,J)\frac{e^{-E(I,J)/T}}{Z'_T} - \frac{1}{T^2}<E>^2_T$$

$$= \frac{1}{T^2}\text{Var}_T E. \qquad \square$$

Lemma 4.2 shows that when $\frac{1}{T^2}\text{Var}_T E$ is large, $<E>_T$ varies very fast—like water freezing into ice. This is where the black and white shapes begin to truly reflect the objects in the image. Note how in the simulation with the cow/tree image in Figure 4.4, the specific heat is maximum around $T=3$ and that at this point J is locating the major shapes in I. For other images I, there may well be several values of T at which certain patterns in I begin to control J.

The case $I=0$ has been a major focus in statistical mechanics. When the underlying lattice gets larger and larger, it is known that the specific heat gets a larger and sharper peak at one particular value of T. Passing to the limit of an infinite lattice, the specific heat becomes infinite at what is called the *critical value* of T or T_{crit}. For $T > T_{\text{crit}}$, there are no long-range correlations, no mostly black or mostly white blobs above some size. For $T < T_{\text{crit}}$, either black or white will predominate on the whole lattice. This behavior has been extensively studied because it is the simplest case of a phase transition—a sudden shift from one type of global behavior to another. The value T_{crit} was calculated explicitly in a major tour de force by Onsager [15, 171]. Recently, a great deal of attention has been given to the study of a certain continuum limit of the Ising model with $I=0, T=T_{\text{crit}}$. It turns out that at the critical temperature, the Ising model has amazing scaling and conformal symmetries when viewed "in the large," so the artifacts of the lattice disappear. It has, in a certain sense, a limit described very precisely by using conformal field theory [75]. We study a simpler example of this in Chapter 6.

4.1.3 Markov Random Fields and Gibbs Fields

The Ising model is a very special case of a large class of stochastic models called *Gibbs fields*. Instead of starting with a finite lattice of pixels α and random variables $J(\alpha)$ at each pixel, we start with an arbitrary finite graph given by a set of vertices V and a set of (undirected) edges E plus a *phase space* Λ. The $\Omega = \Lambda^V$ is called the *state space*, and its elements are denoted by $x, y, \cdots$. Then we say that a random variable $\mathcal{X} = \{\cdots, \mathcal{X}_\alpha, \cdots\}_{\alpha \in V}$ in the state space $\Omega = \Lambda^V$ is a *Markov random field* (MRF) if for any partition $V = V_1 \cup V_2 \cup V_3$ such that there is no edge between vertices in V_1 and V_3,

$$\mathbb{P}(\mathcal{X}_1 = x_1 | \mathcal{X}_2 = x_2, \mathcal{X}_3 = x_3) = \mathbb{P}(\mathcal{X}_1 = x_1 | \mathcal{X}_2 = x_2),$$

$$\text{where } \mathcal{X}_i = \mathcal{X}|_{V_i}, \ i = 1, 2, 3. \quad (4.1)$$

This property, called the Markov property, is also equivalent to saying that $\mathcal{X}_1$ and $\mathcal{X}_3$ are conditionally independent given $\mathcal{X}_2$, or

$$\mathbb{P}(\mathcal{X}_1 = x_1, \mathcal{X}_3 = x_3 | \mathcal{X}_2 = x_2) = \mathbb{P}(\mathcal{X}_1 = x_1 | \mathcal{X}_2 = x_2)\mathbb{P}(\mathcal{X}_3 = x_3 | \mathcal{X}_2 = x_2).$$

Let $\mathcal{C}$ denote the set of *cliques* of the graph (V, E): it is the set of subsets $C \subset V$ such that there is an edge between any two distinct vertices of C. A family $\{U_C\}_{C \in \mathcal{C}}$ of functions on Ω is said to be a family of potentials on cliques if, for each $C \in \mathcal{C}$, $U_C(x)$ depends only on the values of x at the sites in C. A *Gibbs distribution* on Ω is a probability distribution that comes from an energy function deriving from a family of potentials on cliques $\{U_C\}_{C \in \mathcal{C}}$:

$$P(x) = \frac{1}{Z} e^{-E(x)}, \quad \text{where} \quad E(x) = \sum_{C \in \mathcal{C}} U_C(x),$$

and Z is the normalizing constant, $Z = \sum_{x \in \Omega} \exp(-E(x))$, also called the partition function. A *Gibbs field* is a random variable $\mathcal{X}$ taking its values in the configuration space $\Omega = \Lambda^V$ such that its law is a Gibbs distribution.

The main result is the Hammersley-Clifford theorem, which states that Gibbs fields and MRFs are essentially the same thing.

Theorem 4.3. *A Gibbs fields and MRF are equivalent in the following sense:*

1. *If P is a Gibbs distribution then it satisfies the Markov property Equation (4.1).*

2. *If P is the distribution of a MRF such that $P(x) > 0$ for all $x \in \Omega$, then there exists an energy function E deriving from a family of potentials $\{U_C\}_{C \in \mathcal{C}}$ such that*

$$P(x) = \frac{1}{Z} e^{-E(x)}.$$

Part (1) of this theorem is a consequence of the fact that a Gibbs distribution can be written in a product form, which is $P(x) = \frac{1}{Z} \prod_C \exp(-U_C(x))$. A direct consequence of this is that if the set V of sites is partitioned into $V = V_1 \cup V_2 \cup V_3$, where there is no edge between V_1 and V_3, and if for a configuration x, we denote $x_i = x_{/V_i}$, then there exist two functions f_1 and f_2 such that

$$P(x) = P(x_1, x_2, x_3) = f_1(x_1, x_2) f_2(x_2, x_3). \tag{4.2}$$

This comes from the fact that a clique C cannot meet both V_1 and V_3, and that $U_C(x)$ depends only on the values of x at the sites in C. The Markov property is a direct consequence of the factorization of Equation (4.2):

$$P(x_1|x_2, x_3) = P(x_1|x_2).$$

Part (2) of Theorem 4.3 is known as the Hammersley-Clifford theorem, and its proof is mainly based on a combinatorial result called the *Möbius inversion formula*: let f and g be two functions defined on the set of subsets of a finite set V; then

$$\forall A \subset V, \quad f(A) = \sum_{B \subset A} (-1)^{|A-B|} g(B) \iff \forall A \subset V,$$

$$g(A) = \sum_{B \subset A} f(B).$$

Proof: The proof of the Möbius inversion formula is based on the following simple result: let n be an integer; then $\sum_{k=0}^{n} \binom{n}{k}(-1)^k$ is 0 if $n > 0$ (because this is the binomial expansion of $(1-1)^n$) and it is 1 if $n = 0$. Thus, if $f(A) = \sum_{B \subset A} (-1)^{|A-B|} g(B)$, then

$$\sum_{B \subset A} f(B) = \sum_{B \subset A} \sum_{D \subset B} (-1)^{|B-D|} g(D)$$

$$= \sum_{D \subset A} g(D) \times \sum_{k=0}^{|A-D|} \binom{|A-D|}{k} (-1)^k$$

$$= g(A).$$

Conversely, by the same argument, if $g(A) = \sum_{B \subset A} f(B)$, then

$$\sum_{B \subset A} (-1)^{|A-B|} g(B) = \sum_{B \subset A} (-1)^{|A-B|} \sum_{D \subset B} f(D)$$

$$= \sum_{D \subset A} f(D) \times \sum_{k=0}^{|A-D|} \binom{|A-D|}{k} (-1)^k$$

$$= f(A).$$

4.1. Basics IX: Gibbs Fields

We now can prove part (2) of Theorem 4.3.

Proof of (2), the Hammersley-Clifford theorem: We first have to fix some notation. Let 0 denote a fixed value in Λ. For a subset A of V, and a configuration x, we denote by xA the configuration that is the same as x at each site of A and is 0 elsewhere. In particular, we have $xV = x$. We also denote as $x_0 = x\emptyset$ the configuration with 0 for all the sites. Let us also introduce a function E defined on Ω by

$$E(x) = \log \frac{P(x_0)}{P(x)}.$$

The problem is now to show there exists a family $\{U_C\}$ of potentials on cliques such that for all $x \in \Omega$, $E(x) = \sum_C U_C(x)$.

Let x be a given configuration. For all subsets A of V, we define

$$U_A(x) = \sum_{B \subset A} (-1)^{|A-B|} E(xB).$$

By the Möbius inversion formula, we have, for all $A \subset V$,

$$E(xA) = \sum_{B \subset A} U_B(x).$$

In particular, taking $A = V$, we get

$$P(x) = P(x_0)e^{-E(x)} = P(x_0) \exp\left(-\sum_{A \subset V} U_A(x)\right).$$

We first notice that the function U_A depends only on the values of x at the sites in A. The claim now is that $U_A(x) = 0$ unless A is a clique. To prove this, we suppose that A is not a clique, which means that A contains two sites i and j that are not neighbors. Then, the sum over $B \subset A$ in the definition of U_A can be decomposed into four sums, according to the fact that B contains i or j or both or none:

$$\begin{aligned} U_A(x) &= \sum_{\substack{B \subset A \\ i \in B, j \in B}} (-1)^{|A-B|} E(xB) + \sum_{\substack{B \subset A \\ i \notin B, j \notin B}} (-1)^{|A-B|} E(xB) \\ &+ \sum_{\substack{B \subset A \\ i \in B, j \notin B}} (-1)^{|A-B|} E(xB) + \sum_{\substack{B \subset A \\ i \notin B, j \in B}} (-1)^{|A-B|} E(xB) \\ &= \sum_{B \subset A-i-j} (-1)^{|A-B|} E(x(B+i+j)) + \sum_{B \subset A-i-j} (-1)^{|A-B|} E(xB) \\ &- \sum_{B \subset A-i-j} (-1)^{|A-B|} E(x(B+i)) - \sum_{B \subset A-i-j} (-1)^{|A-B|} E(x(B+j)) \\ &= \sum_{B \subset A-i-j} (-1)^{|A-B|} \log \frac{P(x(B+i))P(x(B+j))}{P(xB)P(x(B+i+j))} \\ &= 0, \end{aligned}$$

since by the Markov property of P, we have $P(x(B+i))P(x(B+j)) = P(xB)P(x(B+i+j))$. Indeed, this is a direct consequence of the conditional independence in the case $V_1 = \{i\}$, $V_2 = V - i - j$, and $V_3 = \{j\}$.

In conclusion, we have proved that $U_A(x) = 0$ when A is not a clique, and hence

$$P(x) = P(x_0) \exp\left(-\sum_{C \subset \mathcal{C}} U_C(x)\right). \qquad \square$$

The Hammersley-Clifford theorem is about the existence of a family $\{U_C\}$ of potentials such that the distribution of an MRF is a Gibbs distribution with respect to these potentials. Now, what about unicity? The answer is that the family of potentials is unique if we assume that they are normalized with respect to a given value λ_0 in Λ, which means that $U_C(x) = 0$ as soon as there is an $i \in C$ such that $x(i) = \lambda_0$. The proof of this result is based again on the Möbius inversion formula to obtain an explicit formula of the normalized potentials U_C in terms of the $P(x)$ (the construction being just the exact inverse of the one in the proof of Hammersley-Clifford theorem).

4.2 $(u+v)$-Models for Image Segmentation

4.2.1 The Potts Model

Markov random fields are one of the main techniques for segmenting images. We have seen that if a real-valued image I segments into a foreground and a background with an overall black/white intensity difference, we can segment it by using the most probable state of the Ising model $P_T(J|I)$. We may express this problem as follows: we are given a $[0, 1]$-valued image I and we seek a binary $\{0, 1\}$-valued image u such that the energy $E(u, I)$ given by

$$E(u, I) = \sum_\alpha (I(\alpha) - u(\alpha))^2 + \lambda \sum_{\alpha \sim \beta} (u(\alpha) - u(\beta))^2$$

is minimized in u. (Here, $\lambda = 1/c$ in the notation defined in Section 4.1.3.) This can be interpreted by saying that we want I to be written as the sum $I = u + v$, where u is piecewise constant $\{0, 1\}$-valued, and v is a pure Gaussian noise image. Here v is Gaussian because $I - u = v$; hence e^{-E} has the factor $e^{-\sum v(\alpha)^2}$. Such a decomposition of an image was already used at the beginning of Chapter 3 for character recognition. In that example, I was the image of a character and we wanted to decompose it as $I = u + v$, where u was a binary image with unknown gray levels a for the paper and b for the character. This

4.2. $(u+v)$-Models for Image Segmentation

model extends the Ising model, since it allows u to take on two variable gray values, the mean brightness of the background and that of the characters; hence, it contains two further hidden random variables. The variable u is often called the "cartoon" image associated with I.

In most situations, the image I segment not into two regions but into more regions. Then we want u to take values in a set Λ of q elements called "colors" or "labels" or "states,". In this case, there is a generalization of the Ising model called the q-*state Potts model*. In the physical Potts model, q denotes the number of states of the spins of a lattice. For an image I, we fix the colors of u to be q gray levels and define the energy of a configuration u by

$$E_{\text{q-Potts}}(I, u) = \sum_\alpha (I(\alpha) - u(\alpha))^2 + \lambda \sum_{\alpha \sim \beta} \mathbb{I}_{\{u(\alpha) \neq u(\beta)\}}$$

where $\mathbb{I}_A$ is the indicator function of a statement A with value 1 when it is true and 0 where it is false. In this framework, u and I can be vector-valued images as well. The only difference is that the term $(I(\alpha) - u(\alpha))^2$ is replaced by the norm $\| I(\alpha) - u(\alpha) \|^2$. A typical example of this is the case of color images: they take values in $\mathbb{R}^3$, and many different norms can be considered, depending on the representation used for color (RGB, HSV, Luv, etc.).

If we let the number of colors and the colors themselves be random variables, the q-Potts energy becomes the discrete version of the so-called Mumford-Shah (or MS) segmentation for piecewise constant gray-level or color images [164]. Giving a function u with q values is the same as giving a partition $\pi = \{R_i\}$ of our lattice into q regions (i.e., π is an unordered set whose elements are subsets of the lattice) and an injective map $c : \pi \to \Lambda$, that is a set of distinct constants $\{c_i = c(R_i)\}$, the values of u on each region R_i. Then we may rewrite the energy to be minimized in u as

$$\begin{aligned} E(I, u) = E(I, c, \pi) &= \sum_\alpha (I(\alpha) - u(\alpha))^2 + \lambda \sum_{\text{all pairs } \alpha \sim \beta} \mathbb{I}_{\{u(\alpha) \neq u(\beta)\}} \\ &= \sum_{R_i \in \pi} \sum_{\alpha \in R_i} \left((I(\alpha) - c_i)^2 + \frac{\lambda}{2} \sum_{\text{all } \beta \text{ with } \beta \sim \alpha} \mathbb{I}_{\{\beta \notin R_i\}} \right). \end{aligned}$$

Note that the term with coefficient λ keeps the number of values of u from growing too large. When the values of I and the c_i are constrained to belong to a finite prespecified set of values (for instance, $\{0, 1, \ldots, 255\}$ in the case of a 1-byte gray-level image), there are only a finite number of possible u, and the energy in this discrete form provides a probability distribution on u via

$$P_T(I, u) = \frac{1}{Z} e^{-\sum_\alpha (I(\alpha) - u(\alpha))^2 / T} \cdot e^{-\lambda \sum_{\alpha \sim \beta} \mathbb{I}_{\{u(\alpha) \neq u(\beta)\}} / T}.$$

This probability model can be used in two different ways. First, we may sample from it and obtain synthetic images I, thus checking how much of the characteristics of images these models capture. To sample I from this model, we introduce the probability distribution $\tilde{P}_T(u, u)$ on (u, v) given by

$$\tilde{P}_T(u,v) = \frac{1}{Z} e^{-\sum_\alpha v(\alpha)^2/T} \cdot e^{-\lambda \sum_{\alpha \sim \beta} \mathbb{I}_{\{u(\alpha) \neq u(\beta)\}}/T}.$$

Then sampling from this distribution is just equivalent to sampling u from the q-Potts model (without any external field), and to sampling v independently of u from a Gaussian distribution. Then a sample of I is simply obtained by setting $I = u + v$. A second way to use the probability model $P_T(I, u)$ is to fix I and use the conditional distribution on u given I to analyze the image I and, in particular, assess the qualities of different segmentations of I.

Looking at the first use of the model, Figure 4.5 shows some examples of samples of I from the Potts model with five colors. The most efficient method

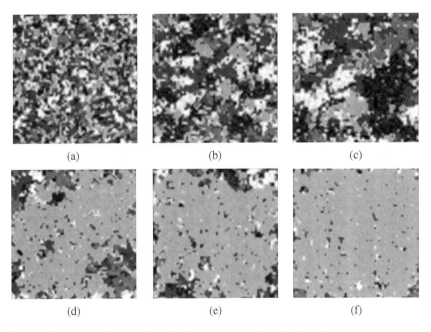

Figure 4.5. Samples I from the five-color Potts model on a 64×64 grid at descending temperatures. The five colors have been chosen as black and white and the primary colors, noise v has been added and the image blown up by resampling and adding a bit of blur to eliminate the blocky artifacts of the small grid. Images (a)–(f) have values $e^{-\lambda/T} = .4, .32, .308, .305, .3, .29$, respectively. Note that at .4 (a), we have essentially random pixel colors. At .305 and less (d)–(f), one color takes over and drives out all others. Image (c), at .308, is roughly at the phase transition where the colors coagulate into larger regions and fight for dominance. (See Color Plate IV.)

4.2. $(u+v)$-Models for Image Segmentation

for sampling the Potts model is the Swendsen-Wang algorithm, which was used to create Figure 4.5 and is described in Exercise 4 at the end of this chapter. Although highly stylized, these samples do capture some of the ways images are segmented, especially for temperatures near the critical point where the colors fight for dominance. To get more realistic samples, we need to incorporate some of idiosyncrasies of real shapes and scenes. This is similar to the situation for music, where constructing models whose samples are more music-like requires that we incorporate musical scales, tempos, and the harmonic composition of the sounds of different instruments. Some of these issues come up in Chapter 5, which discusses the shapes of faces.

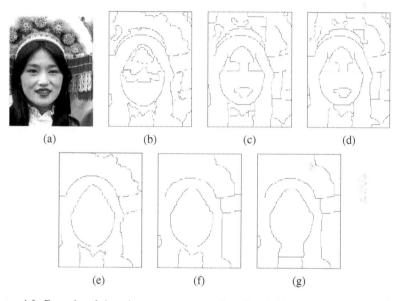

Figure 4.6. Examples of piecewise constant segmentation of a color image by using the Mumford-Shah piecewise constant model for color images. The norm is the Euclidean norm on RGB color space. The implementation is either the greedy merge method developed by Koepfler et al. in [126] (see also Section 4.4) or belief propagation (see Section 4.4.1) followed by the optimal expansion algorithm of Boykov et al. [30] (see also Exercise 2 at the end of this chapter). Image (a) is the original 160×120 image, followed by three segmentations for $\lambda = 0.5$ ((b)–(d)) and three for $\lambda = 2$ ((e)–(g)). The segmentations for $\lambda = 0.5$ have energies $E = 2010, 1736$, and 1733, respectively, and differ in having 22, 32, and 27 regions. respectively. Image (b) is the output of the greedy merge algorithm; (c) is belief propagation followed by optimal expansion; and (d), the last and best, is greedy merge followed by expansion. Between (c) and (d) are clearly multiple trade-offs between the boundary and fidelity terms caused by extra regions—multiple local minima of the energy. The segmentations for $\lambda = 2$ ((e)–(g)) all have eight regions and energies $E = 3747, 3423$, and 3415, respectively. Image (e) is the output of the greedy merge; (f) is the result of following this by optimal expansion, and (g), the third and best, is belief propagation followed by expansion. Images (f) and (g) differ only in the neck area in the trade-off of boundary length with image fidelity. We believe the lowest energy segmentations in each case are very close to the minimum energy segmentations. (See Color Plate V.)

If we fix I, then the probability model defines u as an MRF on the pixel grid V with phase space Λ equal to the value set of u. The second term of the energy is the discrete length of the discontinuities of the cartoon image u. Examples of color image segmentations obtained as the most probable segmentations for such a model are given on Figure 4.6.

A big drawback of the discrete formulation is its bias against diagonal lines. Indeed, if we discretize the square $[0,1]^2$ with a regular $n \times n$ grid (as in Figure 4.7,), then the diagonal of the square is discretized into a staircase with $2n$ segments, and its discrete length is then the "Manhattan" distance between $(0,0)$ and $(1,1)$, that is, $(2n \text{ segments}) \times (\frac{1}{n} = \text{length of segment}) = 2$. Thus, whatever n is, it will never converge to $\sqrt{2}$, the correct length of the diagonal.

A way to avoid this bias of discretization is to use a "tube" around the discontinuity line. This can be done, for instance, by replacing the length term in the energy by

$$E(u) = \sum_{\alpha} \left((I(\alpha) - u(\alpha))^2 + \frac{\lambda}{2r} \mathbb{I}_{\{u|_{(N_r(\alpha))} \text{ not cnst.}\}} \right), \text{ or}$$

$$E(\pi, c) = \sum_{R_i \in \pi} \sum_{\alpha \in R_i} \left((I(\alpha) - c_i)^2 + \frac{\lambda}{2r} \mathbb{I}_{\{N_r(\alpha) \not\subset R_i\}} \right),$$

where $N_r(\alpha)$ denotes the set of pixels at Euclidean distance less than r from the site α. This term provides a much better discrete approximation of length (see more details about this in Exercise 5 at the end of this chapter). See also the paper of Kulkarni et al. [130] about this question. Notice that, as in the previous case, this energy again defines an MRF. To do this, we must enlarge the set of edges so that each neighborhood $N_r(\alpha)$ is a clique. In fact, neurophysiological evidence suggests that vision algorithms in the mammalian cortex involve

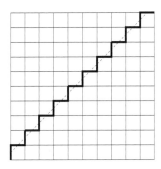

Figure 4.7. Bias against diagonal lines: the length of the line discretized on a grid is 2, whereas its true length is $\sqrt{2}$.

4.2. $(u+v)$-Models for Image Segmentation

integrating evidence over receptive fields that are substantially larger than their nearest neighbors. This chapter later shows many uses for MRFs on a pixel graph with longer range edges.

4.2.2 The Continuum Limit

This energy can also be defined in a continuum setting, giving the MS energy for piecewise constant images. Let I be a real-valued function on a domain $\Omega \subset \mathbb{R}^2$. By fixing I, we can write $E(u)$ instead of $E(I, u)$, and this $E(u)$ is

$$\begin{aligned} E(u) = E(\pi, c) &= \int (I(\alpha) - u(\alpha))^2 + \lambda \cdot \text{len}(\text{discontinuties of } u) \\ &= \sum_{R_i \in \pi} \left(\int_{R_i} (I(\alpha) - c_i)^2 + \frac{\lambda}{2} \cdot \text{len}(\partial R_i) \right). \end{aligned} \quad (4.3)$$

Here "len" stands for the length of a curve. However, it is not clear how to make a stochastic model using the energy in the continuous form because of the term $e^{-\text{len}(\partial R_i)}$.

The length term can be rewritten in another way, which is more mathematically and computationally tractable. The idea is summarized in the rather startling formula

$$\text{len}(\partial R_i) = \int_\Omega \|\nabla(\mathbb{I}_{R_i})(x)\| dx,$$

where in general $\mathbb{I}_R$ is the indicator function of R: 1 on R and 0 elsewhere. However, $\nabla(\mathbb{I}_{R_i})$ is 0 on an open dense set, so this formula doesn't make much sense as it stands. To see what it does mean, we use Proposition 4.4, called the *co-area formula*.

Proposition 4.4 (Co-area formula). *Let f be a real-valued smooth function defined on $\mathbb{R}^n$, and let $h : \mathbb{R} \to \mathbb{R}$ be also a smooth function. Then, for any open domain Ω of $\mathbb{R}^n$, we have*

$$\int_\Omega h(f(x))\|\nabla f(x)\| \, dx = \int_\mathbb{R} h(t)\mathcal{H}^{n-1}(\{x \in \Omega; f(x) = t\})dt, \quad (4.4)$$

where $\mathcal{H}^{n-1}$ denotes the $(n-1)$-dimensional Hausdorff measure.

This result is quite simple for most functions f, namely those for which ∇f vanishes at only isolated points.

Proof (in the generic case): First, throw out the bad points (where ∇f vanishes) and divide what remains into n pieces U_k so that, on U_k, $\frac{\partial f}{\partial x_k}$, which we will write as f_k, is nowhere zero. Now consider the co-area formula on U_1 (all other cases are the same). We may change coordinates via $t = f(\vec{x})$ to $(t, x_2, \cdots, x_n)$. Then the volume forms are related by $dt dx_2 \cdots dx_n = |f_1| dx_1 \cdots dx_n$. The level sets $f = c$ are all given as graphs $x_1 = g(x_2, \cdots, x_n, c)$ and the $(n-1)$-volume on such a graph is given by

$$\sqrt{1 + g_2^2 + \cdots + g_n^2}\, dx_2 \cdots dx_n.$$

But $f(g(x_2, \cdots, x_n, c), x_2, \cdots, x_n) \equiv c$; hence, differentiating with respect to x_k, we get $f_1 \cdot g_k + f_k \equiv 0$. Putting this together, we have

$$\int_{U_1} h(f(x)) |\nabla f(x)| dx_1 \cdots dx_n = \int_{\mathbb{R}} h(t) \left(\int_{f=t} \frac{\|\nabla f\|}{|f_1|} dx_2 \cdots dx_n \right) dt$$
$$= \int_{\mathbb{R}} h(t) \mathcal{H}^{n-1}(f = t) dt. \qquad \square$$

For general f, some technical details are needed: it relies on some arguments from geometric measure theory, which can be found, for instance, in [69].

Now, returning to a region R in the plane domain Ω, we can introduce a level set function $\phi(x)$ that is > 0 if $x \in R$, 0 if $x \in \partial R$, and < 0 for x outside R. A typical example for ϕ is the signed distance to ∂R. Then, if H denotes the Heaviside function, which is 0 for negative arguments and 1 for positive arguments, we have $H \circ \phi = \mathbb{I}_R$, the indicator function of R. This function can be smoothed by convolution of H with the Gaussian kernel $g_\varepsilon(t) = \frac{1}{\sqrt{2\pi\varepsilon^2}} \exp(-t^2/2\varepsilon^2)$ of width ε. We let $H_\varepsilon = g_\varepsilon * H$ and note that $H'_\varepsilon = g_\varepsilon$. Let us denote $f_\varepsilon = (g_\varepsilon * H) \circ \phi$. Then, due to the above proposition (applied with ϕ and g_ε), we have

$$\int_\Omega |\nabla H_\varepsilon(\phi(x))|\, dx = \int_\Omega g_\varepsilon(\phi(x)) |\nabla \phi(x)|\, dx$$
$$= \int_{\mathbb{R}} g_\varepsilon(t) \cdot \mathrm{len}(\{x \in \Omega; \phi(x) = t\})\, dt.$$

As ε goes to 0, this last term goes to $\mathrm{len}(\partial R)$. This gives a limiting sense to the following formula:

$$\mathrm{len}(\partial R) = \int_\Omega \|\nabla(\mathbb{I}_R)(x)\| dx.$$

This equality is the fundamental property used to compute the length of a curve in level-set segmentation methods. See, for instance, the papers of Chan and Vese [47, 222].

4.2.3 The Full MS Model

The correct segmentation of images is not always given by finding regions where the intensity or color is approximately constant. To improve the decomposition $I = u + v$, we need to deal with regions that have:

4.2. $(u+v)$-Models for Image Segmentation

(a) intensity or color that is locally slowly varying, but may have, overall, rather large changes (e.g., images of curved objects that are partly illuminated and partly in shadow);

(b) textures that can have locally large contrasts but give a homogeneous appearance overall (e.g., a checkerboard pattern).

In both of these cases, such regions are likely to be natural segments of the image. To address point (b), we first need to understand and define texture. This is the aim of Section 4.5. For point (a) the full MS model [24, 80, 164] containing a term with $\|\nabla u\|^2$ can be used. Its general formulation in the continuum setting is given by the energy

$$E(u, K) = \int_\Omega (u(x) - I(x))^2 \, dx + \lambda \cdot \text{len}(K) + \mu \int_{\Omega \setminus K} \|\nabla u(x)\|^2 \, dx,$$

where K is the set of discontinuities of u. For this model, the cartoon u is a smooth function with a small derivative within regions and jumps on the boundaries between regions. Mumford and Shah [164] have conjectured that for any continuous I, this E has a minimum with piecewise smooth J and K being a union of a finite number of C^1 pieces, meeting in triple points or ending at "crack tips." But it is not known if this is true. The energy can, of course, be discretized ,but now we need to replace the set of edges by a discrete set of "broken bonds" between adjacent pixels, $K \subset \{(\alpha, \beta) | \alpha \sim \beta\}$. Then we get

$$E(u, K) = \sum_\alpha (u(\alpha) - I(\alpha))^2 + \lambda \cdot \#(K) + \mu \sum_{\alpha \sim \beta, (\alpha,\beta) \notin K} |u(\alpha) - u(\beta)|^2.$$

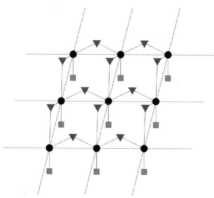

Figure 4.8. The graph of the Markov random field for the MS model. The circles are the vertices for u, the squares for I and the triangles for binary variables to say when the edge is cut by the discontinuity set K.

The resulting probability model, as in the piecewise constant case, defines an MRF and certainly has a minimum. The graph underlying this MRF is shown in Figure 4.8. We will discuss this model further in Exercise 6 at the end of this chapter.

In both the continuous and discrete cases, the discontinuities K of u are difficult to handle analytically and computationally. One striking approach to deal with this is the model of Richardson [186], Richardson and Mitter [185], and Ambrosio and Tortorelli [5]. The idea here is to introduce an auxiliary "line process" l. This line process is a function that takes continuous values in the interval $[0, 1]$, such that $l \simeq 1$ near K and $l \simeq 0$ outside a tubular neighborhood of K. The energy considered by Ambrosio and Tortorelli is of the form

$$E(u,l) = \int_\Omega (u(x) - I(x))^2 + \mu(1 - l(x))^2 \parallel \nabla u(x) \parallel^2$$
$$+ \lambda \left(\varepsilon \parallel \nabla l(x) \parallel^2 + \frac{l^2(x)}{4\varepsilon} \right) dx.$$

Ambrosio and Tortorelli proved [5] that when ε goes to zero this energy converges (in a certain sense called Γ-convergence) to the general MS functional.

Let us illustrate by a simple example the idea of the line process l, and explain why it can be seen as the equivalent of a tube around the discontinuities of u. For this, we assume that the framework is one-dimensional, that $\Omega = \mathbb{R}$, and that u is piecewise constant with a discontinuity at $x = 0$. Then u being fixed, the function l that achieves the minimum of $E(u, l)$ is such that $l(0) = 1$, and $\int_0^1 (\varepsilon l'(x)^2 + l(x)^2/4\varepsilon) dx$ and $\int_{-1}^0 (\varepsilon l'(x)^2 + l(x)^2/4\varepsilon) dx$ are minimal. By a simple calculus of variations, this implies that l is a solution of $l'' = l/4\varepsilon^2$. Since $l \to 0$ as $|x| \to \infty$,

$$l(x) = e^{-|x|/2\varepsilon}.$$

This function l in effect defines a "tube" around the discontinuities of u: it is maximal at the discontinuity and vanishes outside.

In most of the models for image decomposition mentioned so far in this chapter, the norm involved was the L^2 norm. Such a choice is appropriate when v (in the decomposition formula $I = u + v$) is assumed to be Gaussian white noise, but is certainly not the best choice for other types of noise or texture. Many groups have explored other norms. The generic formulation of these models is to find u and v (in some specified functional spaces) such that $I = u + v$ and $E(u, v) = \parallel u \parallel_{N_u} + \parallel v \parallel_{N_v}$ is minimal, where N_u and N_v denote the norms used, respectively for u and v. We list two example models here:

- *Total Variation (TV) model*, using the L^1 norm of the gradient for u and the L^2 norm for the noise v, introduced by Rudin, Osher, and Fatemi [189];

- *The G norm on v* (where G is the norm in a space of "oscillating functions," seen in some sense as the dual of the BV space), to allow v to capture some textures, in the model of Meyer [157]).

This last model has initiated many other studies by Aujol, Vese, Chambolle, Aubert, Osher, Chan and others. See, for instance, [11] and the references therein.

4.3 Sampling Gibbs Fields

Markov random fields pose several difficult computational problems:

1. Compute samples from this distribution.
2. Compute the mode, the most probable state x.
3. Compute the marginal distribution on each component $x_i, i \in V$.

Here we discuss some of the main stochastic approaches to these problems. The simplest method to sample from the field is to use a type of Monte Carlo Markov chain (MCMC) known as the *Metropolis algorithm* [155]. This is a very standard topic for which we will just give a brief introduction in Section 4.3.1. The idea is to define a toy version of a physical system churning around in thermodynamic equilibrium. Running a sampler for a reasonable length of time enables one to estimate the marginals on each component x_i or the mean of other random functions of the state.

Samples at very low temperatures will cluster around the mode of the field. But the Metropolis algorithm takes a very long time to give good samples if the temperature is too low. This leads to the idea of starting the Metropolis algorithm at sufficiently high temperatures and gradually lowering the temperature. This is called *simulated annealing*. We discuss this in Section 4.3.2 and illustrate it by an extended example.

4.3.1 Algorithm III: The Metropolis Algorithm

Let Ω be any finite set called the state-space. In our case Ω is the set of all the possible values of the random variable $\mathcal{X}$ that is finite if Λ and V are finite. Given any energy function $E : \Omega \longrightarrow \mathbb{R}_+$, we define a probability distribution on Ω at all temperatures T via the formula

$$P_T(x) = \frac{1}{Z_T} e^{-E(x)/T},$$
$$\text{where} \quad Z_T = \sum_x e^{-E(x)/T}.$$

What we want to do is to simulate a stochastic dynamical system on Ω via a Markov chain on Ω. Such a process is called a Monte Carlo Markov chain (MCMC) method. We start with some simple Markov chain, given by $q_{x \to y}$, the probability of the transition from a state $x \in \Omega$ to a state $y \in \Omega$. The matrix $Q(x, y) = (q_{x \to y})$ is called the transition matrix, and it has the property that

$$\forall x \in \Omega, \quad \sum_{y \in \Omega} q_{x \to y} = 1.$$

The Markov chain associated with Q is then a discrete stochastic process $\mathcal{X}_t \in \Omega$, the state at time t, such that

$$\mathbb{P}(\mathcal{X}_{t+1} = y \mid \mathcal{X}_t = x) = q_{x \to y}.$$

Chapter 1 presents some basic definitions and properties of Markov chains. We assume in the following that Q is symmetric (i.e., $q_{x \to y} = q_{y \to x}$), irreducible, and aperiodic. We now define a new transition matrix associated with P_T and Q.

Definition 4.5. The Metropolis transition matrix associated with P_T and Q is the matrix P defined by

$$P(x, y) = p_{x \to y} = \begin{cases} q_{x \to y} & \text{if } x \neq y \text{ and } P_T(y) \geq P_T(x), \\ \frac{P_T(y)}{P_T(x)} q_{x \to y} & \text{if } x \neq y \text{ and } P_T(y) \leq P_T(x), \\ 1 - \sum_{z \neq x} p_{x \to z} & \text{if } x = y. \end{cases}$$

Theorem 4.6. *The Metropolis transition matrix $P(x, y) = (p_{x \to y})$ is primitive and aperiodic if $Q(x, y) = q_{x \to y}$ is. The equilibrium probability distribution of $(p_{x \to y})$ is P_T; that is,*

$$\forall x, y, \in \Omega, \ P^n(x, y) \xrightarrow[n \to \infty]{} P_T(y).$$

Proof: The fact that P is primitive and aperiodic comes from the property that, since for all y in Ω, $P_T(y) \neq 0$, if $Q^n(x, y) > 0$, then $P^n(x, y) > 0$.

Now we prove a stronger property, which will imply that P_T is the equilibrium probability distribution of $(p_{x \to y})$. This property (called "detailed balance") is

$$\forall x, y \in \Omega, \quad P_T(x) p_{x \to y} = P_T(y) p_{y \to x}$$

- If $x = y$, it is obvious.
- If $x \neq y$, assume $P_T(x) \geq P_T(y)$ then

$$P_T(x) p_{x \to y} = P_T(x) \frac{P_T(y)}{P_T(x)} q_{x \to y} = P_T(y) q_{x \to y} = P_T(y) q_{y \to x}$$
$$= P_T(y) p_{y \to x}$$

4.3. Sampling Gibbs Fields

So we have that for all x and y, $P_T(x)p_{x\to y} = P_T(y)p_{y\to x}$. Now, summing this on x, we finally get

$$\sum_x P_T(x)p_{x\to y} = P_T(y),$$

which means that P_T is the equilibrium probability distribution of $(p_{x\to y})$. □

Algorithm (Metropolis Algorithm).

Let P_T be the probability distribution on a finite state space Ω from which you want to sample.

1. Let $Q(x, y) = q_{x\to y}$ to be the transition matrix of a primitive and aperiodic Markov chain on Ω. (A simple example of this might be: $q_{x\to y} = 1/|\Omega|$ for all x, y.)

2. Define the Metropolis transition matrix $P(x, y) = p_{x\to y}$ associated with P_T and Q by

$$p_{x\to y} = \begin{cases} q_{x\to y} & \text{if } x \neq y \text{ and } P_T(y) \geq P_T(x), \\ \frac{P_T(y)}{P_T(x)}q_{x\to y} & \text{if } x \neq y \text{ and } P_T(y) \leq P_T(x), \\ 1 - \sum_{z\neq x} p_{x\to z} & \text{if } x = y. \end{cases}$$

3. Choose n large, and let the Markov chain with transition matrix P run n times.

In the Ising model case, we are interested in the following example: Ω is the set of all possible values of $\{J(\alpha)\}_\alpha$, and we define $q_{J\to \tilde{J}} = \frac{1}{\#\alpha} = \frac{1}{N}$ in the case $\tilde{J}(\alpha) = J(\alpha)$ for all but one α and zero otherwise. In other words, under q, you can flip exactly one value of the "field" J at a time, and each flip is equally probable. Then under p, all such flips are possible, but those that increase the energy are less probable by a factor $e^{-\Delta E/T}$. This is how the figures in Section 4.2 were produced.

One of the main problems of the Metropolis algorithm is that many proposed moves $x \to y$ will not be accepted because $P(y) \ll P(x)$, and hence $p_{x\to x} \approx 1$. In particular, because $q_{x\to y} = q_{y\to x}$ is required, we must allow proposed moves from very probable states to very improbable states as often as the other way around. Several ideas suggested to correct this are discussed next.

Gibbs sampler [80]. Let $\Omega = \Lambda^V$ be the Gibbs state space and let $N = |V|$. The Gibbs sampler chooses one site $i_0 \in V$ at random and then picks a new value λ for the state at i_0 with the probability distribution given by P_T, conditioned on the values of the state at all other sites:

$$\forall i_0 \in V, \lambda \in \Lambda, \text{ if } x_2(i) = x_1(i) \text{ for } i \neq i_0 \text{ and } x_2(i_0) = \lambda,$$

$$p_{x_1 \to x_2} = \frac{1}{N} P_T\left(x(i_0) = \lambda \mid \{x(i) = x_1(i)\}_{i \neq i_0}\right);$$

otherwise $p_{x_1 \to x_2} = 0$.

This satisfies the same theorem as the Metropolis algorithm.

Metropolis-Hastings [102]. We start as before with a Markov chain $q_{x \to y}$, but now we drop the symmetry condition and alter the acceptance rule.

$$p_{x \to y} = \begin{cases} q_{x \to y} & \text{if } x \neq y \text{ and } P_T(y)q_{y \to x} \geq P_T(x)q_{x \to y}, \\ \frac{P_T(y) \cdot q_{y \to x}}{P_T(x)} & \text{if } x \neq y \text{ and } P_T(y)q_{y \to x} \leq P_T(x)q_{x \to y}, \\ 1 - \sum_{z \neq x} p_{x \to z} & \text{if } x = y. \end{cases}$$

Theorem 4.6 for the Metropolis algorithm extends to Metropolis-Hastings. (We leave the simple verification to the reader.) The importance of this algorithm is that we construct q precisely to steer the samples toward the places where we expect P_T to have higher values. This has been used with considerable success by Song-Chun Zhu and collaborators [218]. They call this "data-driven Monte Carlo Markov chains' (DDMCMC), and apply this method to the problem of segmentation of images. Prior to their work, Monte Carlo methods were considered as uselessly slow by the computer vision community, but we do not believe this is correct. DDMCMC has turned out to be quite fast and effective when combined with variants of the Swendsen-Wang method due to Barbu and Zhu [12]. This is sketched in Exercise 4 at the end of this chapter.

Langevin equation. This is a version of Metropolis in which the state space is continuous, with a sample $\mathcal{X}$ and the Markov chain replaced by a stochastic differential equation for $\mathcal{X}_t$:

$$d\mathcal{X} = -\nabla E dt + d\mathcal{B}, \quad \mathcal{B} \text{ being Brownian motion.}$$

Particle filter algorithms [112, 201]. Yet another variant, in which, instead of updating *one* sample $\mathcal{X}_t$ from Ω, we update a weighted set of samples $\mathcal{X}_t^{\text{pop}} = \{\mathcal{X}_t^{(n)}, w_n(t)\}$ from Ω, considered to be a "population." There are typically two parts to the updating algorithm: (1) a random perturbation of each sample, as in Metropolis, with an update of the weights using the probability distribution;

4.3. Sampling Gibbs Fields

and (2) a resampling with replacement from the weighted population. The goal is to find weighted samples $\{\mathcal{X}^{(n)}, w_n\}$ such that for some reasonable class of functions f on our state space Ω,

$$\mathbb{E}_{P_T}(f) \approx \sum_n w_n \cdot f(\mathcal{X}^{(n)}).$$

Genetic algorithms [163]. A final variant is the class of algorithms called genetic algorithms'. In this case, particle filtering is combined with a third step, a "splicing" of pairs of samples tailored to the problem at hand (so that fitness in one aspect in one sample can be combined with fitness in another aspect in a second sample). If the set of vertices V of the underlying graph has the property that the states at remote parts of the graph are only loosely coupled with each other, it makes sense to try to combine samples that are good in one part with others that are good elsewhere, as evolution seems to have done with genes.

4.3.2 Algorithm IV: Simulated annealing

In applications of Markov random fields to problems such as segmentation, one usually is primarily interested in the mode of the probability distribution, that is, the minimum of the associated energy function. Note that the mode of P_T doesn't depend on T, but the distribution gets more and more concentrated around its mode as $T \to 0$. To find it, we can run the Metropolis algorithm or the Gibbs sampler at a series of temperatures that decrease: $T_1 \geq T_2 \geq \ldots$ with $T_n \to 0$. This is an inhomogeneous Markov chain on the state space Ω with $\mathcal{X}_{n+1}$ being the result of some steps of a MCMC at temperature T_n starting at $\mathcal{X}_n$. If the temperature decreases slowly enough, we can guarantee that $\mathcal{X}_n$ will converge to $x_{\min}$, the global minimum of E, with probability 1.

Algorithm (Simulated Annealing).
Let E be an energy defined on a finite state space Ω. We want to find the state $x_{\min} \in \Omega$ that has the minimal energy. For each temperature $T > 0$, define a probability distribution on Ω by $P_T(x) = \frac{1}{Z_T} \exp(-E(x)/T)$.

1. Choose a sequence of decreasing temperatures $T_1 \geq T_2 \geq \ldots$ with $T_n \to 0$, and a sequence of times $\tau_1, \tau_2, \ldots$ (Take, for instance, an exponentially decreasing sequence for temperatures and a constant sequence for times.)

2. Run the Metropolis Algorithm during time τ_1 at temperature T_1, then during time τ_2 at temperature T_2, and so on.

Unfortunately, to be *sure* of the convergence of $\mathcal{X}_n$ to $x_{\min}$ requires that we spend enough time at each temperature to "reach equilibrium" at that temperature, that is, to be reasonably sure of sampling all the low energy states. Simple estimates show that if this requires spending time τ at temperatures $T_0 \geq T \geq T_0/2$, then we also need to spend time τ^2 at temperatures $T_0/2 \geq T \geq T_0/4$, time τ^4 at temperatures $T_0/4 \geq T \geq T_0/8$, and so forth. This rapidly gets out of hand! A much better way to use simulated annealing is to accept some probability of error and run the annealing much faster. In fact, a detailed analysis by Catoni [46] of the most effective algorithm given that you have an upper bound N to the number of moves you can make at any temperature and that you seek to find $x_{\min}$ shows the following. If you use exponential cooling schedules of the form

$$\text{for } 1 \leq n \leq N, \quad T_n = A^{-1} \left(\frac{(\log N)^2}{A} \right)^{-n/N},$$

you get nearly optimal performance and explicit bounds:

$$\mathbb{P}(\mathcal{X}_N \neq x_{\min}) = O(N^{-1/D}),$$

where D measures the hilliness of the energy landscape and is defined by

$$D = \max_x (H(x)/(E(x) - E(x_{\min}))),$$
$$\text{where} \quad H(x) = \min_{[\text{paths } \gamma \text{ between } x, x_{\min}]} \max_{x' \in \gamma} (E(x') - E(x)).$$

Example. Let Ω be simply the set $\{1, 2, 3\}$, and the energy function E on Ω be given by $E(1) = 0$, $E(2) = a$, and $E(3) = b$ with $a > b$ (see Figure 4.9).

The constant D controlling the difficulty of finding the minimum energy state 1 in this case is just

$$D = \frac{E(2) - E(3)}{E(3) - E(1)} = \frac{a - b}{b}.$$

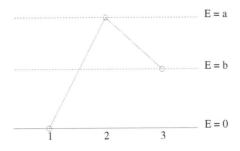

Figure 4.9. Example of simple 3-states space and an energy function E defined on it.

4.3. Sampling Gibbs Fields

The Gibbs probability distribution at temperature T is

$$P_T = \frac{e^{-0/T}}{Z_T}, \frac{e^{-a/T}}{Z_T}, \frac{e^{-b/T}}{Z_T},$$
$$Z_T = 1 + e^{-a/T} + e^{-b/T}.$$

Note that if T goes to $+\infty$, then $P_T \simeq \frac{1}{3}, \frac{1}{3}, \frac{1}{3}$, and if T goes to 0, then $P_T \simeq 1, 0, 0$.

Moreover, we suppose that the probabilities of transition for Q are

$$q_{1\to 2} = q_{2\to 1} = q_{3\to 2} = q_{2\to 3} = 0.5,$$

$$q_{1\to 1} = q_{3\to 3} = 0.5,$$

$$q_{2\to 2} = q_{3\to 1} = q_{1\to 3} = 0.$$

We now compute the Metropolis transition matrix $P^{(T)}$ associated with P_T and the transition matrix $Q(x,y) = q_{x\to y}$. Then, recalling that P_T will be the equilibrium probability distribution of $P^{(T)}(x,y) = p^{(T)}_{x\to y}$, we get

$$P^{(T)} = (p^{(T)}_{x\to y}) = \begin{pmatrix} 1 - 0.5e^{-a/T} & 0.5e^{-a/T} & 0 \\ 0.5 & 0 & 0.5 \\ 0 & 0.5e^{-(a-b)/T} & 1 - 0.5e^{-(a-b)/T} \end{pmatrix}.$$

When T goes to 0 (respectively, ∞), this matrix becomes

$$P^{(0)} = \begin{pmatrix} 1 & 0 & 0 \\ 0.5 & 0 & 0.5 \\ 0 & 0 & 1 \end{pmatrix} \text{ (respectively, } P^{(\infty)} = \begin{pmatrix} 0.5 & 0.5 & 0 \\ 0.5 & 0 & 0.5 \\ 0 & 0.5 & 0.5 \end{pmatrix}).$$

If we run a Markov chain $\mathcal{X}_n$, with the uniform distribution on all states as an initial distribution for $\mathcal{X}_0$, and if we start at a very high temperature (meaning using $P^{(\infty)}$ as a transition matrix from $\mathcal{X}_0$ to $\mathcal{X}_1$), and then freeze directly (i.e., using $P^{(0)}$ as the transition matrix from $\mathcal{X}_1$ to $\mathcal{X}_2$), we get the following probabilities of being in the three states:

	$x=1$	$x=2$	$x=3$
$\mathbb{P}(\mathcal{X}_0 = x)$	$\frac{1}{3}$	$\frac{1}{3}$	$\frac{1}{3}$
$\mathbb{P}(\mathcal{X}_1 = x)$	$\frac{1}{3} + \frac{1}{6}$	0	$\frac{1}{6} + \frac{1}{3}$
$\mathbb{P}(\mathcal{X}_2 = x)$	$\frac{1}{2}$	0	$\frac{1}{2}$

But instead, if in a first step we cook once at a middle temperature, and then in a second step we freeze, we do a bit better:

	$x = 1$	$x = 2$	$x = 3$
$\mathbb{P}(\mathcal{X}_0 = x)$	$\frac{1}{3}$	$\frac{1}{3}$	$\frac{1}{3}$
$\mathbb{P}(\mathcal{X}_1 = x)$	$\frac{1}{3}(1 - \frac{1}{2}e^{-a/T})$ $+\frac{1}{6}$	$\frac{1}{6}e^{-a/T}$ $+\frac{1}{6}e^{-(a-b)/T}$	$\frac{1}{6}$ $+\frac{1}{3}(1 - \frac{1}{2}e^{-(a-b)/T})$
$\mathbb{P}(\mathcal{X}_2 = x)$	$\frac{1}{2}$ $+\frac{1}{12}(e^{-(a-b)/T} - e^{-a/T})$	0	$\frac{1}{2}$ $-\frac{1}{12}(e^{-(a-b)/T} - e^{-a/T})$

If we want to maximize $\mathbb{P}(\mathcal{X}_2 = 1)$, then differentiating with respect to T shows $T = b/\log(\frac{a}{b-a})$ is the best temperature to use. If, for example $a = 2b$ and $T = b/\log 2$, then we get $\frac{1}{2} + \frac{1}{48}, 0, \frac{1}{2} - \frac{1}{48}$.

In all the models (e.g., the cow/tree image shown in Figure 4.4 and samples from the Potts model in Figure 4.5) there are certain critical temperatures or ranges of temperature where the real decisions are made. As we saw in Section 4.1.2, these are characterized by having a high specific heat. At these temperatures, it is essential to anneal slowly to avoid falling into a local but not global minimum of E.

4.4 Deterministic Algorithms to Approximate the Mode of a Gibbs Field

As a method of finding the mode of a Gibbs field, simulated annealing has gotten a bad reputation for being very slow. This is sometimes true, although, when it is enhanced by Metropolis-Hastings with a good choice of proposals, it can be sped up a great deal and can be an excellent technique (see Exercise 4 at the end of this chapter and papers by Zhu's group [12, 218]). However, various deterministic methods are available that in some situations exactly compute the mode, and in others approximate it. A very simple, but surprisingly effective method is a simple greedy merge method introduced by Morel, Koepfler, and collaborators [126, 127]. The latter paper presents the rationale for their method in careful detail. The idea is quite simple: start by taking each pixel as a separate region. Successively merge adjacent regions by seeking pairs of regions whose merger *decreases the energy the most*. Stop when no merger can decrease energy any more. This is especially efficient for models where this energy decrease has a simple

4.4. Deterministic Algorithms to Approximate the Mode of a Gibbs Field

expression. Thus, for the MS locally constant model given by Equation (4.3) with optimal colors $c_i = c(R_i)$, namely, those given by the mean of the image color on the region R_i, it is easy to compare the energy of two segmentations $\pi = \{R_i\}$ and π^* in which R_i and R_j are merged:

$$E(\pi) - E(\pi^*) = -\frac{|R_i| \cdot |R_j|}{|R_i| + |R_j|} \cdot \|c(R_i) - c(R_j)\|^2 + \lambda \cdot \text{len}(\partial R_i \cap \partial R_j).$$

Some results of this algorithm are shown in Figure 4.6 and are compared with more sophisticated methods.

Two main methods have been found to seek the mode of a Gibbs field, especially a Potts model, in a principled and efficient fashion, although neither comes with any guarantees. These are called belief propagation and graph cuts and are be studied in the rest of this section. A third approach can be found in the recent paper of Pock, et al. [178], based on a dual version of the co-area formula (4.4):

$$-\int_\Omega \|\nabla f\|^2 = \inf_{\vec{\xi}:\, \sup \|\vec{\xi}(x)\| \leq 1} \int f \cdot \text{div}(\xi) dx.$$

4.4.1 Algorithm V: Belief Propagation

Let $G = (V, E)$ be a graph, V bethe set of vertices, and E be the set of edges. We assume that the edges are undirected and that two vertices are joined by at most one edge; hence, E is a set of pairs of distinct vertices. Let Λ be a phase space and P be a probability distribution on the configuration space $\Omega = \Lambda^V$. In many problems, lsuch asin the Ising model, the aim is to find the configuration that has the maximal probability. In Section 4.3.2, we described simulated annealing, a method to obtain the mode of Markov random field. An alternative to this is to use the natural generalization of dynamic programming, called belief propagation (BP for short or BBP for the full name Bayesian Belief Propagation). The algorithm computes correctly the marginals and modes of a MRF whenever the graph is a tree, and the bold idea was to use it anyway on an arbitrary graph G! Mathematically, it amounts to working on the universal covering graph $\widetilde{G}$, which is a tree, and hence much simpler, instead of G. (See Figure 4.10.) A drawback is that this tree is infinite. In statistical mechanics, this idea is called the Bethe approximation, introduced by Bethe in the 1930s [20].

The universal covering tree. Let $G = (V, E)$ be a graph (undirected, no multiple edge), and let $i_0 \in V$ be a vertex of the graph. The universal covering tree of G is the graph $\widetilde{G} = (\widetilde{V}, \widetilde{E})$, where $\widetilde{V}$ is the set of all paths $\tilde{i}$ in V starting at i_0, $\tilde{i} = (i_0, i_1, \ldots, i_n)$, such that for all p, $(i_p, i_{p+1}) \in E$ and $i_p \neq i_{p+2}$, and $\widetilde{E}$ is the set of unordered pairs $(\tilde{i}, \tilde{j})$ such that length$(\tilde{j}) = $ length$(\tilde{i}) + 1$ and $\tilde{i} \subset \tilde{j}$, i.e., $\tilde{j} = (\tilde{i}, i_{n+1})$.

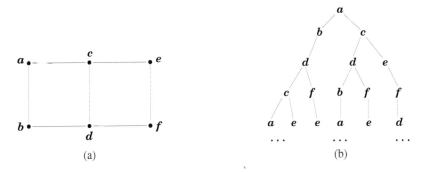

Figure 4.10. An example of (a) a graph and (b) part of its universal covering tree.

Let $\Gamma = \{\tilde{i} \in \tilde{V} | i_n = i_0\}$ be the set of loops in V starting at i_0. Then Γ is a group under concatenation, which is defined as follows: If γ and γ' are two loops in Γ with respective lengths n and n', we define

$$\gamma \cdot \gamma' = (\gamma_0 = i_0, \gamma_1, \ldots, \gamma_{n-1}, \gamma_n = i_0 = \gamma'_0, \gamma'_1, \ldots, \gamma'_{n'-1}, \gamma'_{n'} = i_0)$$

where we erase pairs of points in the middle where the path "doubles back on itself."

In particular, if $\gamma \in \Gamma$, we have $\gamma \cdot \{i_0\} = \gamma$ and $\gamma \cdot \gamma^{-1} = \{i_0\}$, where γ^{-1} is the loop $(\gamma_n = i_0, \gamma_{n-1}, \ldots, \gamma_0 = i_0)$. Thus, Γ is a group, called the *fundamental group of the graph*. It acts on $\tilde{V}$ by $(\gamma, \tilde{i}) \mapsto \gamma \cdot i \in \tilde{V}$. We then have

$$V = \tilde{V}/\Gamma.$$

Indeed, let us consider $\pi : \tilde{V} \mapsto V$ defined by $\tilde{i} \mapsto i_n$. Then $\pi(\tilde{i}) = \pi(\tilde{i}')$ if and only if there exists $\gamma \in \Gamma$ such that $\gamma \cdot \tilde{i} = \tilde{i}'$. To prove this: if two paths $\tilde{i}$ and $\tilde{i}'$ have the same ending point, then take for γ the loop $(i_0, i'_1, \ldots, i'_m = i_n, i_{n-1}, \ldots, i_1, i_0)$, to obtain that $\gamma \cdot \tilde{i} = \tilde{i}'$. Conversely, if $\gamma \cdot \tilde{i} = \tilde{i}'$, then obviously $\pi(\tilde{i}) = \pi(\tilde{i}')$.

To explain the idea of Bethe approximation, start with the mean field approximation. The mean field idea is to find the best approximation of an MRF P by a probability distribution in which the random variables $\mathcal{X}_i$ are all independent. This is formulated as the distribution $\prod_i P_i(x_i)$ that minimizes the Kullback-Leibler distance $D(\prod_i P_i \| P)$ (see Section 1.1). Unlike computing the true marginals of P on each x_i, which is usually difficult, this approximation can be found by solving iteratively a coupled set of nonlinear equations for P_i. But the assumption of independence is much too restrictive. Bethe's idea is instead to approximate P by the closest (for the Kullback-Leibler distance) Γ-invariant MRF on $\tilde{G}$. Next, we explain how this is defined and achieved.

4.4. Deterministic Algorithms to Approximate the Mode of a Gibbs Field

Now, $\widetilde{G}$ is an infinite graph so we must be careful defining probability measures on the state space $\Omega = \Lambda^{\widetilde{V}}$. Typically, the measures of each state $P(x), x \in \Omega$, will be zero, so the measure cannot be given pointwise. Instead, the measures are given by the collection of marginal probabilities $\mathbb{P}(x_{V_1} = a_{V_1}), V_1 \subset V$ with V_1 finite.[3] The Markov property is given by Equation (4.1) but with the constraint that V_1 is finite.

Markov random fields on trees are easy to describe, even when the tree is infinite. First, note that a MRF on a tree is uniquely determined by its marginals $P_i(x_i)$ and $P_e(x_i, x_j)$, for each vertex i and edge $e = (i, j)$. To prove this, we let $V_1 = (i_1, \cdots, i_n)$ be a finite subtree such that each i_k is connected by an edge $e_k = (i_k, i_{f(k)})$ to a vertex $i_{f(k)}$, with $f(k) < k$ for $k \geq 2$. Then

$$\mathbb{P}(x_{V_1} = a_{V_1}) = \prod_{k=2}^{k=n} \mathbb{P}(x_{i_k} = a_{i_k} | x_{i_{f(k)}} = a_{i_{f(k)}}) \cdot \mathbb{P}(x_{i_1} = a_{i_1})$$

$$= \frac{\prod_{k=2}^{k=n} P_{e_k}(a_{i_k}, a_{i_{f(k)}})}{\prod_{k=3}^{k=n} P_{i_{f(k)}}(a_{i_{f(k)}})}.$$

Conversely, if we are given a set of distributions $\{P_e\}_{e \in E}$ consistent on vertices (in the sense that, for all edges (i, j_k) abutting a vertex i, the marginals of $P_{(i,j_k)}$ give distributions on x_i independent of k), they define a MRF on the graph G, when G is a tree.

Notice that this is not true when the graph G is not a tree. For a very simple example of this, we take a triangle $G = (V, E)$ with $V = \{a, b, c\}$, $E = \{(a, b), (b, c), (c, a)\}$, and phase space $\Lambda = \{0, 1\}$. Then we consider the set of distributions on edges given in Figure 4.11. This set of distributions satisfies the consistency requirement but cannot be the set of marginals of a probability P on the triangle. Indeed, if such a distribution P on $\{0, 1\}^3$ exists, then we will have

$$1 - P(0,0,0) - P(1,1,1) = P_a(0) + P_b(0) + P_c(0) - P_{ab}(0,0)$$
$$- P_{bc}(0,0) - P_{ca}(0,0),$$

which is impossible since the left-hand term has to be less than 1 and the value of the right-hand term is 1.2.

So if we start with an MRF on any graph G, we get a Γ-invariant MRF on its universal covering tree $\widetilde{G}$ by making duplicate copies for each random variable x_i, $i \in V$, for each $\widetilde{i} \in \widetilde{V}$ over i and lifting the edge marginals. More generally, if we have *any consistent set of probability distributions* $\{P_e\}_{e \in E}$ on the edges of G, we

---

[3] Technically, the measure is defined on the smallest σ-algebra containing the sets $x_i = a$ for all $i \in V, a \in \Lambda$.

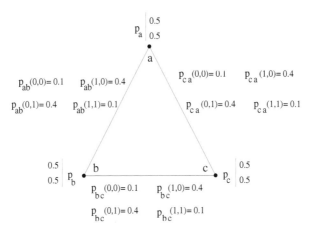

Figure 4.11. Example of a set of distributions on edges, which are consistent on vertices but cannot be the set of marginals of a probability distribution on the whole graph.

also get a Γ-invariant MRF on $\widetilde{G}$. Call such a consistent set a *pseudo-probability distribution*. Then, the Bethe approximation is the pseudo-probability distribution $\{P_e\}_{e \in E}$ that minimizes $D(\{P_e\} \| P)$. This Kullback-Leibler distance is defined by formally putting into the definition for the first argument the function

$$\widetilde{P}(x) = \frac{\prod_{e=(i,j)} P_e(x_i, x_j)}{\prod_i P_i(x_i)^{d(i)-1}}, \quad \text{where } d(i) = \text{degree of vertex } i.$$

As in the mean field case, there is a natural iterative method of solving for this minimum, which remarkably turns out to be identical to the generalization to general graphs G of an algorithm called belief propagation invented originally for trees.

The belief propagation algorithm. The belief propagation algorithm was first introduced by Y. Weiss [226], who generalized to arbitrary graphs the algorithm introduced by Pearl for trees [174]. The aims of the algorithm are (i) to compute the approximate marginal distribution at each node of a graph $G = (V, E)$, and (ii) to find a state close to the mode, the MAP state, for a Gibbs distribution defined only by pairwise potentials (in the sense that $U_C = 0$ for all cliques C such that $|C| > 2$). Such a distribution P is given, for all configurations $x \in \Omega = \Lambda^V$, under the form

$$P(x) = \frac{1}{Z} \prod_{(i,j) \in E} \Psi_{ij}(x_i, x_j) \prod_{i \in V} \Psi_i(x_i), \tag{4.5}$$

There are two versions of this algorithm: the "sum version," which is used to approximate the marginals, and the "max version," which is used to approximate the

4.4. Deterministic Algorithms to Approximate the Mode of a Gibbs Field

mode. For each vertex $i \in V$, we denote as ∂i the set of its neighbors (i.e., $j \in \partial i$ iff $(i,j) \in E$). In the following algorithm, α denotes a generic normalizing constant such that at each step the messages and beliefs are probability distributions.

Algorithm (Belief Propagation Algorithm).

1. Initialization: start with a set of messages $\{m_{ij}^{(0)}\}$, which is a set of probability distributions on Λ, for example, uniform.

2a. In the sum version, update the messages by the iteration of
$$m_{ij}^{(n+1)}(x_j) = \alpha \sum_{x_i} \Psi_{ij}(x_i, x_j)\Psi_i(x_i) \prod_{k \in \partial i \setminus \{j\}} m_{ki}^{(n)}(x_i).$$

2b. In the max version, update the messages by the iteration of
$$m_{ij}^{(n+1)}(x_j) = \alpha \max_{x_i} \Psi_{ij}(x_i, x_j)\Psi_i(x_i) \prod_{k \in \partial i \setminus \{j\}} m_{ki}^{(n)}(x_i).$$

3a. In the sum version, compute the belief after n iterations by
$$b_i^{(n)}(x_i) = \alpha \Psi_i(x_i) \prod_{k \in \partial i} m_{ki}^{(n)}(x_i).$$

3b. In the max version, compute an optimal state by
$$x_i^{(n)} = \arg\max_{x_i} \Psi_i(x_i) \prod_{k \in \partial i} m_{ki}^{(n)}(x_i).$$

Interpretation of messages. *When the graph G is a tree* and the initial messages $\{m_{ij}^{(0)}\}$ all have the uniform distribution on the finite phase space Λ, the messages and beliefs in the sum case can be interpreted as marginals of probability measures on subtrees. More precisely, we let P be an MRF on a tree G given by pairwise potentials, as in Equation ((4.5)). For any subtree T of G, we denote $P_{/T}$ the restriction of P to T, defined by

$$P_{/T}(x_T) = \alpha \prod_{\substack{(i,j) \in E \\ i,j \in T}} \Psi_{ij}(x_i, x_j) \prod_{i \in T} \Psi_i(x_i).$$

Let i be a vertex of G, let m denote the number of neighbors of i, and let $T_1^{(n)}, \ldots, T_m^{(n)}$ denote the subtrees made of the vertices, distinct from i, that are at a distance less than n from i (see Figure 4.12).

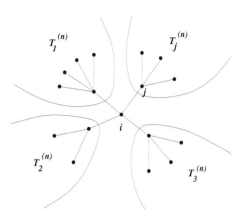

Figure 4.12. A vertex i of a graph G, which is a tree, and the subtrees made of its neighbors, which are at a distance less than n from i.

Then, the message $m_{ji}^{(n)}(x_i)$ is the marginal distribution on x_i of the probability distribution $\Psi_{ij}(x_i, x_j) P_{/T_j^{(n)}}$ on $\{i\} \cup T_j$. The belief on i after n iterations is the marginal on x_i under the subtree $\{i\} \cup T_1^{(n)} \cup \ldots T_m^{(n)}$. Using the Markov property of the Gibbs field, the probability distribution is thus given by

$$b_i^{(n)}(x_i) = P_{/\{i\} \cup T_1^{(n)} \cup \ldots T_m^{(n)}}(x_i)$$
$$= \alpha \, \Psi_i(x_i) \prod_{j=1}^{m} \sum_{x_{T_j^{(n)}}} \Psi_{ij}(x_i, x_j) P_{/T_j^{(n)}}(x_{T_j^{(n)}}).$$

In the max case, the beliefs have an interpretation as MAPs of a conditional distribution. The belief for any value x_i of a node i is the maximum probability of the distribution $P_{/\{i\} \cup T_1^{(n)} \cup \ldots T_m^{(n)}}$ conditioned on that node having value x_i. Therefore, the max-product assignment $\{x_i^{(n)}\}$ is exactly the MAP assignment for $P_{/\{i\} \cup T_1^{(n)} \cup \ldots T_m^{(n)}}$. This is because the max algorithm is a simple variant of dynamic programming. Details can be found, for example, in [174].

If the graph is not a tree, we consider its universal covering tree $\widetilde{G}$. We don't try to put an MRF on $\widetilde{G}$, but instead lift the potentials Ψ at edges and vertices so we can run BP on $\widetilde{G}$. Clearly, it gives the same messages and beliefs except that they are repeated at all the lifts of each edge and vertex. The truncated trees $\widetilde{T}_k^{(n)}$ and $\{\widetilde{i}\} \cup \widetilde{T}_1^{(n)} \cup \ldots \widetilde{T}_m^{(n)}$ in $\widetilde{G}$ are still finite, so the potentials define an MRF on them, and the above results give an interpretation of the messages and beliefs from the BP. In particular, the beliefs in the sum are marginal probabilities at vertices $\widetilde{i} \in \widetilde{G}$ of the probability distribution defined by the potentials on the

4.4. Deterministic Algorithms to Approximate the Mode of a Gibbs Field

subtree of all vertices at a distance $\leq n$ from $\tilde{i}$, and the state $\{x_i^{(n)}\}$ in the max case is the MAP of the same probability distribution.

Convergence of the BP. If the graph G is a finite tree, then for any set of initial messages $\{m_{ij}^{(0)}\}$, both BP algorithms converge in d iterations, where d is the diameter of the graph. This is easily seen because at the leaves of the tree there are no extra "incoming" messages, and only one "outgoing" message, and then the messages progressively stabilize as we move inward from the leaves. In the sum case, the beliefs converge to the true marginals, ot

$$P_i(x_i) = b_i^*(x_i) = \alpha \, \Psi_i(x_i) \prod_{k \in \partial i} m_{ki}^{(n)}(x_i), \quad n \gg 0$$

and the marginals on edges are given by

$$P_{ij}(x_i, x_j) =$$
$$\alpha \Psi_{ij}(x_i, x_j) \Psi_i(x_i) \Psi_j(x_j) \prod_{k \in \partial i \setminus \{j\}} m_{ki}^{(n)}(x_i) \prod_{k \in \partial j \setminus \{i\}} m_{kj}^{(n)}(x_j), \quad n \gg 0.$$

In the max case, the $x_i^{(n)}$ converge to the mode, the state of maximum probability.

On a general graph, neither the sum nor the max BP algorithms need converge and, if they do, the result may be difficult to interpret. Nonetheless, the max BP often gives either the true MAP of the MRF or something quite close; and the sum BP often gives a good estimate of the MRF marginals. For this reason, they are widely used. Yedidia, Freeman, and Weiss have shown in [229] that if the sum-BP algorithm converges, then its limit is a local minimum of the Kullback-Leibler distance $D(\{b_{ij}\}\|p)$. The log of this, incidentally, is also called the *Bethe free energy*. In this case, the limiting beliefs $b_i(x_i)$ are the exact marginal probabilities for a Γ-invariant probability measure on the whole universal covering tree (which is usually a pseudo-probability measure on G).

In their paper entitled "Loopy Belief Propagation and Gibbs Measures" [211], Tatikonda and Jordan related the convergence of the BP algorithm to the existence of a weak limit for a certain sequence of probability measures defined on the corresponding universal covering tree. This connects the convergence of the BP to the phenomenon of "phase transitions" in statistical physics. The idea is that there can be more than one MRF on an infinite graph, all with the same set of potentials. For example, an Ising model with no external magnetic field and at low temperatures defines two different MRFs. For each sufficiently low temperature, there is a probability $p(T) > .5$ such that either the spins are $+1$ with probability $p(T)$, -1 with probability $1 - p(T)$, or vice versa. These two states define two different probability measures on the state space, and two different MRFs. Details can be found in many books, for example, [85, 123].

Another example of this occurs on a grid of pixels (i, j). Call a pixel odd or even depending on $i + j$ mod 2. Then we have a checkerboard pattern, and all messages go from odd pixels to even ones and vice versa. This leads to many situations in which belief propagation settles down to a situation in which two different states alternate, often with actual local checkerboard patterns. This particular problem is solved by using asynchronous updates, or simply selecting randomly some percentage of messages to update at each time step (see [131]).

In addition, Weiss and Freeman [225] have proved a very strong result, which states that the max-product assignment has, on a general graph, a probability greater than all other assignments in a large neighborhood around it. More precisely, they first define the single loops and trees (SLT) neighborhood of an assignment x^* as being the set of all x that can be obtained by changing the values of x^* at any arbitrary subset of nodes of G that consists of a disconnected union of single loops and trees. Their result is that if m^* is a fixed point of the max-product algorithm and x^* is the assignment based on m^*, then $P(x^*) > P(x)$ for all $x \neq x^*$ in the SLT neighborhood of x^*.

Example. We first start with a simple example: the Ising model with state space $\{0, 1\}$ for a 4×4 square made of 16 pixels. Each pixel i, $1 \leq i \leq 4$, has a gray-level $u_i \in [0, 1]$, which we choose randomly. As usual, we set

$$\Psi_{ij}(x_i, x_j) = e^{-(x_i - x_j)^2} \quad \text{and} \quad \Psi_i(x_i) = e^{-c(x_i - u_i)^2},$$

where c is a positive constant representing the relative weight between the regularity term and the data-driven term. We take $c = 4$, as the resulting MAP values for $\{x_i\}$ have interesting variability but do not simply follow the data. Then, strikingly, the max BP converges to the true MAP a bit more than 99 percent of the time. There are two typical ways it can fail. The first and most common is a checkerboard problem: if we divide all the pixels in a checkerboard fashion into even and odd, then the messages from an even pixel to an off one is updated by using messages only from odd pixels to even ones, and vice versa. This sometimes causes the output $\{x_i^{(n)}\}$ to oscillate between two alternatives, both of which are typically very far from the MAP. Less frequently, periods of order 4 occur, and presumably others are possible. The second type of failure is simply that there are competing local minima with very similar energies, and the max BP goes to the wrong one. In yet other cases, even after 1000 iterations of the BP, the max messages were slowly evolving, presumably toward a limit. Two alternate local minima and one 4-cycle are shown in Figure 4.13. A more realistic test of max BP is obtained by considering the Ising model on the cow/tree image we used in Figure 4.4. We can use this same image with various constants c in the Ising term for the external of the external field and then compare the true MAP (obtained by

4.4. Deterministic Algorithms to Approximate the Mode of a Gibbs Field 211

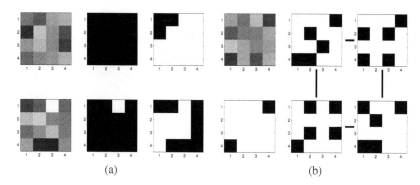

Figure 4.13. The top three images in (a) show the random input $\{u_k\}$, the MAP values of $\{x_k\}$, and the output of max-BP. In this case, the BP energy is 3.4% higher. Below these are three other images, now with the max-BP output energy being only 0.2% higher. In (b), the leftmost two images show a third input (upper image), and the MAP values (lower image) plus a counterclockwise cycle of 4 on the right connected by lines in which BP gets stuck. Note that the MAP has two black squares which are always present in the BP 4-cycle on the right, but each image in this cycle always has three more black squares that move around in checkerboard fashion.

the Greig-Porteous-Seheult method in Exercise 2 at the end of this chapter) with the output of max BP. The max-BP algorithm sometimes finds the MAP exactly, sometimes misses by a few pixels, and sometimes lands in a local energy minimum that is substantially different from the MAP. Thus, for $c = 4$ (the case shown

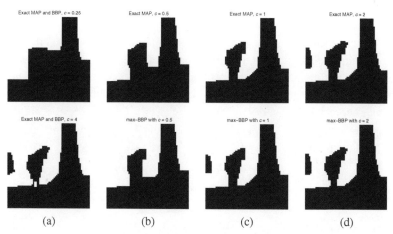

Figure 4.14. MAP and BP estimates (see text). (a) Two values of c where the BP gives the true MAP. (b)–(d) Comparison of the true MAP and the max BP in cases where they differ. The difference is (b) 66 pixels with an energy increase of 0.2% for $c = .5$; (c) 63 pixels with an energy increase of 1.5% for $c = 1$; and (d) 9 pixels for an energy increase of 0.06% for $c = 2$. Case $c = 1$ shows competing nearly equal local minima in the energy landscape.

in Figure 4.4) and for $c = .25$, max BP gives the MAP, even though it must be run for some 200 iterations in the latter case before converging. If $c = 2$, it misses the MAP by a mere 9 pixels, whereas for $c = 1$, it adds a whole black region in the MAP, missing by 63 pixels. These cases are shown in Figure 4.14.

4.4.2 Algorithm VI: Graph Cuts Generalized

We have seen that the exact mode of the Ising model is readily computed, either by direct linear programming (see Theorem 4.1 in Section 4.1.1) or by max-flow graph cut algorithms (see Exercise 2) at the end of this chapter. The ingenious idea of Boykov, Veksler, and Zabih [30] is that we can at least *approximate* the mode of much more general MRF segmentation models by constructing a series of auxiliary Ising models that describe the expansion of one region or pixel swapping between two regions. Then the best expansion or pixel swap is found, and the process is continued until the energy cannot be reduced further by additional expansions or swaps. The authors show that under reasonable assumptions, the resulting energy is at least within a factor of two of the minimum—experimentally, it is usually very close.

The algorithm of Boykov et al. [30] uses the idea of minimum-cost graph cuts: given a connected graph with costs attached to each edge and two distinguished vertices, find the set of edges of minimum total cost that, when cut, results in a graph in which the distinguished vertices are no longer connected. By a famous theorem of Ford and Fulkerson (see Section 3.1 of [26], for example), the cost of this cut is exactly the same as the amount of the maximum "flow" from one of the distinguished vertices to the other; for details see Exercise 2 at the end of this chapter. But max-flow algorithms are special cases of linear programming algorithms, and what we want to present here is a generalization of their methods by using Theorem 4.1.

Consider any graph $G = (V, E)$, a set of labels Λ, a labeling function $u(v)$, $v \in V$, and the following class of energy functions:

$$E(u) = \sum_{v \in V} f_v(u(v)) + \sum_{e=<v_1,v_2> \in E} g_e(u(v_1), u(v_2)).$$

Now take any two labeling functions, u_0, u_1, and any $\{0, 1\}$-valued function J on V, and consider all composite labelings:

$$u^J(v) = \begin{cases} u_0(v), & \text{if } J(v) = 0, \\ u_1(v), & \text{if } J(v) = 1. \end{cases}$$

4.4. Deterministic Algorithms to Approximate the Mode of a Gibbs Field

Then

$$E(u^J) = \sum_{v \in V} (f_v(u_0(v)).(1 - J(v)) + f_v(u_1(v)).J(v))$$

$$+ \sum_{e=<v,w>\in E} \begin{cases} g_e(u_0(v), u_0(w)) & \text{if } J(v) = J(w) = 0, \\ g_e(u_0(v), u_1(w)) & \text{if } J(v) = 0, J(w) = 1, \\ g_e(u_1(v), u_0(w)) & \text{if } J(v) = 1, J(w) = 0, \\ g_e(u_1(v), u_1(w)) & \text{if } J(v) = J(w) = 1. \end{cases}$$

To make sense of the second term here, note the simple result that any function f on $\{0,1\}^2$ can be expressed as follows: If $f(0,0) = a, f(1,0) = b, f(0,1) = c$, and $f(1,1) = d$, then

$$f(x, y) = a + \tfrac{1}{2} \left((b + d - a - c) \cdot x + (c + d - a - b) \cdot y \right.$$
$$\left. + (b + c - a - d) \cdot |x - y| \right).$$

This then allows us to rewrite the energy $E(u^J)$ in the Ising form:

$$E(u^J) = E(u_0) + \sum_{v \in V} (f_v(u_1(v)) - f_v(u_0(v)))$$

$$+ \tfrac{1}{2} \cdot \sum_{w,<w,v>\in E} \left(g_{v,w}^{1,1} + g_{v,w}^{1,0} - g_{v,w}^{0,1} - g_{v,w}^{0,0} \right) J(v)$$

$$+ \tfrac{1}{2} \cdot \sum_{<v,w>\in E} \left(g_{v,w}^{0,1} + g_{v,w}^{1,0} - g_{v,w}^{1,1} - g_{v,w}^{0,0} \right) |J(v) - J(w)|,$$

where $g_{v,w}^{i,j} = g_e(u_i(v), u_j(w))$.

This energy is of the spin-glass form considered in Theorem 4.1. If the inequality

$$g_{v,w}^{0,1} + g_{v,w}^{1,0} \geq g_{v,w}^{1,1} + g_{v,w}^{0,0}$$

holds, then we can find the exact minimum of this energy with respect to J by linear programming, as described at the end of Section 4.1.1.

Boykov et al. [30] study a special case, which they call "optimal expansion." Here, the second function u_1 is always a constant function, and then it is clear that the above inequality holds provided g_e is a *metric*: $g_e(\alpha, \alpha) = 0$ and $g_e(\alpha, \beta) + g_e(\alpha, \gamma) \geq g_e(\alpha, \gamma)$. Their method constructs a sequence of labelings $u^{(n)}$, a sequence of labels λ_n, and, for each n, applies the above method to $u_0 = u^{(n)}, u_1 = $ the constant labeling with value λ_n. Thus, $u^{(n+1)}$ is the labeling obtained from $u^{(n)}$ by expanding the region with label λ_n to whatever size gives minimal energy among such expansions.

Boykov et al. also study the simpler case, which they call "optimal swap": two labels λ_n, μ_n are chosen at each stage, and u_0 is the labeling in which the

λ_n region of $u^{(n)}$ is given the label μ_n, u_1 is the labeling in which the μ_n region of $u^{(n)}$ is given the label λ_n, and the rest of both u_0 and u_1 are the same as $u^{(n)}$. Thus, $u^{(n+1)}$ can be any labeling obtained from $u^{(n)}$ by swapping some of the labels λ_n, μ_n in whatever way gives minimal energy among all swaps. In both cases, they also give an ingenious graph cut interpretation of the linear programming problem associated with the above Ising model. This allows us to use max-flow, min-cut algorithms instead of general purpose linear programming algorithms to solve each optimization problem. Part of their approach is described in Exercise 2 at the end of this chapter.

A point of some dispute involves which of the algorithms presented above is (a) the fastest and/or (b) gets the closest to the exact mode. We can only present our experiences. Simulated annealing, accelerated by suitable importance sampling (such as that described in Exercise 4 at the end of this chapter) can be very good or very bad. For this approach, speed and accuracy play off each other. If it is run with carefully crafted importance sampling or run off line slowly enough, it can give excellent results. This approach has the big advantage that it gives lots of insight into the full energy landscape. Belief propagation's weakness is that it takes a long time for beliefs to propagate across a large pixel grid, and it is very easy for it to fall into local minima. It is best to run it multiple times with different starting messages to see how consistent it is. Graph cuts have the strong advantage that they can handle global labeling changes, and split and merge operations. Graph cuts were found to be best in the review article [209], and we found it gave the best results in the experiments shown in Figure 4.6. Graph-cut methods have a problem when there are many possible labels or a continuum of labels. This occurs, for example, in piecewise-constant segmentation in which each region is fitted with its mean color to get the lowest energy. The set of labels must be the set of *all* possible colors, and this slows the graph cut models down a great deal.

4.5 Texture Models

In many cases, segmentation of natural images must be based not on contrasting black/white intensity but on contrasting textures. In the rest of this chapter, we discuss how segmentation can be performed in this case. We first study a psychophysical approach to texture, then attempt to describe different textures with statistics. A first approach uses autocorrelation (or, equivalently, power spectra) and we show that this is inadequate. A better approach is to use full-filter histograms, and we show how using exponential models to synthesize images with these requirements allows us to synthesize textures quite well. Then we discuss several models for image segmentation in which the segments have contrasting textures.

4.5. Texture Models

4.5.1 What Is Texture?

An attractive idea is to consider texture as an analog of color, where the RGB (red, green, blue) bands are replaced by the output of a specific bank of filters. If this is true, then we can seek to identify these filters by asking when the difference of some specific characteristic of the texture causes one texture to look very clearly different from another. Another central idea is that each texture is constructed out of some simple element, its *texton*, by repeating the texton with some limited range of variations, either regularly or irregularly, over the domain of the texture.

Julesz, Treisman, and many other psychophysicists [117, 214] have approached texture through experiments that use a phenomenon known as "pop out." For example, we take an image with a background made of a certain texture and in the foreground a shape made of another texture. Then the foreground shape often pops out as vividly as though it was drawn in a color different from that of the background. We can use the criterion of pop-out to identify the qualities that make one texture different from another.[4] A set of examples of pop-out (and its failure) are shown in Figure 4.15. Using such experiments, Julesz attempted to codify "pop-out." He started with the idea that every homogeneous texture is made up of recurring elements, its textons. These textons can be measured by assigning them their mean brightness, size and orientation. If the texton is simply a roundish blob of light or dark, it has a size. If the texton is an edgelet or bar, it has an orientation too. But these features alone did not suffice to explain all pop-out and Julesz extended them by introducing the idea of counting "terminators"— for example the letter T has 3 terminators and the letter K has 4 terminators. Although quite suggestive, his theory never reached a definitive stage in which the necessary and sufficient conditions for pop-out were described. But, as we will see in Chapter 6, the essential idea that images are made up of edges, bars and blobs has received considerable statistical support.

The simplest hypothesis for defining the "color" of a texture is that it is given by second-order statistics; that is, you can assign to each texture a vector given by the expectations of all the quadratic functions q of a random image $\mathcal{I}$ of this texture type. Using the standard basis for such quadratic functions, this is the vector of numbers:

$$\mathbb{E}(\mathcal{I}(\alpha_1) \cdot \mathcal{I}(\alpha_2)).$$

---

[4]Another type of pop-out experiment goes like this: try to detect a single small shape such as a cross (called the target) among a set of 5–20 other shapes (e.g., circles) called the distractors. We then judge whether it pops out by measuring whether the time needed to find the cross is independent of the number of distractors. If, instead, the time increases linearly with the number of distractors, then it appears that we must be scanning the distractors to locate the target because it did not pop out right away.

4. Image Texture, Segmentation and Gibbs Models

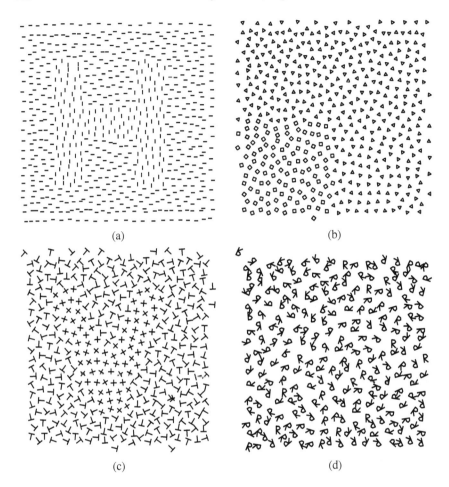

Figure 4.15. Examples of pop-out and non-pop-out. (a) The contrast between vertical and horizontal linelets causes the H image to be perceived immediately. (b) The squares texture elements form a square figure with triangles elsewhere; this is easy to see here, but it works less well if the squares make up a letter shape. (c) The Xs form a large letter V against a background of Ts. Some examination is needed before this is seen. (d) Backwards Rs form a square figure on the top left with the usual R elsewhere. This is invisible unless you make a detailed study.

We assume here, for simplicity, that $\mathcal{I}$ is a random image of this texture type defined on an $N \times M$ grid, and that it wraps around periodically, making the distribution of $\mathcal{I}$ *stationary* with respect to translations on the discrete torus, or

$$\mathbb{P}(\mathcal{I} \in U) = \mathbb{P}(T_\beta \mathcal{I} \in U),$$

4.5. Texture Models

where T_β is a translation: $T_\beta I(\alpha) = I(\alpha - \beta)$. Then, by stationarity, we get that

$$\mathbb{E}(\mathcal{I}(\alpha_1) \cdot \mathcal{I}(\alpha_2)) = \mathbb{E}(\mathcal{I}(0) \cdot \mathcal{I}(\alpha_2 - \alpha_1)),$$

as well as the fact that $\mathbb{E}(\mathcal{I}(\alpha))$ is a constant, the mean brightness of the texture. Then bases for the full set of second-order statistics can be given by any of the following:

- the autocorrelation: $C(\alpha) = \mathbb{E}(\mathcal{I}(0) \cdot \mathcal{I}(\alpha))$;
- the power spectrum: $\mathbb{E}(|\widehat{\mathcal{I}}(\xi)|^2)$;
- the power of responses of any sufficiently large filter bank: $\mathbb{E}((F_\alpha * \mathcal{I})^2)$.

This follows because the power spectrum is the Fourier transform of the autocorrelation. If we assume, for instance, that the image $\mathcal{I}$ is periodic and defined on the unit square, then we have

$$\mathbb{E}(\mathcal{I}(0) \cdot \mathcal{I}(\alpha)) = \mathbb{E}\left(\int \mathcal{I}(x) \cdot \mathcal{I}(x + \alpha) dx\right) = \mathbb{E}((\mathcal{I}^- * \mathcal{I})(\alpha));$$

hence,

$$\int e^{-2\pi i \alpha \cdot \xi} \cdot \mathbb{E}(\mathcal{I}(0) \cdot \mathcal{I}(\alpha)) d\alpha = \mathbb{E}(\widehat{\mathcal{I}^- * \mathcal{I}}(\xi)) = \mathbb{E}(|\widehat{\mathcal{I}}(\xi)|^2).$$

Moreover, the powers of the filter responses are given by $\mathbb{E}((F_\alpha * \mathcal{I})^2) = \mathbb{E}(|\widehat{F_\alpha}|^2) \cdot \mathbb{E}(|\widehat{\mathcal{I}}|^2)$. Some of the most commonly used and useful filters for describing texture are shown in Figure 4.16.

We must be careful and distinguish the second-order statistics and the second-order co-occurrence statistics. The second-order co-occurrence statistics are the full joint probability distribution of two gray values $\mathcal{I}(0)$ and $\mathcal{I}(\alpha)$ whose pixels

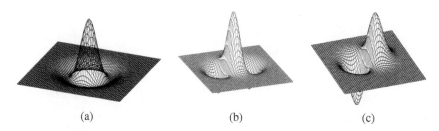

Figure 4.16. Graphs of three types of filters commonly used for finding features and discriminating textures. (a) An isotropic filter given by the Laplacian of a Gaussian. (b) and (c) An approximate Hilbert pair of even and odd filters, respectively, given by the first derivatives of a Gaussian.

differ by a fixed translation and not just their covariance. Except in the case of binary images, they contain more information than the second-order statistics.

In practice, textures often have some characteristic oscillation that we would like to detect first. How can we extract this oscillation most easily from filter responses? Suppose $I(\vec{x}) = \cos(2\pi\vec{\xi}\cdot\vec{x}+a)$ is a pure oscillation. If we filter such textures, the response also oscillates, as is shown in the following proposition, but if you sum the squares of two suitable filter responses, it does not.

Proposition 4.7. *Let I be any pure oscillation $I(\vec{x}) = \cos(2\pi\vec{\xi}\cdot\vec{x}+a)$ and $I'(\vec{x}) = \cos(2\pi\vec{\xi}\cdot\vec{x}+a-\pi/2) = \sin(2\pi\vec{\xi}\cdot\vec{x}+a)$ the phase shifted wave. Then, if F_+ is any even filter and F_- any odd filter,*

$$F_+ * I = \widehat{F_+}(\vec{\xi}) \cdot I,$$
$$F_- * I = i\widehat{F_-}(\vec{\xi}) \cdot I'.$$

F_+, F_- are called a *Hilbert transform pair* if $\widehat{F_-}(\vec{\xi}) = \pm i\widehat{F_+}(\vec{\xi})$ for every $\vec{\xi}$ (usually, $\widehat{F_-}(\vec{\xi}) = \text{sign}(\vec{\xi}\cdot\vec{\gamma})i\widehat{F_+}(\vec{\xi})$ for some $\vec{\gamma}$). In this case, $|F_+ * I|^2 + |F_- * I|^2$ is constant.

The proof follows easily from the fact that $\widehat{F_+}$ is real and even; hence,

$$\begin{aligned}(F_+ * I)(\vec{x}) &= \tfrac{1}{2}\int F_+(\vec{y})\left(e^{2i\pi\vec{\xi}\cdot(\vec{x}-\vec{y})+a} + e^{-2i\pi\vec{\xi}\cdot(\vec{x}-\vec{y})-a}\right)d\vec{y}\\ &= \tfrac{1}{2}\widehat{F_+}(\vec{\xi})\left(e^{2i\pi\vec{\xi}\cdot\vec{x}+a} + e^{-2i\pi\vec{\xi}\cdot\vec{x}-a}\right) = \widehat{F_+}(\vec{\xi})I(\vec{x}).\end{aligned}$$

A similar argument gives the odd case. By summing the squares and using the fact that $|I|^2 + |I'|^2$ is constant, we get the last result.

Notice that due to the hypothesis that the distribution on $\mathcal{I}$ is stationary, we have

$$\mathbb{E}((F_+ * \mathcal{I})^2) = \int (F_+ * I)^2 = \int |\widehat{F_+}(\xi)|^2|\hat{I}(\xi)|^2 = \int |\widehat{F_-}(\xi)|^2|\hat{I}(\xi)|^2$$
$$= \mathbb{E}((F_- * \mathcal{I})^2),$$

but by taking the sum, we detect local areas of constant texture. Most often, we use the approximate Hilbert transform pair called Gabor filters:

$$\begin{cases} F_+(\vec{x}) &= \cos(2\pi\vec{\eta}\cdot\vec{x})\cdot G(\vec{x}),\\ F_-(\vec{x}) &= \sin(2\pi\vec{\eta}\cdot\vec{x})\cdot G(\vec{x}),\end{cases}$$

where G is some Gaussian in two variables. In fact, we can calculate that

$$\widehat{F_+}(\vec{\xi}) = \tfrac{1}{2}\left(\widehat{G}(\vec{\xi}+\vec{\eta}) + \widehat{G}(\vec{\xi}-\vec{\eta})\right),$$
$$\widehat{F_-}(\vec{\xi}) = \tfrac{i}{2}\left(\widehat{G}(\vec{\xi}+\vec{\eta}) - \widehat{G}(\vec{\xi}-\vec{\eta})\right),$$

4.5. Texture Models

so if $\|\vec{\eta}\| \gg \mathrm{supp}(\widehat{G})$, we have something very close to a Hilbert transform pair. In the biological case, the Gaussian is typically anisotropic, with support extended along the line $\vec{\eta} \cdot \vec{x} = 0$ and narrow perpendicular to it. This makes the filter F_- especially sensitive to extended edges in the direction of this line. The filter responses $F_\pm * I$ are reasonable models for the so-called "simple" cells of the primary visual cortex, and the power $|F_+ * I|^2 + |F_- * I|^2$ is a model for the phase insensitive "complex" cells (see Chapter 4 of [173]).

4.5.2 The Limitations of Gaussian Models of Texture

The ideas of the last section are what might be called the "classical" approach to texture using standard electrical engineering tools. The problem is that the expected power of filter responses to any set of textures is identical to their responses to the best fitting Gaussian model for the set of textures. But Gaussian models do not distinguish between textures which are psychophysically very different from each other, nor do samples from them reproduce convincingly realistic images of most textures!

To show this, we use the simple idea of "randomizing phase'. Given a periodic real function f of any number of variables, take its Fourier transform $\widehat{f}$, multiply it by $e^{ih(\xi)}$ with h a random real and odd function and finally take the inverse Fourier transform. The oddness of h makes the final result $\widetilde{f}$ real again. Then f and $\widetilde{f}$ have the same power spectrum but different phases. If f is a random sample from a stationary Gaussian model (so that its covariance is diagonal in the Fourier basis, Chapter 2, Th.2.10), then so is $\widetilde{f}$. So we may ask: if we do this to a textured image, does the result have similar texture?

Let's look first at a one-dimensional example. Consider piecewise constant periodic one-dimensional signals. We generate these with random jumps at random points on its domain. Randomizing the phases of its Fourier transform as above, we can create new functions with the identical power spectrum but new phases. All the second-order statistics will be unchanged. But if we use a simple first derivative filter, the histogram of its response will be totally different for the two functions. For the piecewise constant function, it will have a large peak at 0 and outliers for the jumps, that is, it will have high kurtosis. But the histogram for the phase randomized function will be approximately Gaussian. An example is shown in Figure 4.17.

We can make the same experiment for images. We create a texture which has some simple structure, e.g., take a small image called a texton and place it

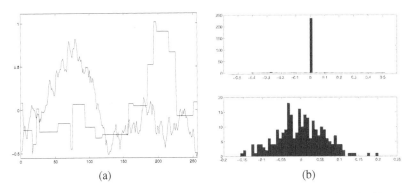

Figure 4.17. (a) A plot showing two functions: a piecewise constant periodic function with 20 random jumps of 20 random sizes and a second function obtained by taking the Fourier transform of the first and randomizing its phases. 9b) The histograms of the first derivative filter applied to each function of (a).

at random places on a white background. Then we randomize its phases[5]. Two examples of this are shown in Figure 4.18. In addition, we show a natural example of the same kind: a sample image of the fur of a cheetah, with characteristic spots. Again randomizing the phase creates an image with no recognizable connection to the original.

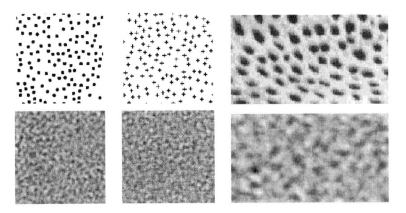

Figure 4.18. The top row shows three textures: one made of small black circles, one made of small crosses and one a detail of the fur of a cheetah. The second row shows what happens when we randomize the phases, keeping the power spectrum fixed, for each of these textures. Clearly, the small elements, the textons, in each texture are lost. (Top row from [237, Figures 6 and 8].)

---

[5]Since the image is not periodic, treating it as periodic creates discontinuities where it wraps around, hence large power for horizontal and vertical frequencies. We prevent this artifact by first reflecting the image across its edges to make a periodic image of twice the size and then randomizing the phases of this.

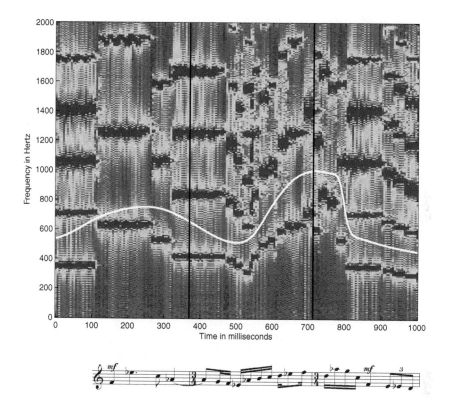

Plate 1. Three bars from an oboe playing *Winter711* by Jan Beran: the spectrogram (top) and the score as humans write it (bottom). The three bars are separated by vertical dotted white lines in the spectrogram and are separated as usual in the human score. The spectrogram shows the distribution of power across frequencies as time progresses. The oboe has a rich set of harmonics. Below the curving white line is the fundamental note, in which the score is easily recognized (although stretched and squeezed a bit). The pixelization of the spectrogram is an unavoidable consequence of Heisenberg's uncertainty: time and frequency cannot be simultaneously made precise. (See Figure 2.1. We thank Chris Raphael, whose playing is shown in the spectrogram and who supplied us with the score.)

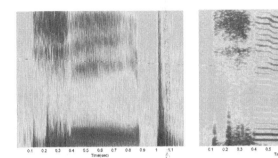

Plate II. Spectral analysis of the word "sheep" (ʃ i p in IPA): On the left, the log-power from a windowed Fourier transform with a window size of 6 msec, giving fine temporal resolution; on the right the window size is 40 msec, giving fine frequency resolution instead. This is a good illustration of Theorem 2.8, which states that we cannot have fine temporal and frequency resolution at the same time. The initial 0.4 sec is the fricative sh, colored noise with power in low and high frequencies. In the middle 0.4–0.9 sec is the vowel ee. It is "voiced"—the vocal chords are vibrating—which creates bursts seen in the thin vertical lines on the left or as the multiple bands of harmonics on the right. Finally, at 0.9–1.1 sec, we have the stop consonant p with a mouth closure and a burst. The horizontal bands seen most clearly on the left during the vowel are the formants—frequencies at which the mouth cavity resonates, reinforcing specific harmonics of the burst. The vertical scales of both spectograms are 0 to 4000 Hertz. (See Figure 2.12.)

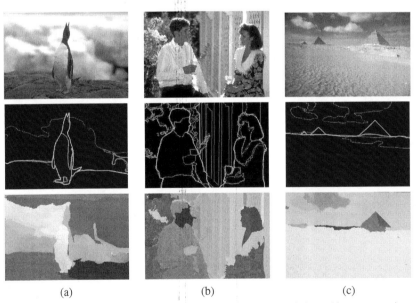

(a)   (b)   (c)

Plate III. More examples of images, their human segmentations and their machine segmentations. Note (a) how the computer algorithm gets confused by artifacts in the jumble of ice flows; (b) how it fails to cluster the panels of the lattice fence; and (c) how it breaks the large pyramid into two segments for its two faces. The algorithm has trouble with *disconnected* regions, with texture gradients, and with discounting lighting artifacts. (See Figure 4.2. Top row courtesy of J. Malik's lab.)

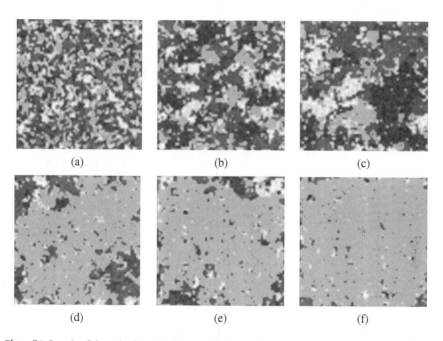

Plate IV. Samples I from the five-color Potts model on a 64×64 grid at descending temperatures. The five colors have been chosen as black and white and the primary colors, noise v has been added and the image blown up by resampling and adding a bit of blur to eliminate the blocky artifacts of the small grid. Images (a)–(f) have values $e^{-\lambda/T} = .4, .32, .308, .305, .3, .29$, respectively. Note that at .4 (a), we have essentially random pixel colors. At .305 and less (d)–(f), one color takes over and drives out all others. Image (c), at .308, is roughly at the phase transition where the colors coagulate into larger regions and fight for dominance. (See Figure 4.5.)

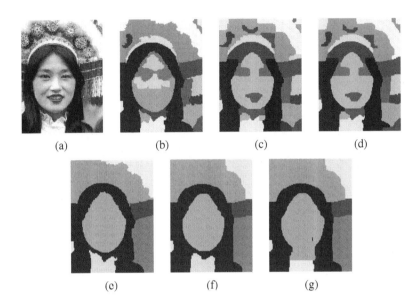

Plate V. Examples of piecewise constant segmentation of a color image by using the Mumford-Shah piecewise constant model for color images. The norm is the Euclidean norm on RGB color space. The implementation is either the greedy merge method developed by Koepfler et al. in [126] (see also Section 4.4) or belief propagation (see Section 4.4.1) followed by the optimal expansion algorithm of Boykov et al. [30] (see also Exercise 2 at the end of this chapter). Image (a) is the original 160×120 image, followed by three segmentations for $\lambda = 0.5$ ((b)–(d)) and three for $\lambda = 2$ ((e)–(g)). The segmentations for $\lambda = 0.5$ have energies $E = 2010, 1736$, and 1733, respectively, and differ in having 22, 32, and 27 regions, respectively. Image (b) is the output of the greedy merge algorithm; (c) is belief propagation followed by optimal expansion; and (d), the last and best, is greedy merge followed by expansion. Between (c) and (d) are clearly multiple trade-offs between the boundary and fidelity terms caused by extra regions—multiple local minima of the energy. The segmentations for $\lambda = 2$ ((e)–(g)) all have eight regions and energies $E = 3747, 3423$, and 3415, respectively. Image (e) is the output of the greedy merge; (f) is the result of following this by optimal expansion, and (g), the third and best, is belief propagation followed by expansion. Images (f) and (g) differ only in the neck area in the trade-off of boundary length with image fidelity. We believe the lowest energy segmentations in each case are very close to the minimum energy segmentations. (See Figure 4.6.)

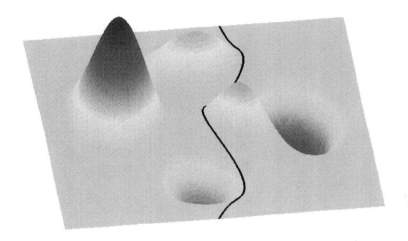

Plate VI. A geodesic on an artificial landscape. (See Figure 5.10.)

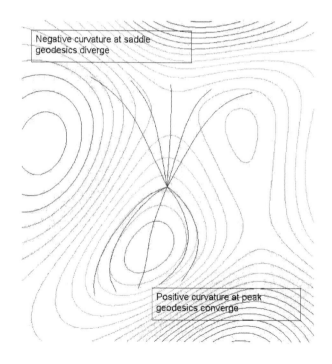

Plate VII. Some geodesics on the landscape used in Figure 5.10, shown here by contour lines. (See Figure 5.11.)

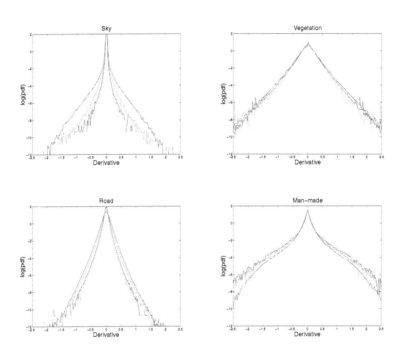

Plate VIII. Log histograms of derivatives from four categories of pixels in the British Aerospace data base: sky, vegetation, roads, and man-made objects. Each plot shows the histograms of derivatives for the original image and for coarser versions in which blocks of pixels are averaged (see text). The four colors red, green, blue, and yellow are data for $n \times n$ block averaged images, $n = 1, 2, 4, 8$, respectively. (From [110, Figure 2.10]. See Figure 6.3.)

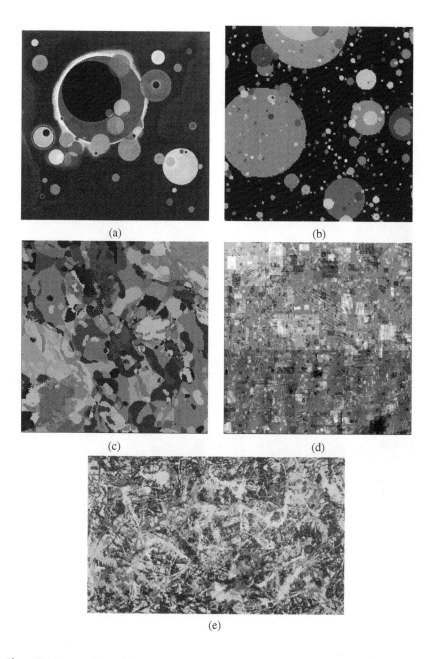

Plate IX. A comparison of abstract art and random wavelet models. (a) A 1926 painting entitled "Quelques Cercles" by Wassily Kandinsky (1926, ©2009 Artists Rights Society (ARS), New York / ADAGP, Paris). (e) Jackson Pollock's "Convergence" (1952, ©2009 The Pollock-Krasner Foundation / Artists Rights Society (ARS), New York). (b)–(d) Random wavelet images from the work of Gidas and Mumford: In (b), the Levy measure is concentrated on simple images of disks with random colors; in (c) the Levy measure is concentrated on snake-like shapes with high clutter; in (d), the Levy measure is on textured rectangles. (See Figure 6.17.)

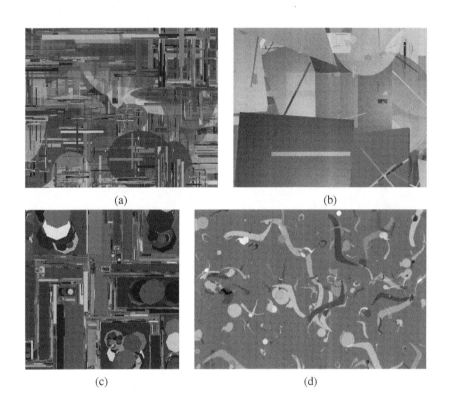

Plate X. Four synthetic images generated by random trees. (a) Trees of depth 4, children constrained to occlude and lie inside parent, top level transparent; (b) top level with uniform sizes but random types and transparency of subobjects; (c) trees of depth 3 with scaling size distribution, disjoint at top level, children lie inside parent; (d) one scale, ν supported at four delta functions, random transparency. (Courtesy of L. Alvarez, Y. Gousseau and J.-M. Morel. See Figure 6.18.)

4.6. Synthesizing Texture via Exponential Models

As in the one-dimensional case, the presence of a specific local characteristic will cause a filter that is "matched" to this texton to have some very large responses and many nearly zero responses. Thus, suppose we take a filter with positive values on a central disk and negative responses in a ring around this. If the size of the disk matches the blobs in the images in the figure, the histogram of this filter's responses will have large values when the filter sits on top of a texton, and small values when it is not so centered. This gives the histogram high non-Gaussian kurtosis. This is an example of the following general observation due to David Field [72]: for almost all images, the statistics of almost all filter responses $F * I$ will have high kurtosis. We discuss this at some length in Chapter 6.

4.6 Synthesizing Texture via Exponential Models

If we seek a stochastic model whose samples will reproduce certain features of an observed set of signals in the real world, a very general method is to use exponential models. We first explain this very useful tool and then apply it to texture synthesis.

4.6.1 Basics X: Exponential Models

We consider a world of signals Ω observed through samples from some probability measure P_{true}. The problem is that we don't know anything about P_{true}. The only thing we can do is measure the marginal distribution on a lot of features

$$f_i : \Omega \longrightarrow \mathbb{R}, \quad \text{for } 1 \leq i \leq n.$$

Assume that we estimate the mean of these features from some measurements $\mathcal{D}$:

$$\mathbb{E}_{P_{\text{true}}}(f_i) \simeq \mathbb{E}_{\mathcal{D}}(f_i) \stackrel{\text{def}}{=} \frac{1}{|\mathcal{D}|} \sum_{x \in \mathcal{D}} f_i(x).$$

We seek the simplest probability distribution on Ω that is consistent with these measurements. We make two weak assumptions:

1. There are enough measurements to distinguish between the features; that is, there is no linear combination $\sum_i a_i f_i$ that is constant on the whole data set.

2. No data point $\bar{x} \in \mathcal{D}$ is infinitely unlikely, in the sense that if we define $\vec{f} = (f_1, \ldots, f_n)$, then for some $\epsilon > 0$,

$$P_{\text{true}}\left(\left\{x \,\Big|\, \|\vec{f}(x) - \vec{f}(\bar{x})\| < \epsilon\right\}\right) = 0.$$

We fix P_{ref}, some mathematically simple measure (possibly not a probability measure) on Ω, and assume P_{true} is absolutely continuous with respect to P_{ref}. If Ω is finite, we just take P_{ref} to be the uniform measure giving each point in Ω measure 1. In this case, we often assume that every element of Ω has nonzero probability, and then assumption 2 is not needed. Then, starting with this reference measure, we can construct a unique *exponential model* for our data:

Theorem 4.8. *Define probability measures P_a by*

$$dP_a(x) = \frac{1}{Z_a}.e^{-\sum a_i f_i(x)} \cdot dP_{\text{ref}}(x),$$

where $\quad Z_a = \int_x e^{-\sum a_i f_i(x)} dP_{ref}(x) \in \mathbb{R}_+ \cup \{\infty\} \quad$ *the "partition function"*

Then, under assumptions 1 and 2 above and with a third mild assumption (to be explained below), there a unique sequence of real numbers $a_1, \ldots a_n$ such that $\mathbb{E}_{P_a}(f_i) = \mathbb{E}_\mathcal{D}(f_i)$.

Theorem 4.8 has a quite elegant, intuitive proof, but it is complicated by convergence problems. Therefore, we give the proof first in the case where Ω is finite, in which case no extra assumption is needed. Then we go on to describe how to beef up the argument for the general case. (Note: in the following proof, we abbreviate $\mathbb{E}_\mathcal{D}(f_i)$ to $\overline{f_i}$.)

Proof (for $|\Omega| = N$ finite): We may assume $P_{\text{ref}}(\omega) = 1$, all $\omega \in \mathcal{D}$. First, we compute the first and second derivatives of $\log Z_a$ seen as a function of $a = (a_1, \ldots, a_n)$. We get the important identities:

$$\frac{\partial}{\partial a_i} \log Z_a = \frac{1}{Z_a} \sum_{x \in \Omega} -f_i(x) e^{-\sum a_k f_k(x)} = -\mathbb{E}_{P_a}(f_i),$$

$$\frac{\partial^2}{\partial a_i \partial a_j} \log Z_a = \frac{1}{Z_a} \sum_{x \in \Omega} f_i(x) f_j(x) e^{-\sum a_k f_k(x)}$$

$$- \frac{1}{Z_a^2} \left(\sum -f_i(x) e^{-\sum a_k f_k(x)} \right) \left(\sum -f_j(x) e^{-\sum a_k f_k(x)} \right)$$

$$= \mathbb{E}_{P_a}(f_i f_j) - \mathbb{E}_{P_a}(f_i) \mathbb{E}_{P_a}(f_j)$$

$$= \mathbb{E}_{P_a}\left((f_i - \mathbb{E}_{P_a}(f_i))(f_j - \mathbb{E}_{P_a}(f_j))\right).$$

This shows that the Hessian matrix of $\log Z_a$ is the covariance matrix of $(f_1, \ldots, f_n)$ for P_a. Since no linear combination of the $\{f_i\}$ is constant on Ω, their covariance matrix must be positive definite, and $\log Z_a(x)$ is strictly convex with respect to $(a_1, \ldots, a_n)$. The convexity of $\log(Z_a)$ implies that $\log(Z_a) + \sum a_i \overline{f_i}$ is also strictly convex; hence, it has at most one minimum. To show the existence of a minimum, we also need to show that it tends to infinity on each ray

4.6. Synthesizing Texture via Exponential Models

$\mathbb{R}_+ \vec{a}_0$. But the scalar feature $\vec{a}_0 \cdot f$ is not constant on our data set $\mathcal{D}$; hence, there is an $x_0 \in \mathcal{D}$ such that $\vec{a}_0 \cdot f(x_0) < \vec{a}_0 \cdot \overline{f}$. Thus, as $t \to \infty$,

$$\log Z_{\vec{a}_0 t} + (\vec{a}_0 \cdot \overline{f}).t \geq \log\left(e^{-\vec{a}_0 \cdot f(x_0) t}\right) + (\vec{a}_0 \cdot \overline{f}).t \longrightarrow \infty.$$

Therefore, $\log(Z_a) + \sum a_i \overline{f_i}$ has a unique minimum. Taking derivatives of $\log(Z_a) + \sum a_i \overline{f_i}$ with respect to a_i, we find that its unique minimum is characterized by $\overline{f_i} = \mathbb{E}_{P_a}(f_i)$ for all i. This proves the theorem. □

In the general case, the integral for Z_a might diverge for some a, and we need to make a third weak assumption:

3. $\mathcal{N} = \{a \mid Z_a < \infty\}$ is an *open subset* of $\mathbb{R}^n$.

Proof (in the general case): First, even if $\mathcal{N} \neq \mathbb{R}^n$, by Fatou's lemma, we have that $Z_a \to \infty$ on the boundary of $\mathcal{N}$. Second, we can use the openness of $\mathcal{N}$ and the bound $(2 + x^2) < e^x + e^{-x}$ to show that all the integrals $\int |f_i| e^{-a \cdot f} dP_{\text{ref}}$ and $\int |f_i||f_j| e^{-a \cdot f} dP_{\text{ref}}$ converge for $a \in \mathcal{N}$. Then the bounded convergence theorem shows that the formulas for the first and second derivatives of $\log Z_a$ still hold. Thus, $\log(Z_a)$ is strictly convex on $\mathcal{N}$.

Consider $\log(Z_a) + \sum a_i \overline{f_i}$ as before; we need to show that it tends to infinity on each ray $\mathbb{R}_+ \vec{a}_0$. But by the second assumption, not only is $\vec{a}_0 \cdot f(x) < \vec{a}_0 \cdot \overline{f}$ at some data point x, but there is a subset $B \subset \Omega$ such that $P_{\text{true}}(B) > 0$ (and hence $P_{\text{ref}}(B) > 0$) and $\vec{a}_0 \cdot f(x) \leq \vec{a}_0 \cdot \overline{f} - c$ on B, for some $c > 0$. Thus, as $t \to \infty$,

$$\log Z_{a_0 t} + (a_0 \cdot \overline{f}).t \geq \log\left(\int_B e^{-a_0 t \cdot f(x)} dP_{ref}(x)\right) + (a_0 \cdot \overline{f}).t$$
$$\geq c.t - a_0 t \cdot \overline{f} + \log(P_{ref}(B)) + (a_0 \cdot \overline{f}).t \longrightarrow \infty.$$

As in the proof above, we find that its unique minimum is characterized by $\overline{f_i} = \mathbb{E}_{P_a}(f_i)$ for all i. □

This exponential model can be characterized in many ways. Taking Ω finite again for simplicity, we can say that among all the distributions P_b, P_a minimizes the Kullback-Leibler distance $D(P_\mathcal{D} \| P_b)$. In fact,

$$D(P_\mathcal{D} \| P_b) = \mathbb{E}_{P_\mathcal{D}}\left(\log\left(\frac{P_\mathcal{D}}{P_b}\right)\right)$$
$$= -H(P_\mathcal{D}) + \mathbb{E}_{P_\mathcal{D}}\left(\log Z_b + \sum b_i f_i\right)$$
$$= -H(P_\mathcal{D}) + \log Z_b + \sum_i b_i \overline{f_i}$$

so this is just a reinterpretation of the proof of the theorem. A second characterization is that among all probability distributions P on Ω such that $\mathbb{E}_P(f_i) = \overline{f_i}$,

P_a is the one with the maximal entropy. We calculate

$$D(P \parallel P_a) = \mathbb{E}_P \left(\log \left(\frac{P(x)}{e^{-\sum a_i f_i(x)}/Z_a} \right) \right) = -H(P) + \log(Z_a) + \sum a_i \overline{f_i};$$

hence, the differential entropy of P is maximized precisely when D is minimized, that is, $P = P_a$.

4.6.2 Applications

We first look at Theorem 4.8 in the Gaussian case. We simply take $\Omega = \mathbb{R}$, $dP_{\text{ref}} = dx$, and $f_1(x) = x$ and $f_2(x) = x^2 = f_1^2(x)$. Then we get exponential models reproducing given means and variances given by the equations

$$dP_a(x) = \frac{1}{Z_a} e^{-a_1 x - a_2 x^2} dx,$$

and the corollary of the main theorem is that there is a unique Gaussian model with given mean and variance. More generally, if $f_k(x) = x^k$ for $1 \leq k \leq n$, we get models of the form $e^{-Q(x)}$, where Q is a polynomial in x of degree n.

First variant of Theorem 4.8 We divide $\mathbb{R}$ in bins: $\mathbb{R} = B_1 \cup \ldots \cup B_n$, and then approximate the full distribution of a function $f : \Omega \longrightarrow \mathbb{R}$ by the binary variables

$$f_i(x) = \mathbb{I}_{B_i}(f(x)),$$

where $\mathbb{I}_{B_i}$ is the characteristic function of B_i. Then,

$$\mathbb{E}(f_i) = \mathbb{E}\left(\mathbb{I}_{B_i}(f)\right) = \mathbb{P}(f(x) \in B_i).$$

Now, taking these f_i's as features, we get the model

$$\frac{1}{Z_a} \cdot e^{-\sum a_i \mathbb{I}_{B_i}(f(x))} dP_{\text{ref}}.$$

If the bins get smaller and smaller, then we get convergence to something such as

$$\frac{1}{Z_\psi} \cdot e^{-\psi(f(x))} dP_{\text{ref}}.$$

for a suitable function $\psi : \mathbb{R} \longrightarrow \mathbb{R}$. So (with appropriate conditions that we ignore) for a unique function ψ, this model has the correct marginal distribution of the scalar feature f.

4.6. Synthesizing Texture via Exponential Models

Second variant of Theorem 4.8 In a groundbreaking article, Heeger and Bergen [103] have suggested that the set of histograms of a big enough filter bank contains enough information on a uniformly textured image I to characterize the texture. So we are led to consider image models of the form

$$dP(I) = \frac{1}{Z} \cdot e^{-\sum_\alpha \sum_{i,j} \psi_\alpha((F_\alpha * I)(i,j))} \cdot \prod_{i,j} dI(i,j),$$

where the F_α are filters, such as nearest neighbors, or more complex ones (a face filte,r for example). Thus, the features are the measurements $(F_\alpha * I)(i,j)$ at all pixels, and we apply to them the above results on exponential models. We want the model to be stationary, and that is why we take the sum over all (i,j). The theorem on the existence and uniqueness of exponential models shows that there is a unique set of functions $\psi_1, \ldots, \psi_n$ such that the above $P(I)$ reproduces the set of empirical histograms of the filter responses $F_\alpha * I$. This method has been used by Zhu, Wu and Munmford [237, 238] for building models that reproduce images with given textures. Some examples are shown in Figures 4.19 and 4.20.

We briefly describe here the algorithm for fitting such a model to a given texture. First, all the F_α are chosen from a big Gabor-type filter bank. At stage k we have chosen k filters, and then we run a simulation of MCMC until we get

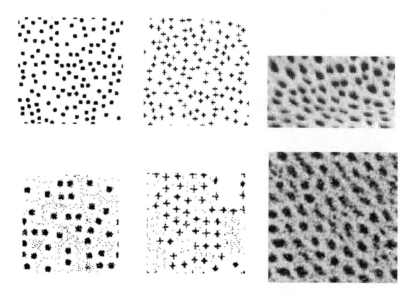

Figure 4.19. The three textures shown on the top row are synthesized on the bottom row by using exponential models and the learning model described in the text to fit the potentials ψ. The first two images use only one filter, the cheetah fur uses six. (Top row from [237, Figures 6 and 8].)

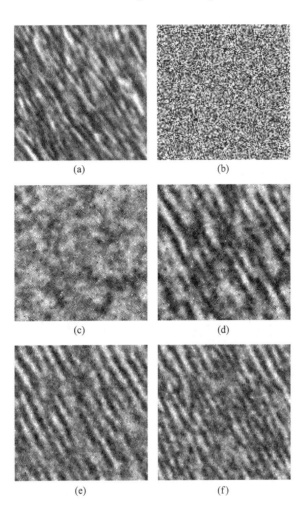

Figure 4.20. The progressive synthesis of a texture by exponential models: (a) is the original fur texture; (b) reproduces the single pixel values randomly; (c) adds an isotropic center-surround filter; and (d), (e), and (f) add even Gabor filters at different orientations and widths. (From [237, Figure 4].)

good samples of the distribution $P_{I,k}$, based on k filters. From the samples, we can estimate the responses of all the other filters. Then, at stage $k+1$, we add the filter whose histogram under $P_{I,k}$ is the furthest from its histogram in the given texture.

To further demonstrate the power of this method for learning exponential models, we next describe a clutter removal application [233]. Suppose that our

4.6. Synthesizing Texture via Exponential Models

Figure 4.21. An image of a building partially obscured by a tree is decomposed into a reconstruction of the whole building and the tree. This is done by modeling the distinctive filter properties of the manmade rectilinear building and the natural tree. (From [233, Figure 17].)

image has a target but also has some other clutter that we want to remove. We first model the clutter with one set of filters, getting a probability distribution of the form $e^{-E_C(I)}$. Then we model the target with a second set of filters, getting $e^{-E_T(I)}$. We assume that our image I can be split into target and clutter $I = I_C + I_T$, and we want to find I_T. To get it, we just have to look for

$$\underset{I_T}{\text{Min}} \left[E_T(I_T) + E_C(I - I_T) \right].$$

This algorithm removed the trees in front of the building as shown in Figure 4.21.

Exponential models are useful not only for modeling image texture but for many other features of signals. A very different example is their use to model the sorts of contours that occur in images. You can look at a lot of data of actual image contours and then find the empirical distribution of their curvature $\kappa(s)$, and even its arc length derivative $\dot{\kappa}(s)$. Song-Chun Zhu did this in [239] by using polygonal approximations Γ to these contours. He got a collection of stochastic polygon models of the form

$$p(\Gamma) \propto e^{-\int_\Gamma (\psi_1(\kappa_\Gamma) + \psi_2(\frac{d\kappa}{ds})) ds_\Gamma},$$

where $\int_\Gamma$ is really a sum over vertices of Γ, s_Γ is arc length along Γ, κ_Γ is a discrete approximation to curvature for polygons, and the density is with respect to a reference measure for polygons given by the coordinates of its vertices. Then he used such models to synthesize curves. The effect of the ψs is to produce models whose sample curves, like actual image contours, have curvature κ, which is mostly small but occasionally very large—instead of hovering around some mean.

4.7 Texture Segmentation

4.7.1 Zhu and Yuille's Models

A stochastic model for segmenting textured images must have hidden variables for the regions $\{R_i\}$ and for the parameters $\{\gamma_i\}$ describing the texture in each region. Zhu and Yuille [236] have considered models of the form

$$p(\{R_i\}|\{\gamma_i\}) = \prod_{i=1}^{M} e^{-c \cdot \text{len}(\partial R_i)} \cdot p\left((I|_{R_i})|\gamma_i\right).$$

This model seeks regions whose intensity values are consistent with having been generated by one of a family of given probability distributions $p(I|\gamma)$ where γ are the parameters of the distribution.

For instance, at the end of their paper [236], Zhu and Yuille consider a family of possible Gaussian distributions on $I|_R$ that are stationary and in which there are interactions only between close neighbors. Reinterpreting a bit their formulas, this means that for each region, we have a mean μ_i and a symmetric matrix $Q^{(i)}(x, y)$, $x, y \in R_i$, depending only on $x - y$ and nonzero only if $\|x - y\| < d$ (for some small d) so that the region-specific probability model is

$$p\left((I|_{R_i})|\mu_i, Q^{(i)}\right) = \sqrt{\det\left(\frac{Q^{(i)}}{2\pi}\right)} \cdot e^{-\frac{1}{2} \sum_{x y \in R_i, \|x-y\|<d} (I(x) - \mu_i) Q^{(i)}(x,y)(I(y) - \mu_i)}.$$

Thus, we get the energy

$$2E(\{R_i\}, \{\mu_i\}, \{Q^{(i)}\}) = \sum_{i=1}^{M} \left(c \cdot \text{len}(\partial R_i) - \log\left(\det\left(\frac{Q^{(i)}}{2\pi}\right)\right) \right.$$

$$\left. + \sum_{x,y \in R_i, \|x-y\|<d} (I(x) - \mu_i) Q^{(i)}(x,y)(I(y) - \mu_i) \right).$$

In this framework the sought segmentation of the image I makes I approximately Gaussian in each region. This energy is quadratic in I, but it is quite complex as

4.7. Texture Segmentation

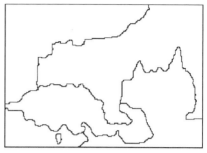

Figure 4.22. An image with three strongly contrasting textures, segmented by the algorithm of Zhu and Yuille. (From [236, Figure 18].)

a function of the parameters $\{\mu_i\}, \{Q^{(i)}\}$. For a fixed region R, minimizing this energy with respect to μ and Q will make the model match $I|_R$. That is, μ is the mean of $I|_R$, and the coefficients of Q are those of the exponential model that reproduce the image's correlations:

$$\left(\text{Average}_{x,x+a\in R} I(x) \cdot I(x+a)\right)$$

for all a with $|a| < d$. We can easily sample from this model if we add to it a hyperprior to give plausible values to the hyperparameters $\mu_i, Q^{(i)}$. We can use Ising type samples to choose regions R_i, then choose $\mu_i, R^{(i)}$, and finally sample from the resulting Gaussian model for I. An example of their algorithm is shown in Figure 4.22.

4.7.2 Lee's Model

The previous approach is designed to apply to segmentation when the texture is uniform in each region. But many textures show a gradient, a slow change in texton orientation or size, for example. A real and an artificial example of such changing textures are used in Figure 4.23. To deal with texture gradients, something like the full MS model is needed. Such a texture segmentation algorithm was investigated by T-S. Lee et al. [136], who proposed a model for the segmentation of an observed texture image with the following hidden random variables:

- $J_0(x, y)$, a textureless, slowly varying gray-level cartoon with discontinuities;

- $J(x, y, \sigma, \theta)$, a slowly varying description of the textural *power* in the channel with scale σ and orientation θ, also with the same discontinuities;

- Γ, the set of discontinuities between regions R_i.

We fix a low-pass filter F_0, which should smooth out all texture, and a pair of even/odd filters F_+, F_-, such as the Gabor filters, which approximately satisfy the Hilbert pair requirement (see Section 4.5.1). Then if we rotate (R_θ) and scale (S_σ) the image, we obtain the local power of the (θ, σ)-channel response:

$$(F_+ * R_\theta S_\sigma I)^2 + (F_- * R_\theta S_\sigma I)^2 .$$

Let us now define the energy function

$$\begin{aligned}E(I, J, \Gamma) &= c_0 \iint_{\mathbb{R}^2 - \Gamma)} (J_0(x,y) - (F_0 * I))^2 \, dx \, dy \\ &+ c_1 \iiiint_{\mathbb{R}^4 - \text{shadow}(\Gamma)} \left(J(x,y,\sigma,\theta) - (F_+ * R_\theta S_\sigma I)^2 \right. \\ &\qquad \left. - (F_- * R_\theta S_\sigma I)^2\right)^2 dx\, dy\, \frac{d\sigma}{\sigma}\, d\theta \\ &+ c_2 \iiiint_{\mathbb{R}^2 \times (\mathbb{R}^2 - \Gamma)} (\|\nabla J_0\|^2 + \|\nabla J\|^2) \, dx\, dy\, \frac{d\sigma}{\sigma}\, d\theta \\ &+ c_3 |\Gamma|\end{aligned}$$

Note that we handle the zero-frequency response $J_0(x,y)$ separately from the nonzero frequencies. We also define the shadow of Γ which is needed to avoid

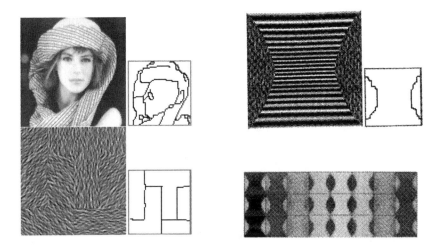

Figure 4.23. Three examples of texture segmentation with Lee's model. Note how the orientation of the stripes of the lady's scarf (top left) changes continuously but is only segmented where there are discontinuities, as in the bottom left. The bottom right image shows the power signal for the upper right image with three scales shown vertically and eight orientations shown horizontally. It shows how distinctly the texture areas are separated by their power. (From [138].)

4.7. Texture Segmentation

looking at filter responses overlapping the discontinuity set Γ. Let $\mathcal{S}$ denote the support of $|F_+| + |F_-|$; then the shadow of Γ is the set of all (x, y, σ, θ) such that

$$((x,y) + \mathcal{S}) \cap R_\theta S_\sigma(\Gamma) \neq \emptyset.$$

Lee applied this to segment textured images I, finding the J_0, J, and Γ minimizing E for fixed I by simulated annealing. Figure 4.23 shows the result of this model on three different textured images.

Remarks.

1. Notice that if we fix I and Γ, then we have a Gaussian model in J. However, if we instead fix J and Γ, the model in I is more complex, having the power 4. This allows I to pick a random phase when implementing the power required by J.

2. This model ahs an analog in the one-dimensional speech case, where J is replaced by the speech spectrogram.

3. The role of the integral on $|\nabla J|^2$ is to penalize jumps and make J vary smoothly in all four dimensions

4. If we make the model discrete by using a grid of pixels and the finite Fourier transform, it is possible to create a probability model from this energy and thus to synthesize sample textured and segmented images I. First, we choose Γ from the Ising-type model given by the last term in E; then, we choose slowly varying Gaussian cartoons J and J_0 from the third term; and finally we choose I from the quartic model given by the first two terms.

4.7.3 Some Other Texture Segmentation Algorithms

Many other texture segmentation algorithms have been proposed. Here we briefly sketch two of these. Many of the ideas that were later developed appeared first in the important work of Geman et al. [83], but we discuss more recent versions of them.

Buhmann's Method. We describe one of several models developed by Joachim Buhmann and collaborators [183]. What distinguishes their approach is that they drop the term $\text{len}(\partial R_i)$ and instead simply put an upper bound M on the number of regions in their segmentation. As an MRF, this is most simply accomplished by using the Potts model approach: introduce a field of labels $\{\rho(x)\}$, one label attached to each vertex with values from a fixed set $\{1, 2, \cdots, M\}$. They follow up an idea thath appeared in [83], which we have seen in texture synthesis:

the whole histogram of filter responses is needed for good texture segmentation. In their case, they take histograms of Gabor filter responses to local patches of the image and they compare these responses from two patches by using the χ^2-statistic.[6] Explicitly, if F_α are the filters, w_α is a function-weighting distance (decreasing from 1 to 0), and χ_{B_k} are the indicator functions for the bins for its range, they measure the difference between the texture statistics at two points x, y by

$$f_{\alpha,k}(x) = \frac{\sum_y w_\alpha(\|x-y\|)\chi_{B_k}(F_\alpha * I(y))}{\sum_y w_\alpha(\|x-y\|)}$$

$$D(x,y) = \sum_{\alpha,k} \frac{(f_{\alpha,k}(x) - f_{\alpha,k}(y))^2}{f_{\alpha,k}(x) + f_{\alpha,k}(y)}.$$

The pixels, as always, lie in a grid, but they introduce larger neighborhood relations by adding edges joining all pairs of vertices within some distance d. In some cases, they also follow an idea from [83] of adding a sparse set of long-distance edges to the graph. Let $x \sim y$ mean that x and y are connected by an edge. Finally, they introduce the energy to be minimized as

$$E(\rho) = \sum_x \underset{\{y|\rho(y)=\rho(x), y \sim x\}}{\text{Average}} (D(x,y)).$$

This energy is clearly a measure of texture inhomogeneity within the regions defined by $R_\ell = \{x|\rho(x) = \ell\}$. Unlike Zhu and Yuille's approach, *no explicit model of the statistics of each texture is made*. Instead distribution-free statistics are used to measure the variation of texture from patch to patch. Thus, it is impossible to synthesize images from this approach—it is a pure bottom-up approach to finding segments without a corresponding generative top-down component.

Shi and Malik's method. The problem of image segmentation for two regions can be seen, in the case of a discrete image made of pixels, as an example of the problem of graph partitioning. Let $G = (V, E)$ be an undirected graph. We assume that to each edge between two vertices i and j is assigned a weight denoted by $w(i, j)$. These weights measure the strength of the bond between two vertices: the larger they are, the harder it is to separate the two vertices. For each partition of the graph in two sets R and S such that $R \cap S = \emptyset$ and $R \cup S = V$, one can define the cost of this partition by

$$\text{cut}(R, S) = \sum_{i \in R, j \in S, (i,j) \in E} w(i,j).$$

---

[6]There are two standard statistics that are used to compare two distributions: the χ^2 statistic used here and the Kolmogorov-Smirnov statistic based on the maximum distance between the cumulative distribution functions. See, for example, [21].

4.7. Texture Segmentation

The number *cut* measures how hard it is to break the graph at this point. The optimal bipartition of the graph is the one that has the minimum cut value. This is a very classic problem in graph theory, and there exist very efficient algorithms to solve this problem. (See Exercise 2 at the end of this chapter on the Greig-Porteous-Seheult min-cut approach [91] for finding the MAP assignment in the Ising model). Looking simply for the minimum cost cut is like taking the $u + v$ energies and concentrating entirely on the discontinuity, ignoring what goes on inside the regions and, in this regard, it is the opposite of the method of Buhmann. A version of this with texture measures as the weights was also introduced in [83].

But such a formulation has a bias: it favors cutting small isolated sets of vertices in the graph. Shi and Malik [196] then proposed to change the cut criterion into a "normalized cut" criterion defined by

$$\text{Ncut}(R, S) = \frac{\text{cut}(R, S)}{\text{assoc}(R)} + \frac{\text{cut}(R, S)}{\text{assoc}(S)},$$

or for M-regions $\{R_i\}$, by

$$\text{Ncut}(\{R_i\}) = \sum_{i=1}^{M} \frac{\text{cut}(R_i, V - R_i)}{\text{assoc}(R_i)},$$

where $\text{assoc}(R) = \sum_{i \in R, j \in V, (i,j) \in E} w(i,j)$ is the total weight of connection from vertices in R to all the other vertices of the graph. Note that we may decompose assoc:

$$\text{assoc}(R) = \text{cut}(R, V - R) + \text{bind}(R),$$

where $\text{bind}(R) = \sum_{i,j \in R, (i,j) \in E} w(i,j)$, with bind giving the internal binding of R to itself. Now, if $\text{cut}(R, S) \ll \min(\text{bind}(R), \text{bind}(S))$ (as we expect), then we can rewrite $\log(\text{Ncut})$ in the two-region case as

$$\begin{aligned}
\log(\text{Ncut}(R, S)) &= \log(\text{cut}(R, S)) + \log(\text{assoc}(R) + \text{assoc}(S)) \\
&\quad - \log(\text{assoc}(R)) - \log(\text{assoc}(S)) \\
&\approx \log(\text{cut}(R, S)) + \log(\text{assoc}(V)) \\
&\quad - \log(\text{bind}(R)) - \log(\text{bind}(S)),
\end{aligned}$$

which now looks very much like our other energies: it has a term measuring how big the discontinuity between R and S is and two terms measuring the homogeneity of R and S.

Unfortunately, the problem of finding the optimal bipartition according to the normalized cut criterion is no longer computationally tractable and only approximate solutions can be found efficiently. The way Shi and Malik approximated it was by embedding the problem in the real-valued domain and by reformulating it as a generalized eigenvalue problem.

There is a wide choice for the weights, depending on the statistics used for the segmentation: Shi and Malik [196] use only gray-levels and distances between pixels, and Malik et al. [144] perform texture segmentation by using the response to filters, also combined with existence of a boundary between the two nodes. Much further work has been done, for example, [213, 232].

4.8 Exercises

1. Monte Carlo Methods with the Ising Model

Let I be a real-valued image of size 64×64 pixels. This exercise explores the probability model for an image $J(i,j)$ whose values are $\{0,1\}$, given by

$$P_T(J|I) = \frac{1}{Z} e^{-\left(c\sum_\alpha (J(\alpha)-I(\alpha))^2 + \sum_{<\alpha,\beta>}(J(\alpha)-J(\beta))^2\right)/T},$$

where $\alpha = (i,j)$ ranges over pixels $1 \leq i,j \leq 64$, and α, β are adjacent ($\alpha \sim \beta$) if $\beta = \alpha + (1,0), \alpha - (1,0), \alpha + (0,1)$ or $\alpha - (0,1)$. This is the Ising Model with an external driving term $I(\alpha) \in \mathbb{R}$. Helpful code for efficiently implementing the Ising model in MATLAB is given below if you wish to use it. The goal is to help you learn to use MATLAB efficiently.

1. The Gibbs sampler section uses two parallel checkerboard-type sweeps to update noninteracting α by computing ΔE, the change in

$$E(J) = c\sum_\alpha (J(\alpha) - I(\alpha))^2 + \sum_{\alpha \sim \beta}(J(\alpha) - J(\beta))^2$$

caused by flipping the α_ith pixel:

$$J(\alpha_i) \to 1 - J(\alpha_i).$$

The flips are accepted with probability $p_i/(p_i+1)$, where $p_i = e^{-(\Delta E)_i/T}$. Explain why this is used with reference to the definition of the Gibbs sampler. Check on paper that the code does this. Why are checkerboards used?

2. The indexing section of the code has as its purpose quick location of adjacent pixels. Thus, ind is a 4×64^2 matrix whose columns correspond to the pixels $\alpha_1, \ldots, \alpha_{64^2}$ suitably ordered, and the ith column contains the indices of $j_1, j_2, j_3,$ and j_4 of the four pixels α_j adjacent to α_i. What have we done at the borders and why?

3. The file nancy.mat is on the book's website. It contains a good 64×64 image I to use for the Ising exercise because it has strong black and white areas. If this is not accessible, choose any image, but reduce it to this small size to make the Gibbs sampler reasonably fast. Check your image with imagesc. To use any image in the Ising model, you must renormalize I to $\lambda I + \mu$ so that a reasonable sample of values of $I(\alpha)$ are greater than 0.5 and another reasonable sample of values are less

4.8. Exercises

than 0.5. For the first part of this exercise, it is easiest to normalize I so that the maximum of the renormalized image is +1 and the minimum is 0.

Make a first set of Gibbs sampler runs to give you a feel of how it works at different temperatures: run the algorithm, starting with $J(\alpha)$ = random 0, 1s, for varying temperatures and report your results. In particular, locate one range of temperatures at which J responds very little to I, another where J quickly freezes under I's control, and a middle range where J responds to I but does not stabilize. Keeping track of E, find out how many iterations of Gibbs are needed before E seems to stabilize. Make three plots of E as a function of iteration number for a high, medium, and low value of T, and compute the variance at each temperature. Finally, run the algorithm at $T = 0$ (i.e., accepting those flips that lower energy and no others) several times with enough iterations for it to stabilize, and examine whether the result depends on the random seed or not. Document your results.

4. Next, run simulated annealing with exponential cooling; that is. set $T = T_0 \sigma^k$ at the kth iteration of Gibbs, where $\sigma < 1$, T_0 is an initial temperature in the high range, and $T_0 \sigma^{-\text{MAXSA}}$ is the final positive temperature in the low range, followed by a series of MAXT0 iterations at $T = 0$. Experiment with different $T_0, \sigma, \text{MAXSA}, \text{MAXT0}$ until you feel you can make a good guess at what the *energy of the ground state* is. We recommend that you document your experiments, making a plot of energy as a function of temperature, which you can think about in your leisure. Print out your presumed ground state and its energy.

5. Finally, try modifying the threshold by replacing I by $I + d$ to get *different ground states*.

Indexing code

This code creates a matrix of neighborhood indices, ind, on the grid of pixels of J. Such arrays of reference indices are one of the nit-picking aspects of MATLAB that cannot be avoided! Note that, in the Gibbs code, we convert both images I and J into one long column so as to index its entries by a single integer: I(1), ..., I(Nv) runs through all entries of I. In this form, it is not clear which are the four nearest neighbors of a pixel. To get these indices, we use ind(:, k), computed below. Figure out how the following code below works, for example, by running it on a small test matrix and examining everything it does. Note that the matrices even and odd are also needed in the Gibbs sampler code below.

```
N = size(I,1); M = size(I,2); Nv = N*M;
[x y] = meshgrid(1:M,1:N);
even = find(rem(x+y,2) == 0); Ne = size(even,1);
odd = find(rem(x+y,2) == 1); No = size(odd,1);
% size 4xNv, x-coord of neighbors
indx = ones(4,1)*x(:)'+[-1 1 0 0]'*ones(1,Nv);
% size 4xNv, y-coord of neighbors
indy = ones(4,1)*y(:)'+[0 0 -1 1]'*ones(1,Nv);
% neighbor coord, between 1, Nv
ind = [(indx-1)*N + indy];
```

```
boundary = [find(indx>M); find(indx<1);
   find(indy>N); find(indy<1)];
% neighbors which are outside grid
ind(boundary) = (Nv+1)*ones(size(boundary));
```

Gibbs sampler code

Note that the code proposes flipping each pixel once for each j.

```
% First initialize J as a random row the size of I(:)
% with values +/-1
% Add a zero at the end so the dummy index Nv+1 doesn't
% give errors
for j = 1:Nsweeps
    adjacent = sum(J(ind(:,odd)))';
    dE = J(odd).*(adjacent+c*I(odd));
    accept = odd([1./(1+exp(dE/T))] > rand(No,1));
    J(accept) = 1-J(accept);
    adjacent = sum(J(ind(:,even)))';
    dE = J(even).*(adjacent+c*I(even));
    accept = even([1./(1+exp(dE/T))] > rand(Ne,1));
    J(accept) = 1-J(accept);
end
```

Remark. You will need to call the indexing code once after you have loaded I (nancy.mat). The Gibbs code should be used as the inner loop in your cooling schedule.

2. Maxflow, the Greig-Porteous-Seheult Algorithm, and Generalizations

One of the main problems for the Ising model is to find the configuration $J(\alpha)$ that has the minimum energy, or equivalently, the maximum probability. Greig, Porteous, and Seheult [91] have shown that the problem of computing this MAP can be reformulated as a minimum cut problem in a directed graph with capacities, and then, due to the classic Ford-Fulkerson theorem/algorithm (see Section 3.1 of [26]), it is equivalent to a maximum flow problem, a special type of linear programming problem. There are special purpose fast algorithms for solving the max-flow problem (e.g., Edmonds-Karp, found in [50]). This enables you to find $J(\alpha)$ quickly and exactly. The aim of this exercise will be to give the details of the result of the Greig-Porteous-Seheult reduction.

We first recall the Ford-Fulkerson theorem: let $G = (V, E)$ be a *directed* graph and let $c(\vec{e}) \geq 0$ be a capacity attached to each edge. A start or source vertex $s \in V$ with all its edges directed outward and an end or sink vertex $t \in V$ with all its edges directed inward are given. A flow f on the graph assigns to each edge a number $f(\vec{e}) \geq 0$ such that for all vertices $x \in V - \{s,t\}$, the total flow over edges ending at x equals the total flow going out:

$$\sum_{y, \vec{yx} \in E} f(\vec{yx}) = \sum_{y, \vec{xy} \in E} f(\vec{xy}).$$

4.8. Exercises

For such a flow, it is easy to check that the total flow out of s equals the total flow into t. The maxflow with given capacities is the max of this flow when $f(\vec{e}) \leq c(\vec{e})$ for all edges $\vec{e}$. In contrast, consider all partitions of the vertices into two sets $V = S \cup T$, with $S \cap T = \emptyset, s \in S, t \in T$. Then the cut defined by this partition is

$$\mathrm{cut}(S,T) = \sum_{\vec{xy} \in E, x \in S, y \in T} c(\vec{xy}).$$

The theorem is that the maxflow equals the capacity of the mincut and, at the maxflow, the flow reaches capacity at all edges in the mincut. Note that maxflow is really a linear programming (LP) problem: flows are points in a convex subset of $\mathbb{R}^{|E|}$, and the maxflow is the point where a linear map on this set attains its maximum. Mincut, in contrast, seems like a difficult combinatorial problem, so this equivalence is most useful.

1. Here is the graph of Greig, Porteous, and Seheult for the Ising problem: Let V be the union of the lattice points α of the Ising model plus the start and ending states s, t. Let E consist in (i) directed edges from s to every lattice point α; (ii) directed edges from every α to t; and (iii) a pair of directed edges, one going each way, between adjacent lattice points $\alpha \sim \beta$. Put capacity $(1 - I(\alpha))^2$ on the edge $\vec{s\alpha}$, $I(\alpha)^2$ on the edge $\vec{\alpha t}$, and capacity 1 on all edges $\vec{\alpha\beta}$. Show that in this case the mincut has value equal to the minimum of the Ising energy.

2. We can simplify a bit by taking away the direct flow $s \to \alpha \to t$. Let $\lambda_\alpha = 1 - 2I(\alpha)$, and if $\lambda_\alpha > 0$, put capacity λ_α on $\vec{s\alpha}$, 0 on $\vec{\alpha t}$; if $\lambda_\alpha < 0$, put capacity 0 on $\vec{s\alpha}$, capacity $-\lambda_\alpha$ on $\vec{\alpha t}$. Show that this graph has the same mincut and its value is

$$\min_J E - \sum_\alpha \min(I(\alpha)^2, (1 - I(\alpha))^2).$$

3. MATLAB has a handy built-in function `linprog`, which solves linear programming problems. Take a simple black-and-white image, not too large (e.g., 64×64) and normalize its min and max to be 0 and 1. Using `linprog` and the above graph, solve for the mode of the Ising model with this image as external field I. This method was used to compute the mode in Figure 4.4. You can also download the original cow/tree image from our website.

The original algorithm of Greig-Porteous-Seheult applies the Ford-Fulkerson mincut-maxflow theorem to finding the minimum energy state for the Ising model. But, as described in Section 4.4.2, Boykov, Veksler, and Zabih [30] extended it to very good approximations for multiple-label problems. One of their results is the following: suppose $G = (V, E)$ is any graph, Λ is a set of labels, and $u : V \to \Lambda$ is a labeling. Suppose we have an energy attached to each labeling of the form,

$$E(u) = \sum_{v \in V} f_v(u(x)) + \sum_{e=<v,w> \in E} g_e(u(v), u(w)),$$

where g_e is a *metric* on the set Λ. Take any $\alpha \in \Lambda$. Then an α-expansion of u is a new labeling $\tilde{u}$ such that

$$\tilde{u}(v) \neq u(v) \implies \tilde{u}(v) = \alpha.$$

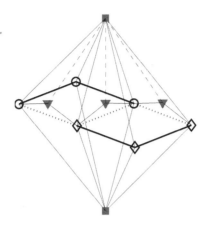

Figure 4.24. The graph used by Boykov, Veksler, and Zabih in their algorithm. Here we have a 2×3 image with two labels: three pixels marked with circles have one label and the other three marked with diamond shapes have another. A source and sink node are added (marked with squares). The sink is at the top for better visibility. The edges between distinct labels are erased (shown by dotted lines) and are replaced by blue triangle-shaped vertices connected as shown.

That is, an α-expansion simply increases the set of vertices with label α. The main result is a way to find the α-expansion that decreases E the most by using a maxflow graph cut. Then one can cycle through all labels and use this to decrease E step by step until no α-expansion can improve the labeling.

4. Their construction is a carefully crafted graph, illustrated in Figure 4.24. Take the Greig-Porteous-Seheult graph but, for every edge e connecting vertices with distinct labels in u, delete the edge e, add to the graph one new vertex v_e, and connect this vertex to the two former ends of e and to the sink. Then put weights (i.e., capacities) on the edges as follows:

   a. weight $f_v(\alpha)$ on edge from source to $v \in V$,

   b. weight $f_v(u(v))$ on edge from $v \in V$ to sink *except* weight ∞ if $u(v) = \alpha$,

   c. weight $g_e(u(v), \alpha)$ on edge $e =< v, w >$ if $u(v) = u(w)$,

   d. weight $g_e(u(v), \alpha)$ on edge from v to new vertex v_e,

   e. weight $g_e(u(v), u(w))$ on edge from v_e to sink if $e =< v, w >, u(v) \neq u(w)$.

   Prove that the mincut of this graph is the minimum of the energy E over all α-expansions $\tilde{u}$ of u.

4.8. Exercises

3. Propp and Wilson's Algorithm for Exactly Sampling the Ising Model

Section 4.3.1 showed that the Metropolis algorithm was one way to obtain "approximate" samples from the Ising probability distribution P. In [182], Propp and Wilson have proposed a method for exactly sampling from P. Their method called, "coupling from the past," is a protocol that runs two Gibbs samplers starting in the past that are coupled in the sense that the same random numbers are used to drive each. The main point is that, if one starts the chains from a time far enough in the past, then at time 0 the two chains become equal and the obtained configuration is distributed according to P. Here we provide more details about this.

1. We first define a partial order on the set of configurations: we say that $J_1 \leq J_2$ if for all α, one has $J_1(\alpha) \leq J_2(\alpha)$. Let $\widehat{B}$ and $\widehat{T}$ denote the "bottom" and "top" configurations where all states are 0 and $+1$, respectively. Notice that we have, then, for any configuration J, $\widehat{B} \leq J \leq \widehat{T}$. For a site α and a real number $u \in [0,1]$, we define the Gibbs sampler: $J \mapsto \text{Gibbs}(J, \alpha, u) \in \Omega$ by

$$\text{Gibbs}(J, \alpha, u) = \begin{cases} J_{B_\alpha} & \text{if } u < P(J_{B_\alpha})/(P(J_{B_\alpha}) + P(J_{T_\alpha})), \\ J_{T_\alpha} & \text{if } u \geq P(J_{B_\alpha})/(P(J_{B_\alpha}) + P(J_{T_\alpha})), \end{cases}$$

where J_{B_α} and J_{T_α} are the configurations K such that $K_\alpha = 0$ and $+1$, respectively, and $K_\beta = J_\beta$ for all $\beta \neq \alpha$. Prove that if two configurations J_1 and J_2 are such that $J_1 \leq J_2$, then $\text{Gibbs}(J_1, \alpha, u) \leq \text{Gibbs}(J_2, \alpha, u)$. Prove that, given the external field I, there is always some $\epsilon > 0$ such that if $u < \epsilon$, then the value of $\text{Gibbs}(J, \alpha, u)$ at α is 0 for all J; and if $u > 1 - \epsilon$, then it is always 1.

2. We assume in all the following that $\mathcal{V} = (\mathbf{i}, \mathcal{U})$ is a random variable, where $\mathbf{i}$ is uniformly among all sites, and $\mathcal{U}$ is uniform on $[0, 1]$. Prove that

$$\forall J_2, \quad \sum_{J_1} P(J_1) \, \mathbb{P}(Gibbs(J_1, \mathcal{V}) = J_2) = P(J_2).$$

3. Let $\ldots, \mathcal{V}_{-3}, \mathcal{V}_{-2}, \mathcal{V}_{-1}$ be iid random variables (with distribution as above). For $t < 0$, we define Gibbs_t and F_t by

$$F_t = \text{Gibbs}_{-1} \circ \text{Gibbs}_{-2} \circ \ldots \circ \text{Gibbs}_t,$$

where $\text{Gibbs}_t(J) = \text{Gibbs}(J, \mathcal{V}_t)$. Prove that with probability 1, there exists τ for which F_τ is constant (i.e., $F_\tau(J_1) = F_\tau(J_2)$ for all J_1, J_2); and that for all $t < \tau$, $F_t = F_\tau$. This constant, which is thus defined with probability 1, is a random variable denoted by $\mathcal{Z}$. Prove that its probability distribution is P.

Finally, the algorithm of Propp and Wilson for an exact sampling from P is summarized in the following pseudocode:

$T \longleftarrow 1$
randomly generate v_{-1}
repeat
    upper $\longleftarrow \widehat{T}$
    lower $\longleftarrow \widehat{B}$

```
        for t = -T to -1
            upper ⟵ f(upper, v_t)
            lower ⟵ f(lower, v_t)
        T ⟵ 2T
        randomly generate v_{-2T}, v_{-2T+1}, ..., v_{-T-1}
until upper = lower
return upper
```

4. Sampling the Set of Partitions of a Graph with Swendsen-Wang

One of the main problems in finding the best image segmentations is that we need to explore all possible partitions of the lattice of pixels into disjoint regions. The lattice of pixels is a particular graph, and there is a large literature on finding optimal partitions of graphs into a finite sets of disjoint full subgraphs. This means partitioning the set of vertices V into disjoint subsets V_n, cutting the edges between distinct V_n but leaving all edges in each partition, making it a full subgraph. Remarkably, an idea due to Swendsen and Wang [208] was adapted to this problem by Barbu and Zhu [12], and this has provided a fast and efficient method of sampling all possible partitions. The usual problem is that most Monte Carlo algorithms propose *small* moves from one state to another. In the case of partitions, we need *large* moves—moves that create new partitions by splitting an existing one into two or by merging two existing partitions into a larger one. It is not easy to combine split-and-merge ideas with Monte Carlo algorithms constrained to have a fixed equilibrium distribution, but it can be done, as we see here.

Let $G = (V, E)$ be any finite graph and let $\{q_e\}$ be any set of probabilities (just numbers in $[0, 1]$), one for each edge. Consider the set Ω of all possible (unordered) partitions $\pi = \{V_n\}$ of V into disjoint subsets (i.e., $V = \coprod V_n$). Suppose $P(\pi)$ is *any* probability distribution on Ω. Then consider the following Markov chain $q_{\pi \to \pi'}$ on Ω:

1. Starting with a partition $\pi = \{V_n\}$, "turn off" all edges $e \in E$ with probability q_e.

2. Using only the remaining edges, find all the connected components $\{V_{n,i}\}$ of each V_n.

3. Choose with a uniform probability *one* of these connected components $W = V_{n,i}$. Let it be part of V_n.

4. Consider the new partitions obtained by

    a. merging W with one of the other partitions $V_m, m \neq n$:
    $$\pi_m = \{W \cup V_m, V_n - W, \{V_k\}|_{k \neq n, m}\}$$

    b. making a new partition splitting V_n into W and $V_n - W$:
    $$\pi_0 = \{W, V_n - W, \{V_k\}|_{k \neq n}\}$$

    c. merging W back into $V_n - W$, getting back π.

    If $W = V_n$, then options (b) and (c) are the same and are not considered separately.

4.8. Exercises

5. Calculate the ratios of the probabilities $P(\pi), P(\pi_m), P(\pi_0)$.

6. Move to one of these new partitions with relative probabilities:

   probability of move to π_m $\quad \propto \quad P(\pi_m) \cdot \prod_{\{\text{edges } e \text{ between } W, V_m\}} q_e,$
   probability of move to π_0 $\quad \propto \quad P(\pi_0) \cdot \prod_{\{\text{edges } e \text{ between } W, V_n - W\}} q_e,$
   probability of move to π $\quad \propto \quad P(\pi).$

Note that we can have $W = V_n$; then option (a), merging, decreases the number of partitions.

Prove that this algorithm satisfies detailed balance:

$$q_{\pi \to \pi'} P(\pi) = q_{\pi'' \to \pi} P(\pi');$$

hence, the stationary distribution of this Markov chain is P.

In the applications to vision, we take G to be the lattice of pixels with nearest-neighbor edges. Consider the MS model for locally constant approximations of an image I on G:

$$E(\pi, c) = \sum_{V \in \pi} \sum_{x \in V} (I(x) - c(V))^2 + \lambda \sum_{V, W \in \pi, V \neq W} |E_{V,W}|$$

$$P_T(\pi, c) \propto e^{-E(\pi,c)/T}, \quad E_{V,W} = \text{edges between } V, W.$$

This gives us a probability model on partitions π either by substituting the most likely value for c or by taking the marginal on π. In the first case, we set $c(R) = \text{mean}(I|_R)$, giving:

$$P_T(\pi) \propto \prod_{R \in \pi} e^{-(|R| \cdot \text{variance}(I|_R))/T} \cdot \prod_{V, W \in \pi, V \neq W} e^{-\lambda \cdot |E_{V,W}|/T}.$$

If we let the set of possible values of c be $\mathbb{R}$, then the marginal works out to be:

$$P_T(\pi) \propto \prod_{R \in \pi} \frac{e^{-(|R| \cdot \text{variance}(I|_R))/T}}{\sqrt{|R|}} \cdot \prod_{V, W \in \pi, V \neq W} e^{-\lambda \cdot |E_{V,W}|/T}.$$

Prove this.

The big advantage of this in the image application is the possibility of using importance sampling: we can take the q_e to reflect our guess as to whether e is an edge between partitions, making it larger for edges that look likely to separate partitions, and we can simply take $q_e = e^{-a|I(x) - I(y)|}$ for a suitable constant a. Or, being a bit more sophisticated, take neighborhoods of x and y, *displaced away from each other a small amount*, compute means of I on each, and divide this by the square root of the variance of I on their union. Let $t(e)$ be the result: this is a sort of t-test for whether both want to be in the same partition. Then $q_e = 1 - e^{-at(e)}$, for a suitable constant a, is perhaps a better choice (compare the method in [12] that uses the KL-distance of histogrammed filter values).

This gives algorithms for segmenting images that are very suited for experimentation. MATLAB has a subroutine `graphconncomp` in the `bioinformatics` toolbox to work out connected components, and the energy above is readily calculated if we keep track of the sums $\overline{I_n} = \sum_{x \in V_n} I(x)$ and $\overline{I_n^2} = \sum_{x \in V_n} I(x)^2$ on each partition because

$$\sum_{x \in V_n} (I - \mathrm{mean}_{V_n}(I))^2 = \overline{I_n^2} - (\overline{I_n})^2/|V_n|.$$

The project is to implement the Swendsen-Wang-Barbu-Zhu algorithm and see (a) how fast it is and (b) by combining it with annealing, how well it finds the true minimum of E. Although the true minimum may be difficult to determine, we can run it many times with various parameters and compare the resulting minima. Also, the α-expansion algorithm described in Section 4.4.2 can be used to see which finds lower values of E.

5. Approximating the Length of a Curve

In Section 4.2, about the $(u + v)$-decomposition of an image, we saw that one often needs to get a good discretized version of the length term. A typical example of this was the MS model discretized on a grid. For this, we approximated the length of a curve by the area of a "tube" centered on the curve divided by the width of this tube. The aim of this exercise is to try to get an idea of how good this approximation is. Let us assume that in the continuum setting, the region R is a connected closed subset of the square $[0,1]^2$ and that its boundary is described by a C^2 curve γ.

1. Let n be an integer. We assume that the square $[0,1]^2$ is discretized into a regular $n \times n$ grid, and the set of vertices, denoted by V_n, of this lattice is the set of points with coordinates $(k/n, l/n)$ with $0 \leq k, l \leq n-1$. For $\alpha \in V_n$, and for $r > 0$ real, we also denote by $N_r(\alpha)$ the set of vertices $\beta \in V_n$, which are at a Euclidean distance less than r from α. Let

    $$A_n(r) = \#\{\alpha | N_r(\alpha) \not\subset R \text{ and } N_r(\alpha) \not\subset R^c\}/n^2.$$

    Prove that as n goes to infinity then

    $$A_n(r) \xrightarrow[n \to +\infty]{} A(r) = \mathrm{Area}(\{x | d(x, \gamma) \leq r\}).$$

2. We assume that $s \mapsto \gamma(s)$ is a parameterization of γ by arc length, and that $r < 1/\kappa_{\max}$, where $\kappa_{\max}$ is the maximal curvature of γ, that is, $\kappa_{\max} = \max_s |\gamma''(s)|$. Let $S_\gamma(r)$ be the set of points defined by

    $$S_\gamma(r) = \{\gamma(s) + t\gamma'^\perp(s) | s \in [0, \mathrm{len}(\gamma)] \text{ and } t \in [-r, r]\}.$$

    Prove that

    $$S_\gamma(r) = \{x | d(x, \gamma) \leq r\} \text{ and that } \mathrm{Area}(S_\gamma(r)) = A(r) = 2r \cdot \mathrm{len}(\gamma).$$

3. Let us consider here bias at corner points—what happens when γ has two straight branches meeting in a "corner" with an angle $\theta \in (0, \pi)$. We moreover assume that the two straight branches have a length $\geq r \tan(\frac{\pi}{2} - \frac{\theta}{2})$. Prove that

    $$A(r) = 2r \cdot \mathrm{len}(\gamma) + r^2 \left(\frac{\pi}{2} - \frac{\theta}{2} - \tan(\frac{\pi}{2} - \frac{\theta}{2}) \right),$$

    which shows that the method of approximating the length by the area of a tube around the curve underestimates the length at corner points.

4.8. Exercises

Exercise 6. Segmentation via the Blake-Zisserman and Geman-Yang Energies

Recall the energy of the MS model on a discrete lattice. Let $\mathbb{E}$ denote the set of edges $\{(\alpha,\beta)|\alpha \sim \beta\}$ in the lattice and $K \subset \mathbb{E}$ the boundary edges. Then

$$E(u,K) = \sum_\alpha (u(\alpha) - I(\alpha))^2 + \lambda \cdot \#(K) + \mu \sum_{(\alpha,\beta) \in \mathbb{E}-K} |u(\alpha) - u(\beta)|^2.$$

Define an auxiliary function

$$\Phi(x) = \min(x^2, \lambda/\mu).$$

Prove the result on which much of the analysis in Blake and Zisserman's book [24] is based:

$$\min_K E(u,K) = \sum_\alpha (u(\alpha) - I(\alpha))^2 + \mu \sum_{\alpha \sim \beta} \Phi(u(\alpha) - u(\beta)).$$

Functions such as Φ also arise in the context of robust statistics. Suppose we have some data that are corrupted by a few wild values. Then it makes sense to discard some outlying values because they are likely to include the wild ones. Instead of minimizing $\sum_i (a_i - \bar{a})^2$ to find the true center of the data $\{a_i\}$, we minimize $\sum_i \Phi(a_i - \bar{a})$. Show that the result will be a number $\bar{a}$, which is the mean of all the data within a distance $\sqrt{\lambda/\mu}$ of $\bar{a}$.

Yet another rearrangement, due to D. Geman and C. Yang [82], is very effective for calculations. Define another auxiliary function

$$\Psi(x) = \begin{cases} 2\sqrt{\frac{\lambda}{\mu}}|x| - x^2 & \text{if } |x| \leq \sqrt{\frac{\lambda}{\mu}}, \\ \frac{\lambda}{\mu} & \text{if } |x| \geq \sqrt{\frac{\lambda}{\mu}}. \end{cases}$$

Let ℓ denote a real-valued function on $\mathbb{E}$. Graph these two auxiliary functions. Then prove that

$$\min_K E(u,K) =$$

$$\min_\ell \left[\sum_\alpha (u(\alpha) - I(\alpha))^2 + \mu \sum_{\alpha \sim \beta} \left((u(\alpha) - u(\beta) - \ell(\alpha,\beta))^2 + \Psi(\ell(\alpha,\beta)) \right) \right].$$

(The point is that $x^2 - \Phi(x)$ and $x^2 + \Psi(x)$ are Legendre transformations of each other.) The function ℓ is similar to the line process introduced in Section 4.2.3. The main reason why this is useful is that it is now quadratic in u for fixed ℓ and convex in ℓ for fixed u; show this. It is relatively easy to minimize, and we can always find the local minima of E by alternately minimizing the Geman-Yang form in u and ℓ.

7. Fourier Transforms of Images

Let $I = I(x,y)_{1 \leq x,y \leq N}$ be a discrete gray-level image made of $N \times N$ pixels. The Fourier transform of I is defined by

$$\forall 1 \leq k,l \leq N, \quad \widehat{I}(k,l) = \sum_{x=1}^N \sum_{y=1}^N I(x,y) e^{-2i\pi(kx+ly)/N}.$$

244   4. Image Texture, Segmentation and Gibbs Models

This transform can be computed in MATLAB using the function fft2; its inverse transform is the function ifft2.

1. *Visualize the Fourier transform.* Load an image I in MATLAB. Take one of your own or one from the list of images that come with MATLAB (check them via the command imageext). Use the function fft2 to compute the Fourier transform of this image. We then obtain a *complex valued* image $\widehat{I}$, which can be written as

$$\forall (k,l), \quad \widehat{I}(k,l) = R(k,l)e^{i\Phi(k,l)}.$$

Visualize separately the image of the phase Φ and the one of the amplitude R. A better visualization usually results from plotting $\log(R+1)$ instead of R and also using the command fftshift, which puts the zero frequency in the middle of the plot. Check that the inverse Fourier transform (function ifft2) allows us to recover I from R and Φ.

```
load clown;     %the image is now in the variable X
subplot(2,2,1);
imagesc(X); colormap(gray); axis image
FX = fft2(X); R = abs(FX); Phi = angle(FX);
subplot(2,2,3);
imagesc(log(fftshift(R)+1));
colormap(gray); axis image
subplot(2,2,4);
imagesc(fftshift(Phi)); colormap(gray); axis image
X2 = real(ifft2(R.*exp(i*Phi)));
subplot(2,2,2);
imagesc(X2); colormap(gray); axis image
```

2. *Procedure for avoiding border artifacts.* You may have noticed vertical and horizontal lines through the center in the Fourier transform plots, which come from the fact that the finite Fourier transform assumes that the image wraps around at its edges, top with bottom, left with right. This causes discontinuities on the "seams," and hence large power for pure horizontal and pure vertical frequencies. The best way to get a picture of the frequency makeup of the image itself is to gray out the borders. If X is your image, choose the size of border to have n1 gray pixels (e.g., n1 = 2) and then n2 blending pixels between the gray and the true image (e.g., n2 = 8). Then you can put this frame on your image by using code such as:

```
[Ncol, Nrow] = size(X);
edge = [zeros(1,n1), sin((1:n2)*pi/(2*n2)).^2];
rowwin = [edge; ones(1,Nrow-2*(n1+n2));
    fliplr(edge)];
colwin = [edge'; ones(Ncol-2*(n1+n2),1);
    flipud(edge')];
Xbar = mean(X(:));
Xwin = Xbar + (X-Xbar).*(colwin*rowwin);
```

Redo the tests in part 1 after this procedure and note the difference.

4.8. Exercises

3. *Does the power or the phase help more for recognition?* Take two images I_1 and I_2 of same size and compute their respective Fourier transforms to get two amplitude images R_1 and R_2, and two phase images Φ_1 and Φ_2. Then, apply the inverse Fourier transform to $R_1 \exp(i\Phi_2)$ and to $R_2 \exp(i\Phi_1)$. Visualize the two obtained images. What happens? This shows that for a "natural" image, the most important part of its geometric information is contained in the phase of its Fourier transform. Use this code:

```
load clown; X1=X;
load mandrill;  X2=X(1:200,1:320);
subplot(2,2,1);
imagesc(X1); colormap(gray); axis image
subplot(2,2,2);
imagesc(X2); colormap(gray); axis image
FX1=fft2(X1);   R1=abs(FX1);   Phi1=angle(FX1);
FX2=fft2(X2);   R2=abs(FX2);   Phi2=angle(FX2);
X3 = real(ifft2(R1.*exp(i*Phi2)));
X4 = real(ifft2(R2.*exp(i*Phi1)));
subplot(2,2,3);
imagesc(X3); colormap(gray); axis image
subplot(2,2,4);
imagesc(X4); colormap(gray); axis image
```

4. *Texture synthesis from the power spectrum.* Take a very simple image (for instance a black square on a white background), compute its Fourier transform, and *replace the phase by random values, uniform on $[0, 2\pi]$*. Use the inverse Fourier transform and visualize the texture image thus obtained. Try this with other shapes, such as a disk or stripes. Some aspects of texture are well preserved; others are not. Explain this. Also, what happens if, in the code below, you omit real in the last line? What shortcut are we taking by randomizing all phases?

```
% the image is completely white
X=ones(256,256);
% this creates a black square
X(110:140,110:140)=0;
% OR make a black disk or stripes like this:
% x = [-127:128]; r = 20;
% disk
% X = (ones(256,1)*x.^2 + x'.^2*ones(1,256) > r^2);
% stripes
% X = 1+sign(sin(0.2*(ones(256,1)*x
     + x'*ones(1,256)))));
[n,m]=size(X);
FX=fft2(X); R=abs(FX); Phi_alea=2*pi*rand(n,m);
X2=ifft2(R.*exp(i*Phi_alea));
subplot(2,1,1); imagesc(X); colormap(gray);
    axis image
subplot(2,1,2); imagesc(real(X2)); colormap(gray);
    axis image
```

246   4. Image Texture, Segmentation and Gibbs Models

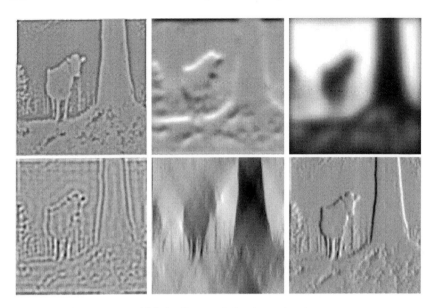

Figure 4.25. Six filters applied to the cow/tree image.

8. Identify the Filter!

This is an easy game. In Figure 4.25, the image of a cow next to a tree used earlier in Figure 4.4 has been convolved with six filters (or, equivalently, its Fourier transform has been multiplied by some function). The six filters are:

1. smoothing by a Gaussian;
2. possible smoothing followed by discrete x-derivative (i.e., horizontal first difference);
3. possible smoothing followed by discrete y-derivative;
4. possible smoothing followed by discrete Laplacian;
5. band-pass filtering (i.e., taking spatial frequencies (ξ, η) with $a \leq ||(\xi, \eta)|| \leq b$);
6. sum of Fourier terms with frequencies $|\eta| \leq 0.5|\xi|$.

Which filters were used in which images?

9. Discriminating Textures

Figure 4.26 shows images of six textures. We see plainly the differences between them. The problem is to devise a MATLAB program to distinguish them and (this is what makes it tough) *if you are given an arbitrary* 32×32 *patch from any of them, to identify from which of the large images the patch came.* This actually can be done with very small error rate, but it is not easy. (This problem has been used as a contest in several of our classes.)

There are many ways of approaching this. For instance, you may use in any order any of the following operations:

4.8. Exercises

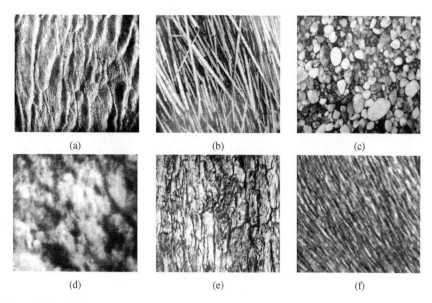

Figure 4.26. Six images of textures: (a) water (rotated to resemble (b) and (f)), (b) straw, (c) pebbles, (d) clouds, (e) bark, and (f) fur. (From [35].)

1. linear operations on I, meaning convolution with any kernel you want, such as Gaussian smoothing, Laplacian ($I(\alpha) - 0.25 \sum_{\beta \sim \alpha} I(\beta)$), difference of Gaussians with two different variances, or gradient along x and y. In general, the idea is to find a "matched filter," an auxiliary image of small support that mimics a structure found in one of the textures: and then convolution with this so-called filter will take some quite large values for this image compared to the others;

2. nonlinear operations, such as taking the square, the absolute value of some derived image, or pixel-wise maxs, mins (or more complex operations) of two images.

3. Fourier transform to obtain the power spectrum or the phase; multiplying the power spectrum with a predetermined mask to select part of the power spectrum;

4. computing means, medians, variances, or higher moments; quartiles of various computed auxiliary images.

To decide on your final criteria, the following strategy may be useful: having chosen one, two, or three features that seem useful to discriminate the textures, run them on your own random selection of 32×32 subimages and make a scatterplot in one-, two-, or three dimensions of the clusters of feature values. You will want to use distinctive colors or marks for the values arising from different textures, and in three dimensions, you will want to rotate the display on the screen. Seek linear or more complex combinations of the feature values that separate the clusters. If you have taken a statistics course, you may know techniques such as Fisher's optimal linear discriminant; tree-based classifiers such as "CART" (classification and regression trees) [32]; kth nearest neighbors, finding clusters by k-means or fitting them by multidimensional Gaussians, followed by maximum-likelihood tests. Any of these can be used.

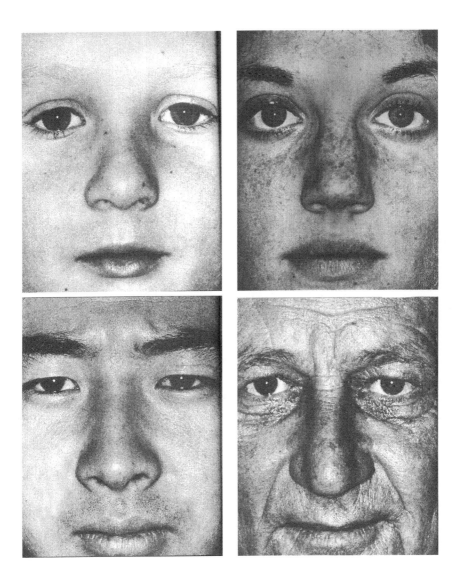

Figure 5.1. Photos of three faces taken at 57th St. and 7th Ave. in New York City. Here the illumination, viewing angle, and expression have been more or less fixed; hair, beards, ears, and glasses are absent so only individual central face appearance varies. An extraordinary variation remains. The faces are part of a "phonebook" of faces shot by Ken Ohara, originally published with the title *One* in 1970 and reprinted by Taschen GmbH in 1997 and reproduced with Ohara's permission. The book is recommended for anyone studying face recognition. (From [169].)

– 5 –
Faces and Flexible Templates

FACES ARE a very interesting application of pattern theory ideas. The difficulty in modeling faces is the extraordinary number of variables affecting their appearance: (a) changes in illumination; (b) changes in viewing angle; (c) varying expressions, (d) varying appearance between individuals; and (e) partial occlusion by hair, glasses, or other factors. The variations (a), (b), (c), and (e) are eliminated in Figure 5.1, so you can see how great the remaining factor, (d), is.

The goal of this chapter will be to seek a stochastic model of faces that will incorporate at least some of these factors. Our model is very crude, but it illustrates the challenges of this task. We model a face by a two-dimensional signal $I(x, y)$. The first point is that changing illumination causes huge changes in the actual signal $I(x, y)$. This is what we have been calling a value pattern and, measured by mean square pixel differences, varying illumination dwarfs all other sources of variation. Fortunately, it is also relatively easy to model, as we see in Section 5.1. To illustrate the degree to which lighting interferes with the recognition of shape, Figure 5.2 shows an example of three different images displayed by their *graphs* rather than as black/white images. Two of them are faces and one is not a face. The image in Figure 5.2(a) is obviously a face, but which of the others is a face? Note that Figures 5.2(b) and (c) have similar structure and are certainly closer as functions. Nonetheless, one is a face and *perceptually*, when presented as an image, is much closer to image 5.2(a). Which is which? Figure 5.3 shows the same

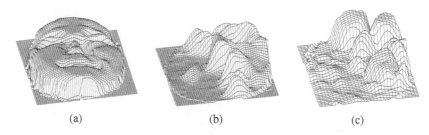

Figure 5.2. Three black-and-white images, shown as graphs $z = I(x, y)$ rather than as images. Image (a) is clearly a face; but (b) and (c) are deeply shadowed scenes (light to the top right). One is a face and the other is a jumble of pebbles: but which is which? See the text for the answer. (From [99], [97, Figure 1.6].)

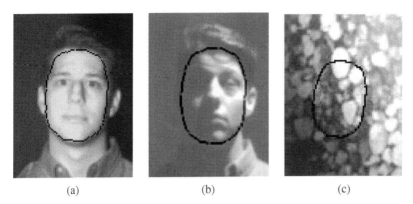

(a)   (b)   (c)

Figure 5.3. The images from Figure 5.2; who would now group (b) and (c)? (From [97, Figure 1.7].)

three images in the usual way and there is no doubt which is the face, and which is a jumble of pebbles. Exercise 1, at the end of the chapter, has been included just to show how difficult it can be to recognize very familiar scenes from the graphs of their images.

However, the good news is that, although lighting variations produce huge changes in the image, they are *additive*. If you have two ways of illuminating a scene, the image produced by both lights at once is the sum of the images produced by each separately. We see that this leads to an additive decomposition of the face image into "eigenfaces" which —in the example in Figure 5.3—easily separates true faces from pebbles (see Section 5.1.1).

Looking at the remaining sources of variations, we may group them into three types. (1) Small changes in viewing angle from the frontal, expression changes, and individual face variations can be seen as *warpings* of the image, that is, replacing $I(x,y)$ by $I(\phi_1(x,y), \phi_2(x,y))$ where $(x,y) \mapsto (\phi_1(x,y), \phi_2(x,y))$ is a diffeomorphism of the image domain. Warpings are one of the most common geometric patterns found in all types of signals and, in general, are much more difficult to model than value patterns. (2) Going further, large viewpoint changes, beards, bangs or hair combed forward, and glasses produce occlusions. These cover up part of the face proper and are best modeled by the grammatical models discussed in Chapter 3. (3) Wrinkles are another source of variation. They might be considered three-dimensional warpings of the physical face, but no two-dimensional warping of the smooth skin of the child's face in Figure 5.1 will create the wrinkles in the older man's face. To a large extent, moreover, wrinkles are independent of illumination: a concave part of a shape almost always receives less illumination than the convex parts. We will not attempt to model occlusions or wrinkles in this chapter. Our main focus is to model warpings, which is not simple. An example of a pure warping is shown in Figure 5.4.

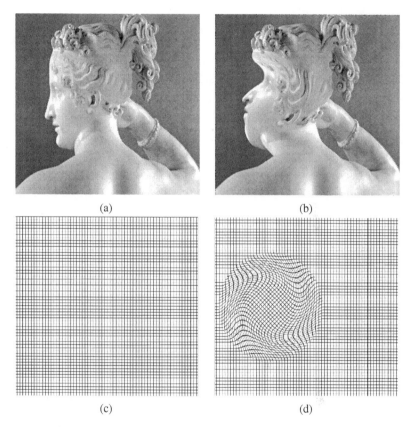

Figure 5.4. (a) An image $I(x, y)$ of the statue of the beautiful Paolina. (b) A warping $I \circ \phi$, producing a startlingly different figure. (c) and (d) The warping shown by its effect on a grid.

We may put lighting variations and warpings together and say that we are modeling a face image I on a domain $\mathcal{D}$ by

$$I(x,y) = \bar{I}(\phi(x,y)) + \sum_\alpha c_\alpha F^{(\alpha)}(\phi(x,y)) + \text{noise},$$

where $\bar{I}$ is a frontally viewed average face on a domain $\mathcal{D}_0$, under uniform ambient lighting and without expression, $F^{(\alpha)}$ are eigenfaces whose sums synthesize all possible lighting conditions, and $\phi : \mathcal{D} \to \mathcal{D}_0$ is the warping that takes the sample face to the average frontally viewed face. The noise term may be considered as allowing for effects such as wrinkles: we might put slight wrinkles in the mean face model, and then the noise term can both eliminate or enhance them.

The challenge of this model is that it introduces hidden variables $\{c_\alpha\}$ for the value pattern operating on the *range* of the image at the same time as hidden variables ϕ for the geometric pattern operating on the *domain*. Solving for both sets of hidden variables involves minimizing an energy of the form

$$E(c_\alpha, \psi) = \int_{\mathcal{D}_0} \left(I(\psi(x,y)) - \bar{I}(x,y) - \sum_\alpha c_\alpha F^{(\alpha)}(x,y) \right)^2 dxdy$$
$$+ E_1(c_\alpha) + E_2(\psi),$$

where E_1 is some penalty of large values of the lighting coefficients and E_2 is a penalty on large distortions of the domain. (Here $\psi = \phi^{-1} : \mathcal{D}_0 \to \mathcal{D}$.)

What we focus on in this chapter is what are good expressions for E_1 and E_2. E_2 is especially difficult, and Sections 5.2–5.5 are devoted to expressions for this term. Some of the more technical parts of this discussion have been included in an appendix, Section 5.8. Whereas the coefficients $\{c_\alpha\}$ lie naturally in a vector space, ψ is a diffeomorphism, an invertible map, and does not naturally belong in a linear space. The variable ψ can be treated as a particular smooth map from $\mathcal{D}_0$ to $\mathbb{R}^2$ but addition in this vector space has no natural meaning for a diffeomorphism ψ. For *small* deformations of the identity (i.e., $\psi \approx \text{Id}$), it is reasonable to treat $f(\vec{x}) = \psi(\vec{x}) - \vec{x}$ as an arbitrary smooth function with small sup-norm, which then lies in a vector space. But for large deformations, the natural operation in the group $\text{Diff}(\mathcal{D}_0)$ is the composition of two maps, not their point-wise sum. This creates the need for new tools to deal with such variables. In particular, data sets in vector spaces have a mean and covariance matrix. Can we define a mean and covariance matrix for data sets $\{\psi_\alpha\} \subset \text{Diff}(\mathcal{D}_0)$? One approach here is to linearize, somehow, all or a large part of the group of diffeomorphisms.

As in previous chapters, if we discretize the problem, energies such as the above functions E_i are equivalent to probability distributions. For example, in the face images in Figure 5.3, the face circumscribed by the oval is represented by about 1500 pixels, giving a vector of values of I on these pixels $\vec{I} \in \mathbb{R}^{1500}$. We can define a finite-dimensional stochastic model for faces as

$$p(\vec{I}, \vec{c}, \vec{\psi}) = \frac{1}{Z} e^{-\sum_i \left| I(\psi_i) - \bar{I}(p_i) - \sum_\alpha c_\alpha F^{(\alpha)}(p_i) \right|^2 / 2\sigma^2} \cdot e^{-E_1(c_\alpha)} \cdot e^{-E_2(\psi_i)},$$

where p_i are the pixels, ψ_i are the values $\psi(p_i) \in \mathbb{R}^2$ (the now finite set of hidden variables for the warping), and $I(\psi_i)$ is the linear interpolation of the discrete image I to the points ψ_i. Using a finite set of points p_i instead of a continuum is known as the *landmark point method* in the statistics literature. The landmark points need not be the points of a grid, but may be some significant points in an image, or equally spaced points along a contour. Defining directly a probability distribution on warpings—that is, on the space of all diffeomorphisms on all or part of $\mathbb{R}^n$—is an active research area that we will not pursue in this book.

5.1. Modeling Lighting Variations

The general problem of face recognition in the literature is often divided into two different steps:

- First, we locate all faces in a cluttered scene and normalize them with translation and scale.

- Once the faces are located and normalized, we get a vector, such as $\vec{I}$ in $\mathbb{R}^{1500}$, and then determine the hidden variables $\vec{c}$ and find the best match of our template with $\vec{\psi}$. We then identify the person, if possible.

Our stochastic model can be used for the second step. For the first step, many faster ways exist based on looking for quickly extracted features suggestive of a face (see, for example, the work of Amit and Geman [6–8]; Ullman et al. [2, 190]; and Viola and Jones [115, 223, 224]). But all such fast algorithms also locate a spectrum of "near faces," and we need a closer look to decide exactly which are real faces.[1] That is where an accurate stochastic model is essential: if the model doesn't fit at all, you don't have a face, and if it fits partially, we can analyze why not—we may have an outlier, such as a hockey mask or a face with paint.

5.1 Modeling Lighting Variations

5.1.1 Eigenfaces

Suppose we look at all images of a fixed object in a fixed pose but under all possible lighting conditions. The set of these images will form a convex cone $\mathcal{C}$ in a huge space $\mathbb{R}^n$ (e.g., $n = 1500$ if, as above, 1500 pixels are used). The extreme points in this cone are given by the images $I_{\theta,\varphi}$ of the object lit by a single spotlight from the angle with latitude θ and longitude φ, $-\pi/2 \leq \theta, \varphi \leq \pi/2$. Normalizing the total illumination by fixing the mean $\sum_i I(p_i)/n$, we have a convex set in a hyperplane in $\mathbb{R}^n$. What sort of convex set do we get? This question was first investigated for frontally viewed, size-, and position-normalized face images by Kirby and Sirovich in 1987 [125, 200]. They used the Gaussian method, also known as *Karhunen-Loeve (K-L) expansions* or *principal components analysis (PCA)*.

The general concept of the Gaussian approach is this: given N data points $\vec{x}^{(\alpha)} = (x_1^{(\alpha)}, \cdots, x_n^{(\alpha)})$, we get a Gaussian approximation to the distribution of this cluster of data by

$$p(\vec{x}) = \frac{1}{Z} e^{-(\vec{x}-\overline{x})^t C^{-1} (\vec{x}-\overline{x})/2},$$

---

[1] More precisely, the algorithms have an ROC curve (or receiver operating characteristic). There is a threshold to set, and depending on where you set it, you either miss more true faces or include more false faces.

where
$$\bar{x} = \frac{1}{N}\sum_{\alpha=1}^{N} \vec{x}^{(\alpha)}$$
is the mean of the data, and
$$C_{ij} = \frac{1}{N}\sum_{\alpha=1}^{N}(x_i^{(\alpha)} - \bar{x}_i)(x_j^{(\alpha)} - \bar{x}_j)$$
is the covariance matrix of the data. Since C is a symmetric positive matrix, we can diagonalize it. Let, for $1 \leq k \leq n$, $\vec{e}^{(k)}$ be an orthonormal basis of eigenvectors of C, and λ_k be the associated eigenvalues. In the work of Kirby and Sirovich [125, 200], the $\vec{x}^{(\alpha)}$ are images of faces, the indices i are pixels, and the vectors $\vec{e}^{(k)}$ are called eigenfaces. Moreover, we suppose that the eigenvalues are ordered such that
$$\lambda_1 \geq \lambda_2 \geq \cdots \geq \lambda_n.$$
If we define
$$\vec{x} = \bar{x} + \sum_{k=1}^{n} \sqrt{\lambda_k}\, \xi_k\, \vec{e}^{(k)},$$
where the ξ_k are independent normal random variables with mean 0 and standard deviation 1, then $\vec{x}$ is a random sample from the Gaussian model p.

In particular, for faces, this model gives
$$\forall \text{ pixels } i,j, \quad I(i,j) = \bar{I}(i,j) + \sum_k c_k F^{(k)}(i,j),$$
where
$$\bar{I} = \text{ the mean face,}$$
$$F^{(k)} = \text{ the eigenfaces,}$$
$$\{c_k\} = \text{ independent Gaussian random variables, mean 0, variance } \lambda_k,$$
that is, with energy $E_1(\{c_k\}) = \sum_k \frac{c_k^2}{2\lambda_k}$.

The next observation is that we can truncate this sum because λ_k is small for large k and the corresponding term $F^{(k)}$ usually represents noise and/or the overfitting of the data. In fact, note that
$$\mathbb{E}\left(\|\vec{x} - \bar{x}\|^2\right) = \mathbb{E}\left(\sum_{k=1}^{n} \lambda_k \xi_k^2\right) = \sum_{k=1}^{n} \lambda_k = \text{tr}(C).$$

5.1. Modeling Lighting Variations

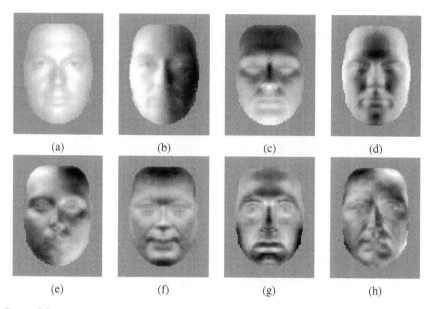

Figure 5.5. An adult male subject was photographed from a frontal view under illumination from a spotlight in 130 positions. Mirror-image faces were added. The faces were cropped and normalized for position. Image (a) is their mean, and (b)–(h) show the top seven eigenfaces. Note that eigenfaces (b) and (c) represent averages of left minus right illumination and bottom minus top illumination, respectively. Images (d)–(f) are second derivatives, and (g) and (h) seem to be dealing with specularities and shadowing. Because it is a single face, no warping is needed, and crisp—not blurred—eigenfaces are obtained. (From [97, Figure 3.5], [99].)

Thus, the variance accounted for (VAF) by the first m components is defined by

$$\text{VAF}(m) = \frac{\sum_{k=1}^{m} \lambda_k}{\sum_{k=1}^{n} \lambda_k}.$$

Kirby and Sirovich [125] get, for $m = 50$, $\text{VAF}(m) = 0.964$.

Kirby and Sirovich used a large database of faces of different people. If instead we use only one face but vary the illumination, a large amount of the variance is accounted for by a really small number of eigenimages. Typically five or six eigenimages account for 90–95 percent of the variance! This means that if we take a face illuminated from a certain direction (θ, φ), then we can write

$$I_{\theta,\varphi} \approx \overline{I} + \sum_{k=1}^{m} c_{\theta,\varphi}^{(k)} F^{(k)},$$

where m is just 5 or 6. Hallinan conducted extensive experiments on eigenfaces [67, 97–99], an example of which is shown in Figure 5.5. Here, if $I_{\theta,\varphi}$ denotes

256   5. Faces and Flexible Templates

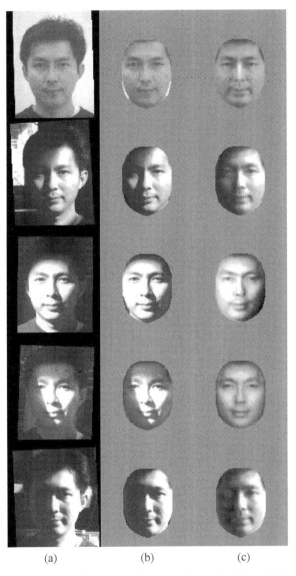

(a)   (b)   (c)

Figure 5.6. Column (a) shows five photographs of an adult male subject, the bottom four under extreme shadowing. These faces are cropped columns (b) and (c). The subject was photographed under normalized lighting, as in Figure 5.5, in order to compute eigenfaces for this subject. Column (c) shows the projection of these shadowed faces onto the span of the top five eigenfaces. Except for the case of the face where a triangular patch is brightly illuminated, the five eigenfaces in column (c) give a softened but generally accurate version of the face. (From [97, Figure 5.6], [99].)

5.1. Modeling Lighting Variations

the face lit from angle (θ, ϕ), then the mean is

$$\bar{I} = \iint_{[-\pi/2,\pi/2]^2} I_{\theta,\varphi} \cdot \cos\varphi \, d\theta \, d\varphi.$$

(Recall that the uniform measure on the sphere parameterized by spherical coordinates (θ, φ) is given by $\cos\varphi \, d\theta \, d\varphi$). In all of Hallinan's experiments, the low eigenfaces turned out to be approximately moments, that is,

$$\iint \theta \, I_{\theta,\varphi} \cos\varphi \, d\theta \, d\varphi, \qquad \iint \varphi \, I_{\theta,\varphi} \cos\varphi \, d\theta \, d\varphi,$$

$$\iint (\theta^2 - \overline{\theta^2}) \, I_{\theta,\varphi} \cos\varphi \, d\theta \, d\varphi, \qquad \iint (\varphi^2 - \overline{\varphi^2}) \, I_{\theta,\varphi} \cos\varphi \, d\theta \, d\varphi,$$

$$\iint \theta \, \varphi \, I_{\theta,\phi} \cos\varphi \, d\theta \, d\varphi,$$

where $\overline{\theta^2}$ and $\overline{\varphi^2}$ are the mean of the square of the lighting angles, or $\frac{2}{\pi} \int_0^{\pi/2} x^2 dx = \pi^2/12$. The meaning of these moments is the following: the first two represent mean left versus right illumination and top versus bottom illumination, respectively. The next three are like second derivatives in the space of lighting directions (e.g., extreme top plus extreme bottom lighting minus frontal lighting). A problem with using small sets of eigenfaces is that they don't model specularities and sharp shadows. But this may sometimes be an advantage: recall the deeply shadowed face in Figure 5 in Chapter 0. One possible way to explain our ability to reconstruct the missing features in such faces is that we mentally project onto a low-dimensional eigenspace, and this projection reconstructs the missing parts of the face. See the examples in Figure 5.6.

In conclusion, even though changing illumination accounts for huge changes in the image (considered in the L^2 norm), the images are remarkably well modeled in a very simple way by superposition of eigenfaces and by Gaussian models for the resulting lighting coefficients $\{c^{(k)}\}$.

5.1.2 Why Gaussian Models Fail for Warpings

The K-L expansion does not work well if more than one face is used. The eigenfaces $F^{(k)}$ all turn out to be blurred, and adding up a few terms does not reproduce sharp edges on eyes, mouth, and other features. This stems from the fact that *black/white lighting patterns and the geometric patterns are getting mixed up in the Gaussian model*. To explain this, we present a very simple but instructive example illustrating the failure of the principal components analysis when the patterns in a stochastic signal are purely geometric.

Example (Delta Signals). Define a random signal $\vec{s} = (s_1, s_2, \cdots, s_n)$ to be a delta function at some value l, where l has equal probability of being any integer between 1 and n, which means that $s_k = \delta_{k,l}$ for all k. The resulting probability distribution is invariant under all permutations of the axes. The only invariant subspaces are $(1, 1, \cdots, 1)$ and its orthogonal complement, the set V_0 of $\vec{x}$ such that $\sum_i x_i = 0$. In particular, the mean signal is $\bar{s} = (1/n, \cdots, 1/n)$, and the covariance matrix has only these two eigenspaces; that is, by subtracting the mean, all eigenvalues in V_0 are equal. In fact, we can calculate the covariance matrix to be

$$C = \begin{pmatrix} 1 - \frac{1}{n} & -\frac{1}{n} & \cdots & -\frac{1}{n} \\ -\frac{1}{n} & 1 - \frac{1}{n} & \cdots & -\frac{1}{n} \\ \cdots & \cdots & \cdots & \cdots \\ -\frac{1}{n} & -\frac{1}{n} & \cdots & 1 - \frac{1}{n} \end{pmatrix}.$$

The eigenvalues of the covariance matrix are $\lambda_1 = \lambda_2 = \ldots = \lambda_{n-1} = 1$ on V_0 and $\lambda_n = 0$ on $(1, 1, \cdots, 1)$. Thus, *no patterns are detected by PCA*: the fact that all our signals are localized at a single point, but at different points, cannot be expressed in a Gaussian model.

Let's make the model a bit more realistic in order to see what may be expected from PCA whenever there are geometric variations in a pattern. Suppose the signal is a function $s : [0, 1] \to \mathbb{R}$. Instead of localizing the signal at an integer l, let's choose $0 \leq a \leq 1$ at random, and then define the signal as $s(t) = \delta_a(t)$. Now we observe the signal with n sensors at points $\frac{k}{n}$. We suppose that the response of the sensor w is of the form shown in Figure 5.7.

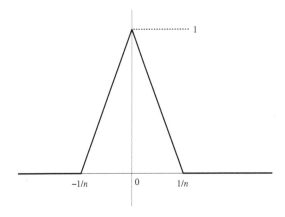

Figure 5.7. Form of the response of the sensor w.

If $\frac{k_0}{n} \leq a \leq \frac{k_0+1}{n}$, then the n sensors respond to s by

$$s_k = \langle s, w_{k/n}\rangle = w(a - \frac{k}{n}) = \begin{cases} a - \frac{k_0}{n} & \text{if } k = k_0 + 1, \\ \frac{k_0+1}{n} - a & \text{if } k = k_0, \\ 0 & \text{otherwise.} \end{cases}$$

The set of all possible responses to these signals forms a polygon in $\mathbb{R}^n$, which passes through all the unit vectors. If we allow $\vec{s}$ to be cyclic (this has only a small effect on the covariance matrix C), then C is diagonalized by Fourier transform and the "eigensignals" for this situation are just periodic signals—revealing nothing about the fact that the true signals are localized! An analysis similar to that in Chapter 2 (on music) shows that projecting one of the true signals onto the low dimensional eigensignals gives a sinc-like function, which rings strongly on each side of the true location of the signal.

These examples make it quite clear that the only way to model geometric patterns is to explicitly identify the deformations or warpings that create them. Thus delta signals are all *translations* of one basic delta signal, and this should be used to model them. In general, if we need to linearize not the space of signals but the space of warpings in some natural way, and then we can apply Gaussian methods to find, for instance, the mean warp and eigenwarps and expansions of a general warp as a sum of the mean plus combinations of the eigenwarps. We see an example of this in Section 5.6.

5.2 Modeling Geometric Variations by Elasticity

A much more difficult question is how to associate an energy $E(\phi)$ to a deformation ϕ that measures how far it is from the identity. We first seek to adapt physical models based on ideas from continuum mechanics. In particular, if ϕ is a deformation, $E(\phi)$ might represent the actual physical energy required to deform some realization of the object by the diffeomorphism ϕ of some elastic material. For example, a spring model was introduced by Fischler and Elschlager [73] in 1973, and a similar rubber mask model was introduced by Widrow [228] in the same year. A good introduction to elasticity theory is found in [96].

The setup for elasticity theory is this: we have a body at rest occupying the subset $\mathcal{D} \subset \mathbb{R}^n$, whose points $\vec{x} \in \mathcal{D}$ are called *material points*. The deformed body can be modeled as $\phi(\mathcal{D}) \subset \mathbb{R}^n$, where ϕ is a diffeomorphic mapping whose points $\vec{y} \in \phi(\mathcal{D})$ are called *spatial points*. Then we define the *strain matrix* in material coordinates ($\vec{x}$) or in spatial coordinates ($\vec{y}$) by

$$(D\phi)^t \circ (D\phi) : T_{\vec{x}}\mathcal{D} \longrightarrow T_{\vec{x}}\mathcal{D}$$
$$(D\phi) \circ (D\phi)^t : T_{\vec{y}}\phi(\mathcal{D}) \longrightarrow T_{\vec{y}}\phi(\mathcal{D}),$$

where $D\phi$ denotes the differential of ϕ, and $T_{\vec{x}}\mathcal{D}$ is the tangent space to $\mathcal{D}$ at point $\vec{x}$, and similarly for $\vec{y}$.

If we assume that no energy is dissipated in heat during the deformations, the total energy stored in the deformation is given by a function of the form

$$E_s(\phi) = \int_{\mathcal{D}} e_s(D\phi(\vec{x}), \vec{x}) d\vec{x},$$

where e_s is called the *strain energy density*. Such objects are called *hyperelastic*. The energy should satisfy the following properties:

- e_s is a nonnegative C^1 function in the matrix J, zero only at $J = I_n$, the identity in $\mathbb{R}^n$, and such that $e_s(J, \vec{x}) \to +\infty$ if $\| J \| \to \infty$ or $\| J^{-1} \| \to \infty$.

- If R is a rotation in spatial coordinates, $e_s(R \cdot J, \vec{x}) = e_s(J, \vec{x})$ (called "frame indifference"). This implies that e_s depends only on the strain matrix $J^t \circ J$.

- If the object is isotropic (i.e., has the same properties in each direction), then for all rotations R in material space, $e_s(J \cdot R, \vec{x}) = e_s(J, \vec{x})$. In this case, e_s depends only on the *singular values* of J or, equivalently, on the eigenvalues of the strain matrix $J^t \circ J$.

- If the object is homogeneous, $e_s(J, \vec{x})$ is independent of $\vec{x}$.

The partial derivatives

$$S(\vec{x}) = \frac{\partial}{\partial J} e_s(J, \vec{x}) \Big|_{J = D\phi(\vec{x})}$$

describe the forces caused by the deformation and are called the *stress* (in this form, the Piola-Kirchhoff stress). Newton's laws give

$$\rho(\vec{x}) \cdot \frac{d^2\phi}{dt^2}(\vec{x}) = \mathrm{div} S(\vec{x}),$$

where ρ is the density of the undeformed object, and div is the divergence in material coordinates $\vec{x}$. Equations written in terms of functions of the material coordinates $\vec{x}$ are referred to as Lagrangian, whereas equations in the spatial coordinates $\vec{y}$ are called Eulerian, so we have written Newton's law in its Lagrangian form. Another way of saying this is that if there is an additional *body force* $b(\vec{x})$ acting on the body, then the deformed body $\phi(\mathcal{D})$ will be in equilibrium if

$$b(\vec{x}) \equiv -\mathrm{div} S(\vec{x}).$$

5.2. Modeling Geometric Variations by Elasticity

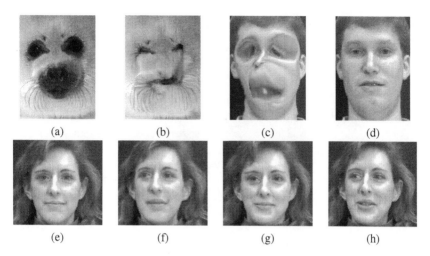

Figure 5.8. Four examples of warpings that minimize a sum of elastic strain energy and an L^2 image matching term. (a)–(d) A student is warped toward a harp seal and vice versa. Although clearly unsuccessful, it shows an honest effort to shrink the harp seal's generous features to human size and to put a rubber mask on the student to make him seal-like. (e)–(h) A more serious experiment in which another student is shown on the far left looking frontally with a neutral expression; and on the right looking left with a smile. Second from left, the left-most image is warped by moving the pupils left and stretching the lips to match the right-most image. Note that warping cannot create teeth, so they are missing. Second from right, the right-most image is warped by moving the pupils to the center of the eyes and closing the lips in order to match the left-most image. (From [97, Figures 4.12 and 4.15], [99].)

As explained in the introduction to this chapter, we can add $E_s(\phi)$ to a term such as $\iint |I_1(\phi(x,y)) - I_2(x,y)|^2 dx dy$ and solve for an optimal warping between two images. Some examples of such optimal warpings from the work of Peter Hallinan are shown in Figure 5.8. Elasticity may also used to put more natural linear coordinates on the space of ϕ. For example, we might use, as a coordinate for ϕ mod translations, the body force $b(x,y)$ that balances the stress of ϕ. This approach to linearization has not, to our knowledge, been explored.

Here are three expressions for energy.

1. In linear elasticity theory, we assume J is close to the identity $J = I_n + \delta J$, where δJ is small. Then the strain matrix $J^t \circ J \approx I_n + \delta J + \delta J^t$ and energies that are isotropic, quadratic in δJ, and depend only on the strain are all of the form

$$e_s(J) = \lambda \text{tr}\left((\delta J + \delta J^t)^2\right) + \mu (\text{tr}(\delta J))^2.$$

In this formula, λ and μ are called the *moduli of elasticity*. The stress is given by

$$S = \frac{\partial e_s}{\partial(\delta J)} = 2\lambda(\delta J + \delta J^t) + 2\mu \cdot \text{tr}(\delta J) \cdot I_n.$$

Closely related to this is the simple expression $e_s(J) = \text{tr}(\delta J \cdot \delta J^t) = \|\delta J\|^2$, which was the first expression researchers in computer vision studied. However, it does not satisfy the infinitesimal form of frame indifference.

2. A common type of expression in nonlinear elasticity is the so-called *Hadamard strain energies*,

$$e_s(J, \vec{x}) = \lambda \cdot \text{tr}(J^t \cdot J) + h(\det(J)),$$

where h is convex and goes to infinity when $\det(J)$ goes to 0 or ∞. We need to assume that $h'(1) = -2\lambda$ in order that this has a minimum at $J = I_n$. This has all the assumed properties. If we let $J = I_n + \delta J$, then it is not difficult to check that this approaches the classical linear expression for energy, provided $h''(1) = 2(\lambda + \mu)$.

3. Another type of expression is the so-called *Mooney-Rivlin strain energies*:

$$e_s(J, \vec{x}) = \lambda \cdot \left(\text{tr}(J^t \cdot J) + \text{tr}((J^t \cdot J)^{-1})\det(J)\right) + h(\det(J)).$$

This has the extra property that $e_s(J) = e_s(J^{-1}) \cdot \det(J)$, which, when integrated, gives us the equality of global energy for a diffeomorphism ϕ and its inverse:

$$E_s(\phi) = E_s(\phi^{-1}).$$

In Hallinan's experiments shown in Figure 5.8, he used the similar

$$e_s(J, \vec{x}) = \text{tr}(J^t \cdot J)(1 + (\det(J))^{-2}),$$

which kept his algorithm more strongly from letting $\det(J)$ go to 0.

5.3 Basics XI: Manifolds, Lie Groups, and Lie Algebras

The set Diff of diffeomorphisms of a domain $\mathcal{D} \subset \mathbb{R}^n$ is a *group*: composition of two diffeomorphisms is the natural operation on Diff. We want to devote a substantial part of this chapter to exploring what can be done by using the group

5.3. Basics XI: Manifolds, Lie Groups, and Lie Algebras

operation. Grenander [93] has often emphasized that when we want to model some collections of patterns, it is very important to consider the symmetries of the situation—whether there is an underlying group. For warpings of domains of signals, the natural group is the group of diffeomorphisms. Elasticity theory is, in some sense, the study of diffeomorphisms but in a very asymmetric way: the domain of the diffeomorphism is the material body at rest, and the range is the deformed body. After developing in this section and in Section 5.4 some theory of groups, and the distances between elements of groups and of the specific group Diff, we come back and discuss the relationship of the new approach to elasticity theory. Diff is a special kind of group, a group that is also a manifold, albeit an infinite-dimensional one. Groups that are manifolds are called Lie groups (after Sophus Lie, who introduced them). In this section, we sketch the basic ideas of manifolds and Lie groups and we do this in both finite and infinite dimensions. This is a huge topic, and we insert here an outline of the basic definitions of these theories. Many books on these subjects are available, such as [44, 116, 132, 193, 202].

5.3.1 Finite-Dimensional Manifolds

A n-dimensional manifold M is a set that "locally looks like an open subset of $\mathbb{R}^n$." In this section and in the following ones, we focus on manifolds, so we must first give a more precise definition of them.

Definition 5.1. Let M be a set. We say that M is an n-dimensional smooth manifold if we endow it with an atlas, that is, a collection of pairs (U_a, φ_a) (called "charts") such that

- Each U_a is a subset of M, and the U_a cover M.

- Each φ_a is a bijection of U_a onto an open subset V_a of $\mathbb{R}^n$.

- For any a, b, $\varphi_a(U_a \cap U_b)$ and $\varphi_b(U_a \cap U_b)$ are open in $\mathbb{R}^n$, and the map

$$\varphi_b \varphi_a^{-1} : \varphi_a(U_a \cap U_b) \to \varphi_b(U_a \cap U_b)$$

is a C^∞ diffeomorphism.

The most basic examples of finite-dimensional manifolds are any open subset of $\mathbb{R}^n$ (it only needs one chart with the identity map), a smooth curve or a smooth surface in 3-space, the sphere S^n (such as the sphere S^2 of Figure 5.9), the torus $\mathbb{T}^n$, the set of real $n \times n$ matrices, and so on. We can define smooth maps from one manifold to another and we can associate with each point $x \in M$ a vector space, the *tangent space* $T_x M$ of M at x (e.g., as the equivalence classes of 1-jets

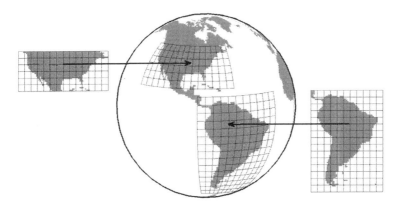

Figure 5.9. Some charts in an atlas for the sphere S^2.

of smooth maps $f : \mathbb{R} \to M$ with $f(0) = x$, the equivalence relation being that two maps agree in a suitable chart to the first order). The equivalence class of such a smooth map f is simply denoted by $f'(0)$. The tangent space $T_x M$ is then the set of all these $f'(0)$. If $x \in U_a$, then to any curve γ such that $\gamma(0) = x$, one can associate the n-tuple of derivatives $\frac{d}{dt}(\varphi_a \circ \gamma)\big|_{t=0} \in \mathbb{R}^n$ and, in this way, identify $T_x M$ with $\mathbb{R}^n$. The union of all tangent spaces $\{T_x M | x \in M\}$ forms another manifold TM whose dimension is twice that of M and which is called the *tangent bundle* to M. For a smooth function F defined on M, we denote by $D_x F$ its differential at point $x \in M$.

We introduce distances on manifolds using Riemannian metrics described in Definition 5.2.

Definition 5.2. A Riemannian structure on an n-dimensional manifold M is a set of positive definite inner products $\langle u, v \rangle_x$ on all tangent spaces $T_x M$ to M that are given, in local charts, by expressions

$$\langle u, v \rangle_x = \sum_{1 \leq i,j \leq n} g_{ij}(x) u^i v^j, \quad \forall u, v \in T_x M,$$

where g is an $n \times n$ symmetric matrix depending smoothly on x. (It is usual to denote tangent vectors in local charts by n-tuples with superscripted indices.)

Given these inner products, a norm is then defined on each $T_x M$ by $\| u \|_x = \sqrt{\langle u, u \rangle_x}$.

For a smooth path $\gamma : [0, 1] \to M$ with tangent vectors $\gamma'(t)$, its length is defined by

$$\ell(\gamma) = \int_0^1 \sqrt{\langle \gamma'(t), \gamma'(t) \rangle_{\gamma(t)}} dt.$$

5.3. Basics XI: Manifolds, Lie Groups, and Lie Algebras

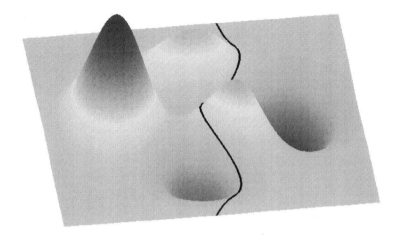

Figure 5.10. A geodesic on an artificial landscape. (See Color Plate VI.)

The distance $d(x, y)$ between two points x and y of M is then given by

$$d(x, y) = \text{Inf}\{\ell(\gamma); \gamma \text{ smooth path such that } \gamma(0) = x \text{ and } \gamma(1) = y\}.$$

A path γ that minimizes distance, at least locally, is called a *geodesic*. It is standard theorem that on any manifold, any two points x, y that are sufficiently close are connected by a unique geodesic. If the manifold is compact, then any two points, even when far apart, are connected by at least one geodesic, and one of them has length $d(x, y)$. An example of a geodesic on a surface in 3-space is shown in Figure 5.10. When solving numerically for geodesics, it is usually better to minimize a sum of squares, viz. the *energy* of a path

$$E(\gamma) = \int_0^1 \langle \gamma'(t), \gamma'(t) \rangle_{\gamma(t)} dt.$$

This has the same minimizing paths in the manifold as the length ℓ. To see this, note how both behave if we reparametrize γ by setting $t = f(\tau)$ or $\tau = g(t)$ with f, g mutually inverse increasing functions from $[0, 1]$ to itself. Let $\tilde{\gamma} = \gamma \circ f$. Then

$$\ell(\tilde{\gamma}) = \int_0^1 \| \gamma' \circ f(\tau) \| f'(\tau) d\tau = \int_0^1 \| \gamma'(t) \| dt = \ell(\gamma), \quad \text{but}$$

$$E(\tilde{\gamma}) = \int_0^1 \| \gamma' \circ f(\tau) \|^2 f'(\tau)^2 d\tau = \int_0^1 \frac{\| \gamma'(t) \|^2}{g'(t)} dt.$$

Taking the first variation in g of the last integral above, it is easy to see that $E(\widetilde{\gamma})$ is minimized over all choices of g when $g'(t) \equiv C \cdot \| \gamma'(t) \|$. That is, g is the scaled arc length along the curve—$g(t) = C \cdot \int_0^t \| \gamma'(s) \| \, ds$, C being chosen so that $g(1) = 1$. After this substitution, $\| \widetilde{\gamma}' \|$ is a constant and $E(\widetilde{\gamma}) = \ell(\widetilde{\gamma})^2 = \ell(\gamma)$. Thus minimizing ℓ chooses the best path but ignores the parameterization whereas minimizing E chooses both the best path and the simplest parameterization.

Setting the first variation of E equal to zero, one finds an elegant ordinary differential equation of second order for geodesics on any manifold M. We sketch this result in Exercise 2 at the end of this chapter. It is a standard result that for all $x \in M, u \in T_x M$, there is a unique geodesic γ with $\gamma(0) = x, \gamma'(0) = u$. This defines the *Riemannian exponential map*:

$$\exp : T_x M \to M, \quad \exp(u) = \gamma(1), \quad \text{where } \gamma(0) = x, \gamma'(0) = u.$$

By the discussion above, the map exp is always a bijection between some neighborhood of $0 \in T_x M$ and some neighborhood of $x \in M$. If M is compact, it is always surjective.

5.3.2 Lie Groups

Some finite-dimensional manifolds, moreover, have the property of being groups:

Definition 5.3. A group G is a Lie group if it is a smooth (C^∞) manifold such that the group operations

$$(g, h) \mapsto gh \quad \text{and} \quad g \mapsto g^{-1}$$

are smooth.

For all $g \in G$, we can define the left and right translations on G, respectively, by the maps

$$L_g : x \mapsto gx \quad \text{and} \quad R_g : x \mapsto xg.$$

Let e denote the identity element of G. The left translations L_g give a way to map any tangent space $T_g G$ into the tangent space $T_e G$ via $(D_e L_g)^{-1}$. This shows in particular that the tangent bundle TG is a trivial vector bundle since it is isomorphic to $G \times T_e G$. The tangent space $T_e G$ to G at the identity is also called the *Lie algebra* of the group G, and, as is usually done, we denote it by $\mathfrak{g}$ (it is a time-honored tradition to use a German Fraktur g for the Lie algebra).

A vector field X on G is a function from G to TG: its value at a point $g \in G$ is denoted by X_g, and it belongs to $T_g G$. The vector field X is called left-invariant (or right-invariant) if

$$\forall g, h \in G \quad D_h L_g \cdot X_h = X_{gh} \quad (\text{or } D_h R_g \cdot X_h = X_{hg}), \quad \text{respectively}.$$

5.3. Basics XI: Manifolds, Lie Groups, and Lie Algebras

Then, the main property of a left-invariant (resp. right-invariant) vector field X is that the value of X at any point g can be recovered from the value X_e of X at e by $D_e L_g \cdot X_e = X_g$ (resp. $D_e R_g \cdot X_e = X_g$). This shows that the vector space of all left- or right-invariant vector fields on G is isomorphic to the tangent space $T_e G$ or to $\mathfrak{g}$.

The group exponential map. Let X be a vector of $T_e G$. We are interested in curves $g : t \mapsto g(t)$ starting at e at time $t = 0$, with initial tangent vector X. This curve is called a *1-parameter subgroup* if

$$\forall t, s, \quad g(t+s) = g(t)g(s).$$

This condition implies that the tangent vector $g'(t)$ at each time t is the left- and right-translated image of the initial vector $X = g'(0)$; that is,

$$\forall t, \quad g'(t) = D_e L_{g(t)} X = D_e R_{g(t)} X.$$

Given X, such a curve exists and is unique. It is called the exponential map and it is denoted by

$$g(t) = \exp(tX).$$

The exponential map is always locally a diffeomorphism, although globally it may be neither injective nor surjective. However, it does define a canonical coordinate system on G near the identity.

The bracket product. Let X and Y be two vectors of the Lie algebra $\mathfrak{g}$ of G. We want to "measure" how far X and Y are from commuting in the following sense. We have the two exponential maps $t \mapsto \exp(tX)$ and $t \mapsto \exp(tY)$, and we thus can compare $t \mapsto \exp(tX)\exp(tY)$ and $t \mapsto \exp(tY)\exp(tX)$. For this, we look at $t \mapsto \exp(tX)\exp(tY)\exp(-tX)\exp(-tY)$ when t goes to 0. At the first order, we have

$$\left.\frac{d}{dt} \exp(tX)\exp(tY)\exp(-tX)\exp(-tY)\right|_{t=0} = 0.$$

The bracket product $[X, Y]$ is then the element of $\mathfrak{g}$ defined using the second-order derivative:

$$[X, Y] = \frac{1}{2}\frac{d^2}{dt^2} \exp(tX)\exp(tY)\exp(-tX)\exp(-tY)\bigg|_{t=0}.$$

The group law in terms of exponential coordinates can be written out explicitly by using the Lie bracket; this is known as the *Campbell-Hausdorff formula* (see [193], Section 4.8).

The adjoint representation. Let g be an element of G; it then defines an inner automorphism of G by the map $h \mapsto ghg^{-1} = L_g R_{g^{-1}} h$. Considering the map it defines on the tangent space to G at the identity element e, we get a map, denoted by Ad_g in the Lie algebra $\mathfrak{g}$. For any vector X in $\mathfrak{g}$, it is given by

$$\mathrm{Ad}_g X = (D_g R_{g^{-1}}) \cdot (D_e L_g) X.$$

The map $\mathrm{Ad}: G \to \mathrm{GL}(\mathfrak{g}, \mathfrak{g})$ is called the adjoint representation of G. Finally, the derivative $D_e \mathrm{Ad}$ of the map Ad at point e is denoted by $\mathrm{ad}: \mathfrak{g} \to L(\mathfrak{g}, \mathfrak{g})$ (L being the space of linear maps), and it is called the adjoint representation of the Lie algebra $\mathfrak{g}$. The adjoint representation ad is related to the Lie bracket by

$$\forall X, Y \in \mathfrak{g},\ \mathrm{ad}(X)Y = [X, Y].$$

Example (The Linear Group). Let $G = GL(n, \mathbb{R})$ be the linear group of $n \times n$ real matrices with nonzero determinant. It is a smooth manifold of dimension n^2, since it is an open subset of the set $M(n, \mathbb{R}) \simeq \mathbb{R}^{n^2}$ of all $n \times n$ matrices. Thus, its tangent space at $e = I_n$ is simply $M(n, \mathbb{R})$. The Lie algebra $\mathfrak{g}$ of G is the algebra of all real $n \times n$ matrices with the usual bracket product on matrices given by

$$[A, B] = AB - BA.$$

The exponential map is the usual exponential map for matrices:

$$\exp(A) = I_n + A + \frac{A^2}{2!} + \ldots + \frac{A^k}{k!} + \ldots.$$

Example (The Orthogonal Group). Let $G = O(n)$ be the orthogonal group, that is, the set of $n \times n$ real matrices H such that $HH^t = I_n$. If we consider the map

$$F: M(n, \mathbb{R}) \to \mathrm{Sym}(n, \mathbb{R}) \simeq \mathbb{R}^{n(n+1)/2}, \quad F(M) = MM^t,$$

then $O(n) = F^{-1}(I_n)$. Moreover, for all $A \in M(n, \mathbb{R})$, the derivative $D_A F$ of F is given by $D_A F(H) = AH^t + HA^t$. It is surjective for all $A \in O(n)$ since for $S \in \mathrm{Sym}(n)$, we have $D_A F \cdot \frac{SA}{2} = S$. Thus, $O(n)$ is a manifold of dimension $n^2 - \frac{n(n+1)}{2} = \frac{n(n-1)}{2}$. Although it is not difficult to write down *local* charts for the orthogonal group, unlike the case of the general linear group, there is no global chart that covers the whole orthogonal group. The tangent space at a point $H \in O(n)$ is

$$\begin{aligned} T_H G &= \{M \in M(n, \mathbb{R}); D_H F \cdot M = 0\} \\ &= \{M \in M(n, \mathbb{R}); HM^t + MH^t = 0\}. \end{aligned}$$

5.3. Basics XI: Manifolds, Lie Groups, and Lie Algebras

In particular, the tangent space at $e = I_n$ is the set of matrices M such that $M + M^t = 0$, that is, the space of real $n \times n$ skew symmetric matrices. The Lie algebra of $O(n)$ is the tangent space at $e = I_n$ with the same bracket product as in the case of $GL(n)$. Notice that if $A^t = -A$ and $B^t = -B$, then $[A, B]^t = B^t A^t - A^t B^t = BA - AB = -[A, B]$, so that the bracket $[A, B]$ belongs to the same Lie algebra. Notice also that if $A^t = -A$, then for all t real, $\exp(tA)^t = \exp(-tA) = \exp(tA)^{-1}$, and it thus belongs to the group.

Metrics on Lie groups. Let G be a Lie group. Given an inner product on its Lie algebra $T_e G$, denoted by $\langle .,. \rangle_e$, we can then, by using left translations, define an inner product on $T_g G$ for any $g \in G$ by setting

$$\forall u, v \in T_g G, \quad \langle u, v \rangle_g = \langle D_g L_{g^{-1}} u, D_g L_{g^{-1}} v \rangle_e.$$

This defines a Riemannian structure and hence a metric on the Lie group G. Computationally, if g^* is a group element very close to g, then we left-translate the vector u from g to g^* to the vector from $g^{-1} \circ g^*$ to e. Writing $g^{-1} \circ g^* = \exp(\widetilde{u})$, for $\widetilde{u} \in \mathfrak{g}$, the distance from g to g^* is approximated by $\| \widetilde{u} \|_e$. The length of a path is obtained by integrating along the path the norm of its tangent vectors. If $\mathcal{P} = \{g_i\}$ is a polygon in G with small edges, we can approximate its length by

$$\ell(\mathcal{P}) \approx \sum_i \| a_i \|_e, \quad \text{where } g_i^{-1} \circ g_{i+1} = \exp(a_i).$$

Since the inner product on each $T_g G$ was defined using left translations, for a smooth path γ on G and for $g \in G$, we have $\ell(g\gamma) = \ell(\gamma)$, and the metric thus obtained on G is left-invariant (i.e., $d(x, y) = d(gx, gy)$). There is no particular reason for choosing left translations instead of right translations. If we used right translations, then the length of a path would instead be

$$\ell(\mathcal{P}) \approx \sum_i \| a_i \|_e, \quad \text{where } g_{i+1} \circ g_i^{-1} = \exp(a_i).$$

Note, however, that usually the metrics constructed by using left and right translations will not coincide.

Example. Given two real $n \times n$ matrices $A = (a_{ij})_{i,j}$ and $B = (b_{ij})_{i,j}$, we can define their inner product by

$$\langle A, B \rangle_e = \sum_{i,j} a_{ij} b_{ij} = \operatorname{tr}(AB^t).$$

We let $G = O(n)$ and $h \in G$. The inner product on $T_h G$ is then given by

$$\langle A, B \rangle_h = \langle h^{-1} A, h^{-1} B \rangle_e = \operatorname{tr}(h^{-1} A B^t h^{-t}) = \operatorname{tr}(AB^t)$$

because $hh^t = I$. The distance thus obtained on the group $G = O(n)$ is then both left- and right-invariant.

If we have a path $g(t) \in G$ in a Lie group, and a left invariant metric, then to measure its length we left translate the tangent vectors to the path back to e, take their norms, and integrate:

$$\ell(g) = \int_0^1 \left\| D_{g(t)} L_{g(t)^{-1}} \left(\frac{\partial g}{\partial t} \right) \right\|_e dt.$$

There is a very elegant theory of geodesic paths on Lie groups with respect to left- or right-invariant metrics. This theory gives a simple explicit differential equation for $u(t) = D_{g(t)} L_{g(t)^{-1}} \left(\frac{\partial g}{\partial t} \right)$, which characterizes geodesics. We describe this in the appendix to this chapter (Section 5.8).

5.3.3 Statistics on Manifolds

Classical statistics is based on the analysis of data that are scalars or, more generally, points in a vector space. Some of the key tools of data analysis are (a) taking means and covariance matrices, (b) principal components analysis and Gaussian fits to the data and (c) breaking the data into clusters by techniques such as k-means, nearest neighbor, and so forth. In many applications, however, the data may lie not in a linear space at all, but in some nonlinear space—a sphere, a Lie group, the Grassmannian of subspaces of a vector space, or any other manifold. How can we do data analysis for sets in such nonlinear spaces? This topic is still under active investigation, but we can sketch here some approaches to it by using differential geometry.

We assume we are given a data set $\{x_\alpha \in M\}$, where M is a Riemannian manifold. We first need an effective way of computing distances $d(x, y)$ between any two points $x, y \in M$, and, better, a geodesic between them realizing this distance. Of course, in very bad cases these may not exist, but in many cases, such as when M is compact, they always do. There are simple formulae giving an ODE for geodesics (see any differential geometry text, e.g., [116, 140]) and, by using these, geodesics are often found by the shooting method. We start with an initial guess for the tangent vector $t \in T_x M$ pointing to y, integrate the geodesic $\exp_x(\lambda t)$ in this direction, find how close to y it comes, and make a correction to t. Iterating, we converge to the desired geodesic.

We thus get a matrix of distances $D_{\alpha\beta} = d(x_\alpha, x_\beta)$. There is a class of rather ad hoc techniques called multidimensional scaling (MDS), which seeks a set of points p_α in a low-dimensional Euclidean space whose Euclidean distances are close to the entries in this matrix (see, e.g., [129]). This is good for getting a rough idea of what information is contained in the matrix of distances.

5.3. Basics XI: Manifolds, Lie Groups, and Lie Algebras

More fundamental is Definition 5.4 for a Karcher mean.

Definition 5.4. A Karcher mean of a set of points $\{x_\alpha\} \subset M$ is a point $\bar{x} \in M$ that achieves the infimum of

$$\sum_\alpha d(\bar{x}, x_\alpha)^2.$$

Clearly, for Euclidean spaces, this is the usual mean. If we have a Karcher mean, we can go further using distance-minimizing geodesics $g_\alpha(t)$ from $\bar{x}$ to x_α. We assume g_α has unit speed, so that $g_\alpha(0) = \bar{x}, g_\alpha(d_\alpha) = x_\alpha$, where $d_\alpha = d(\bar{x}, x_\alpha)$ is the distance of the αth data point from the mean. Let $t_\alpha \in T_{\bar{x}}M$ be the unit vector that is the tangent to the path g_α at $\bar{x}$, so that it points from the mean to x_α. Then

$$\exp_{\bar{x}}(d_\alpha \cdot t_\alpha) = g_\alpha(d_\alpha) = x_\alpha.$$

Then replacing the cluster of points $\{x_\alpha\}$ in M by the cluster of tangent vectors $\{d_\alpha \cdot t_\alpha\}$ in $T_{\bar{x}}M$ *linearizes the data set*. This construction is a key reason for using differential geometry in statistics. In particular, it allows us to apply the classical techniques of data analysis to the x_α. The set of vectors $\{t_\alpha\}$ now has a covariance matrix, has principal components, and can be split into clusters, and so forth.

This approach is a reasonable first approximation, but it also leads us directly to the main problem in nonlinear data analysis: curvature. We do not develop the theory of curvature here, but just indicate how it applies to the problem of linearization. A full exposition can be found in many places, such as [116, 140]. The problem is that although exp is an isometry between $T_{\bar{x}}M$ and M to *first order*, it begins to distort distances at second order. The correction term is

$$d(\exp_{\bar{x}}(t_1), \exp_{\bar{x}}(t_2))^2 = \| t_1 - t_2 \|^2 - \tfrac{1}{3}R(t_1, t_2, t_1, t_2) \\ + o(\max(\| t_1 \|, \| t_2 \|)^4), \qquad (5.1)$$

where $R(s, t, u, v) = \sum_{ijkl} R_{ijkl} s_i t_j u_k v_l$ is the Riemann curvature tensor. R looks complicated, but it makes a bit more sense if we bring in its symmetries:

$$R_{ijkl} = -R_{jikl} = -R_{ijlk} = R_{klij}.$$

In other words, R is skew symmetric in each of the pairs (i, j) and (k, l), and symmetric if these two pairs are interchanged. This means that if $s \wedge t$ is the 2-form, $(s \wedge t)_{ij} = s_i t_j - s_j t_i$, then we can write

$$\sum_{ijkl} R_{ijkl} s_i t_j u_k v_l = \tfrac{1}{4} \sum_{ijkl} R_{ijkl}(s \wedge t)_{ij}(u \wedge v)_{kl}$$

$$= \text{a symmetric inner product } \langle s \wedge t, u \wedge v \rangle_R,$$

where $\langle,\rangle_R$ is the curvature tensor inner product. So we can rewrite Equation (5.1) as

$$d(\exp_{\overline{x}}(t_1), \exp_{\overline{x}}(t_2))^2 = \| t_1 - t_2 \|^2 - \tfrac{1}{3} \| t_1 \wedge t_2 \|_R^2 \\ + o(\max(\| t_1 \|, \| t_2 \|)^4). \qquad (5.2)$$

Note that if t_1 and t_2 are multiples of each other, the correction vanishes and, in fact, exp preserves distances along individual geodesics. This is why the correction term involves only $t_1 \wedge t_2$. Since it must also be symmetric in t_1 and t_2, a quadratic function of $t_1 \wedge t_2$ is needed. The best known cases of curvature are the sphere and the hyperbolic plane. On a sphere, geodesics converge at the antipodal point—an extreme example of distances being less on the manifold than on a tangent plane. In the hyperbolic plane, any two geodesics starting at the same point diverge exponentially, much faster than in the tangent plane. On surfaces in 3-space, the curvature is positive where the surface is concave or convex and negative where there is a saddle point; this is illustrated in Figure 5.11. Regarding statistics, the main thing to keep in mind is that some tight clusters of data in a

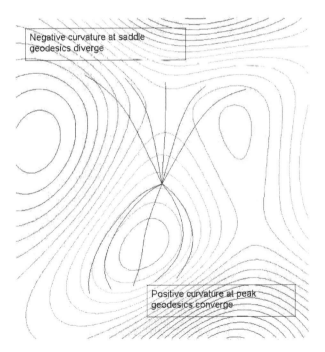

Figure 5.11. Some geodesics on the landscape used in Figure 5.10, shown here by contour lines. (See Color Plate VII.)

5.3. Basics XI: Manifolds, Lie Groups, and Lie Algebras

manifold can be pulled apart when viewed in the tangent space if the curvature is positive, and some quite distant sets of points may appear as a cluster in the tangent space when the curvature is negative. An example of this problem is given in Exercise 2 at the end of this chapter.

5.3.4 Infinite-Dimensional Manifolds

Ideally, the definition of infinite-dimensional manifolds would be exactly like the one in the finite-dimensional case, except that we would have to replace $\mathbb{R}^n$ by an infinite-dimensional vector space, for example, a Banach space E (that is, a topological vector space whose topology is given by a norm, and is moreover complete). This is the definition that we can usually find in books [132]. Unfortunately, however, things are not so simple. All too often, the maps between two coordinate systems on the overlap of two charts is not differentiable in these Banach norms. As described in Kriegl and Michor's foundational book [128], to get a good theory, we need to use charts based on vector spaces $C^\infty(\mathbb{R}^n, \mathbb{R}^m)$ of smooth functions of $\mathbb{R}^n$ to $\mathbb{R}^m$. This is not a Banach space, and neither is the space $C_c^\infty(\mathbb{R}^n, \mathbb{R}^m)$ of smooth functions with compact support. So, instead of modeling a manifold on a Banach space E, we need to use more general topological vector spaces such as these. We do not give any general theory in this book, but rather explain the idea through a simple example, the space of smooth curves. This example shows clearly what charts and an atlas can look like in the infinite-dimensional setting. More details can be found in [159].

Let us consider the space $\mathcal{S}$ of smooth simple closed plane curves. To show that this space is a manifold, we denote by $[C] \in \mathcal{S}$ the point defined by a curve C, and we take some point $[C] \in \mathcal{S}$ and represent its curve by a parameterization $s \mapsto C(s), s \in S^1$. Then all points of $\mathcal{S}$ near $[C]$ are assumed to be given by curves that lie in a band around C, which may be parameterized by $s, t \mapsto C(s) + t \cdot \vec{n}(s)$, where $\vec{n}(s)$ is the unit normal to C at s. There is some $\tau > 0$ such that if $|t| < \tau$, this is injective (i.e., $C \times [-\tau, \tau] \subset \mathbb{R}^2$); this is the band we want. Then we can define a local chart on the infinite-dimensional object $\mathcal{S}$ near the point $[C]$ as follows. First, we define curves C_a by

$$s \mapsto C_a(s) = C(s) + a(s) \cdot \vec{n}(s),$$

where $\vec{n}(s)$ is the unit normal to C and a is any smooth function on the circle with sup less than τ (see Figure 5.12). Let $U_C \subset C^\infty(S^1, \mathbb{R})$ denote this set of functions a. Then $(U_C, a \mapsto [C_a])$ is a local chart near C, and the $a(s)$ are the local linear coordinates on the space $\mathcal{S}$. Thus, the charts for $\mathcal{S}$ are all based on the topological vector space $C^\infty(S^1, \mathbb{R})$. The collection of such charts for all $[C] \in \mathcal{S}$ gives the atlas for $\mathcal{S}$.

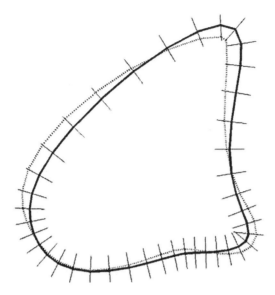

Figure 5.12. A local chart in the space of smooth curves. We start with the dark curve and lay out its unit normals (the perpendicular "hairs"). The nearby curves, such as the light one, meet each normal in a unique point; hence, they are of the form C_a (see text).

Note that $T_C \mathcal{S}$ is naturally isomorphic to the vector space of normal vector fields $a\vec{n}_C$ along the curve C: if a is a smooth function on C, it defines the element of $T_C \mathcal{S}$ given by the infinitesimal perturbation of C:

$$C_\epsilon : s \mapsto C(s) + \epsilon a(s)\vec{n}(s),$$

where $\vec{n}(s)$ is the unit normal to C. The simplest inner product on this tangent space is the L^2 one: $\langle a, b \rangle = \int_C a \cdot b \cdot ds$, where s is arc length along C. This and many variants have been studied in [158, 159]. But very surprisingly, in the L^2 metric, paths can be found of arbitrarily small length joining any two simple closed curves. Strange things sometimes happen in infinite dimensions.

In order to make the geometry in infinite-dimensional manifolds seem more accessible, we now describe a simple example of a gradient flow on the space $\mathcal{S}$. First, we recall what a gradient "really" is in a manifold setting.[2] Let f be a real-valued function defined on a smooth manifold. The derivative $D_x f$ of f at a point $x \in M$ is by definition the linear approximation to f, that is, a linear mapping from $T_x M$ into $\mathbb{R}$ whose values are the directional derivatives of f. Equivalently,

---

[2]It is a common mistake, found even in multivariable calculus textbooks, to draw contour plots of functions f of two variables that have different dimensions, such as space and time, and then to call vectors perpendicular to these level lines gradients of f. The result depends on the choice of units for two independent variables.

5.3. Basics XI: Manifolds, Lie Groups, and Lie Algebras

it is a co-vector, an element of T_xM^*, the dual of T_xM. If we want to convert it into a vector, we need an inner product on T_xM. If $\langle .,. \rangle$ denotes such an inner product, then we can define the gradient $\nabla f(x)$ of f at point x as the vector of T_xM such that

$$\forall h \in T_xM, \quad D_x f(h) = \langle \nabla f(x), h \rangle.$$

We want here to compute the gradient of the length function $C \mapsto \ell(C) =$ length of C on $\mathcal{S}$. We need an inner product on $T_C \mathcal{S}$. For this, we take the simplest choice, $\langle a, b \rangle = \int_C a \cdot b \cdot ds$, where ds is arc length along C. Next, if $s \mapsto C(s)$ is a parameterization of C by arc length, then recall that the ordinary curvature of the plane curve C is given by $\kappa(s) = \langle \vec{n}'(s), C'(s) \rangle$. Then (using the fact that $\langle C'(s), \vec{n}(s) \rangle = 0$):

$$\begin{aligned}
\ell(C_\epsilon) &= \int_0^{l(C)} \| C'_\epsilon(s) \| \, ds \\
&= \int_0^{l(C)} \| C'(s) + \epsilon a'(s)\vec{n}(s) + \epsilon a(s)\vec{n}'(s) \| \, ds \\
&= \int_0^{l(C)} (1 + \epsilon \langle C'(s), a'(s)\vec{n}(s) + a(s)\vec{n}'(s) \rangle) \, ds + O(\epsilon^2) \\
&= \ell(C) + \epsilon \int_0^{l(C)} a(s)\kappa(s) ds + O(\epsilon^2).
\end{aligned}$$

This means that the derivative of ℓ at the point C in the direction a is $\int a(s)\kappa(s) ds$, which is the inner product of the two tangent vectors $a, \kappa \in T_C \mathcal{S}$. Therefore, this computation shows the nice fact that the gradient of the length on the space of smooth plane curves is the curvature: $\nabla \ell(C) = s \mapsto \kappa(s)$!

Now, $-\nabla \ell$ is a vector field on $\mathcal{S}$. What are its flow lines? Starting from a smooth closed curve C_0, the flow line of this vector field will be a family of curves C_t that satisfy the following partial differential equation called the *geometric heat equation*:

$$\frac{\partial C_t(s)}{\partial t} = -\kappa_t(s)\vec{n}_t(s).$$

This equation was analyzed in an important series of papers by Gage and Hamilton [77] and Grayson [90]. This evolution $t \mapsto C_t$ will make the length of the curves decrease: it "pushes" the concave parts outside while "pulling" the convex parts inside (see Figure 5.13 for an example of this). Eventually, it produces curves that are nearer and nearer to circles which shrink down in finite time to a point—and then, like a black hole, the equation becomes meaningless.

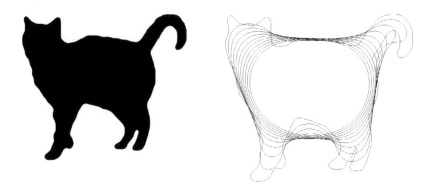

Figure 5.13. A cat silhouette evolving by the negative gradient of length, that is, the geometric heat equation. (Figure obtained by the free software MegaWave2, intended for image processing.)

5.4 Modeling Geometric Variations by Metrics on Diff

Finally, we come to the application of differential geometry to warpings. We consider the group $\mathcal{G} = \text{Diff}(\mathbb{R}^n)$ of all smooth diffeomorphisms of $\mathbb{R}^n$ that are the identity outside a compact subset of $\mathbb{R}^n$. As in elasticity theory, we call coordinates of points $\vec{x}$ in the domain of ϕ material coordinates, and coordinates of the values $\vec{y} = \phi(\vec{x})$ spatial coordinates. Then, $\mathcal{G}$ is a manifold simply using the single global chart, which subtracts material coordinates from spatial ones. That is, it associates to a diffeomorphism $\phi : \mathbb{R}^n \to \mathbb{R}^n$ the infinite-dimensional "coordinate" $\vec{x} \mapsto \phi(\vec{x}) - \vec{x}$, a point in the vector space $C_c^\infty(\mathbb{R}^n, \mathbb{R}^n)$ of smooth functions with compact support. For a convenient strong topology (which we don't discuss), the group operations of composition and inversion are smooth (have infinitely many derivatives). Thus, the group $\mathcal{G} = \text{Diff}(\mathbb{R}^n)$ can be seen as a manifold—in fact, an infinite-dimensional Lie group—modeled on the space of functions with compact support. Mathematical details on this example can be found in [128, 160].

The Lie algebra $\mathfrak{g}$ of $\mathcal{G}$ is the tangent space $T_e\mathcal{G}$ at point $e = \text{Id}$. It is given by the vector space of smooth vector fields $\vec{v}$ with compact support on $\mathbb{R}^n$ via the infinitesimal diffeomorphism $\vec{x} \longrightarrow \vec{x} + \varepsilon \vec{v}(\vec{x})$. Similarly, a tangent vector at the point $\phi \in \mathcal{G}$ is given by $\phi(\vec{x}) + \varepsilon \vec{v}(\vec{x})$. But we are in a Lie group, so it is much more relevant to translate this tangent vector back to a tangent vector at the identity diffeomorphism. If we use right translation, this means we write it instead as the infinitesimal deformation of ϕ:

$$\phi(\vec{x}) + \varepsilon \vec{v}(\vec{x}) = (\text{Id} + \varepsilon \vec{w}_1) \circ \phi = \phi(\vec{x}) + \varepsilon \vec{w}_1(\phi(\vec{x}))$$

5.4. Modeling Geometric Variations by Metrics on Diff

for the vector field $w_1 = v \circ \phi^{-1}$ on $\mathbb{R}^n$. We could have used left translation, in which case we would write the infinitesimal deformation as

$$\phi(\vec{x}) + \varepsilon \vec{v}(\vec{x}) = \phi \circ (\mathrm{Id} + \varepsilon \vec{w}_2(\vec{x})) = \phi(\vec{x}) + \varepsilon D_{\vec{x}}\phi \cdot \vec{w}_2(\vec{x}).$$

Unlike the case for finite-dimensional Lie groups such as rotations, using left or right translation gives quite different formulas. We will use the right translation, which is perhaps more natural. Notice that, in this case, left and right translations do not provide the same description of the tangent space at a point ϕ of the group $\mathcal{G}$. For the left translation, we need the derivative $D\phi$, which shows that we "lose" one degree of regularity. In our case, we avoid all these questions by working directly in the C^∞ setting.

All of the previous setting remains valid if we replace the space $\mathbb{R}^n$ by a domain $\mathcal{D}$ of $\mathbb{R}^n$. In all the following, we denote by $\mathcal{G}$ either the group $\mathrm{Diff}(\mathbb{R}^n)$ of smooth diffeomorphisms of $\mathbb{R}^n$, which constitute the identity outside a compact subset of $\mathbb{R}^n$, or the group $\mathrm{Diff}(\mathcal{D})$ of smooth diffeomorphisms of a domain $\mathcal{D} \subset \mathbb{R}^n$, which equal the identity outside a compact subset of $\mathcal{D}$.

What kind of metrics can we put on the group $\mathcal{G}$? The simplest inner product on the vector space of vector fields is again the L^2 one:

$$\langle \vec{v}, \vec{w} \rangle = \int \langle \vec{v}(\vec{x}), \vec{w}(\vec{x}) \rangle dx_1 \cdots dx_n.$$

Unfortunately, as in the case of simple closed curves, there are paths joining any two diffeomorphisms of arbitrarily small length; see Exercise 5 at the end of this chapter. What we need to do is to construct the norm by using derivatives of the vector fields as well as their values. These are the so-called *Sobolev norms* and their generalizations. Fix some $k > 0$ and look at all sequences $D^{i(\cdot)} = D_{x_{i(1)}} \circ \cdots \circ D_{x_{i(p)}}$ of at most k derivatives. Then the simplest higher-order norms are given by adding all these up:

$$\langle \vec{v}, \vec{w} \rangle_k = \int_\mathcal{D} \sum_{p=0}^{k} \left(\sum_{1 \leq i(1), \ldots, i(p), q \leq n} D^{i(\cdot)}(v_q)(\vec{x}) \cdot D^{i(\cdot)}(w_q)(\vec{x}) \right) \cdot dx_1 \cdots dx_n.$$

It is often convenient to integrate by parts and rewrite this inner product as

$$\langle \vec{v}, \vec{w} \rangle_L = \int_\mathcal{D} \langle L\vec{v}(\vec{x}), \vec{w}(\vec{x}) \rangle d\vec{x} \quad \text{and} \quad \|\vec{v}\|_L = \int_\mathcal{D} \langle L\vec{v}(\vec{x}), \vec{v}(\vec{x}) \rangle d\vec{x},$$

where L is a positive self-adjoint differential operator, which in our case equals

$$L_k = (I - D_{x_1}^2 - \cdots - D_{x_n}^2)^k = (I - \Delta)^k.$$

Other operators are possible, and they can incorporate things such as div (divergence) and curl (also called the rotational operator); we call the resulting norms $\langle \vec{v}, \vec{w} \rangle_L$.

Let $\Phi : [0,1] \to \mathcal{G}$ be a path on $\mathcal{G}$, that is, for each t, the map $\vec{x} \mapsto \Phi(\vec{x}, t)$ is a diffeomorphism of $\mathcal{D}$. Then, for each t, the map $\vec{x} \mapsto \frac{\partial \Phi}{\partial t}(\vec{x}, t)$ belongs to $T_{\Phi(\cdot, t)} \mathcal{G}$. With *right translations*, it corresponds to the vector field $\vec{x} \mapsto \frac{\partial \Phi}{\partial t}(\Phi^{-1}(\vec{x}), t)$ on the Lie algebra $\mathfrak{g}$. By using the above inner product, we can define (as we did it for finite-dimensional Lie groups) the length and the energy of the path Φ by

$$\ell(\Phi) = \int_0^1 \|\vec{v}(\cdot, t)\|_L \, dt \quad \text{and}$$

$$E(\Phi) = \int_0^1 \|\vec{v}(\cdot, t)\|_L^2 \, dt = \int_0^1 \int_{\mathcal{D}} \langle L\vec{v}(x, t), \vec{v}(x, t) \rangle \, dx \, dt,$$

where $\vec{v}(x, t) = \frac{\partial \Phi}{\partial t}(\Phi^{-1}(x, t), t)$.

Computationally, if we have a polygon $\mathcal{P} = \{\phi_i\}$ in the diffeomorphism group, then we can approximate its length by a finite sum:

$$\ell(\mathcal{P}) \approx \sum_i \|v_i\|_L, \quad \text{where } \mathrm{Id} + v_i = \phi_{i+1} \circ \phi_i^{-1}.$$

This shows why translating vectors back to the origin is the natural thing to do. As in the finite-dimensional case, we can define a metric on $\mathcal{G}$ as the infimum of lengths of paths joining two points. As previously mentioned, this distance is zero for the simplest case, $L = \mathrm{Id}$. But for-higher order L, it gives a perfectly good distance, and the corresponding energy $E(\phi)$ is finally the diffeomorphism-based measure of the degree of warping (see Exercises 5 and 8at the end of this chapter).

5.4.1 The Geodesic Equation on Diff($\mathcal{D}$)

To avoid too big a digression into differential geometry, we have not worked out the equations for geodesics on general manifolds. But the geodesic equation on Diff($\mathcal{D}$) is central to the application of this method to shape analysis, and it can be done by a direct application of the techniques of variational calculus. We work this out in this section. The appendix to this chapter (Section 5.8) gives a second derivation of this equation by using a more abstract point of view.

We have to calculate the first variation of the energy $E(\Phi)$ of a path Φ:

$$E(\Phi) = \int_0^1 \left(\int_{\mathcal{D}} \left\| \frac{\partial \Phi}{\partial t}(\Phi^{-1}(\vec{y}, t), t) \right\|_L^2 d\vec{y} \right) dt.$$

5.4. Modeling Geometric Variations by Metrics on Diff

We consider the variation $\Phi \longrightarrow \Phi + \varepsilon \vec{f}(\vec{x}, t)$, where $\vec{f}$ is a function with compact support in $\mathcal{D} \times [0, 1]$. We write the norm $\| \vec{v} \|^2 = L\vec{v} \cdot \vec{v}$, where L is a self-adjoint operator, and we use the notation $\vec{v}(\vec{y}, t) = \frac{\partial \Phi}{\partial t}(\Phi^{-1}(\vec{y}, t), t)$ for the velocity and $\vec{g}(\vec{y}, t) = \vec{f}(\Phi^{-1}(\vec{y}, t), t)$.

Step 1. We first obtain some formulas that will be useful in what follows. Since $\Phi(\Phi^{-1}(\vec{y}, t), t) = \vec{y}$ we get, by differentiation with respect to t:

$$\frac{\partial (\Phi^{-1})}{\partial t}(\vec{y}, t) = -(D_{\vec{x}}\Phi)^{-1}(\Phi^{-1}(\vec{y}, t), t) \cdot \frac{\partial \Phi}{\partial t}(\Phi^{-1}(\vec{y}, t), t).$$

We need the first variation of Φ^{-1}. Expanding the identity $(\Phi + \varepsilon \vec{f}) \circ (\Phi^{-1} + \varepsilon \delta \Phi^{-1})(\vec{y}) = \vec{y}$, we get

$$\delta \Phi^{-1}(\vec{y}) = -(D_{\vec{x}}\Phi)^{-1}(\Phi^{-1}(\vec{y}, t), t) \cdot \vec{f}(\Phi^{-1}(\vec{y}, t), t).$$

We also need

$$\frac{\partial \vec{g}}{\partial t} = \frac{\partial}{\partial t} \vec{f}(\Phi^{-1}(\vec{y}, t), t) = \frac{\partial \vec{f}}{\partial t} - D_{\vec{x}}\vec{f} \cdot (D_{\vec{x}}\Phi)^{-1} \cdot \frac{\partial \Phi}{\partial t}.$$

Step 2. We now can calculate the first variation $\delta \vec{v}$ of $\vec{v}$ by using the results of Step 1.

$$\begin{aligned}
\delta \vec{v} &= \frac{\partial}{\partial \varepsilon} \left[\frac{\partial}{\partial t}(\Phi + \varepsilon \vec{f})((\Phi + \varepsilon \vec{f})^{-1}(\vec{y}, t), t) \right]_{\varepsilon = 0} \\
&= \frac{\partial \vec{f}}{\partial t}(\Phi^{-1}(\vec{y}, t), t) - D_{\vec{x}}\frac{\partial \Phi}{\partial t}(\Phi^{-1}(\vec{y}, t), t) \cdot (D_{\vec{x}}\Phi)^{-1}(\Phi^{-1}(\vec{y}, t), t) \\
&\qquad \cdot \vec{f}(\Phi^{-1}(\vec{y}, t), t) \\
&= \frac{\partial \vec{g}}{\partial t}(\vec{y}, t) + D_{\vec{x}}\vec{f} \cdot (D_{\vec{x}}\Phi)^{-1} \cdot \frac{\partial \Phi}{\partial t} - D_{\vec{x}}\frac{\partial \Phi}{\partial t} \cdot D_{\vec{x}}\Phi \cdot \vec{g}.
\end{aligned}$$

Then, since

$$D_{\vec{x}}\frac{\partial \Phi}{\partial t}(\Phi^{-1}(\vec{y}, t), t) \cdot D_{\vec{x}}\Phi(\Phi^{-1}(\vec{y}, t), t) = D_{\vec{y}}\vec{v}(\vec{y}, t)$$

and

$$D_{\vec{x}}\vec{f}(\Phi^{-1}(\vec{y}, t), t) \cdot (D_{\vec{x}}\Phi)^{-1}(\Phi^{-1}(\vec{y}, t), t) = D_{\vec{y}}\vec{g}(\vec{y}, t),$$

we finally have

$$\delta \vec{v} = \frac{\partial \vec{g}}{\partial t} + D_{\vec{y}}\vec{g} \cdot \vec{v} - D_{\vec{y}}\vec{v} \cdot \vec{g}.$$

Step 3. So the first variation of E is

$$\delta E = -2 \int_0^1 \int_{\mathcal{D}} (L\vec{v} \cdot \delta\vec{v}) d\vec{y} dt$$

$$= -2 \int_0^1 \int_{\mathcal{D}} \left(L\vec{v} \cdot \frac{\partial \vec{g}}{\partial t} + D_{\vec{y}}\vec{g} \cdot \vec{v} - D_{\vec{y}}\vec{v} \cdot \vec{g} \right) d\vec{y} dt.$$

Integrating by parts, as usual, we get

$$\delta E = -2 \int_0^1 \int_{\mathcal{D}} \sum_j \left(\frac{\partial}{\partial t}(Lv_j) + \sum_i Lv_i \cdot \frac{\partial v_i}{\partial y_j} + \sum_i \frac{\partial}{\partial y_i}(Lv_j \cdot v_i) \right) g_j \, d\vec{y} dt.$$

So if we want $\delta E = 0$ for all $\vec{g}$, we get that $\vec{v}$ is a solution of the following equation, called *EPDiff* (which stands for Euler-Poincaré equation for diffeomorphisms):

$$\frac{\partial}{\partial t}(L\vec{v}) + (\vec{v} \cdot \vec{\nabla})(L\vec{v}) + (\text{div}\vec{v}) \cdot L\vec{v} + L\vec{v} \cdot D\vec{v} = 0. \tag{5.3}$$

Example (The Case of Dimension 1). In dimension 1, the vector $\vec{v}$ reduces to a scalar function $u(x,t)$, and Equation (5.3) simplifies to

$$(Lu)_t + u \cdot (Lu)_x + 2u_x \cdot Lu = 0.$$

In particular, if $L = \text{Id}$, then u satisfies Burger's equation,

$$u_t + C \cdot u \cdot u_x = 0, \quad \text{for } C = 3.$$

Then Φ itself satisfies the equation

$$\Phi_{tt}\Phi_x + 2\Phi_{tx}\Phi_t = 0.$$

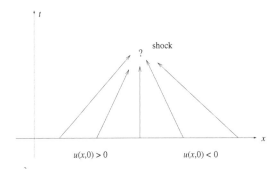

Figure 5.14. Burger's equation $u_t + C \cdot u \cdot u_x = 0$ is the simplest example of a shock–wave-producing equation. The level lines of u are straight lines whose slope equals C times their value. We can easily check that if $u(x_0, 0) = a$, then $u(x_0 + C \cdot a \cdot t, t) \equiv a$.

5.4. Modeling Geometric Variations by Metrics on Diff

This should not be smoothly integrable, and, in fact, it is the simplest example of a shock–wave-producing equation; see Figure 5.14.

Example (The Group of Volume-Preserving Diffeomorphisms).

Theorem 5.5 (V. Arnold [9]). *If $\mathcal{G}$ is the group of volume-preserving diffeomorphisms and $L = \mathrm{Id}$, giving the L^2 metric*

$$\| \vec{v} \|^2 = \int_D \| \vec{v}(\vec{y}) \|^2 d\vec{y},$$

then the geodesics are given by Euler's equation for inviscid, incompressible fluid.

If $\mathcal{G}$ is the group of volume-preserving diffeomorphisms, then in order to define a variation in the volume-preserving group, $\vec{v}$ and $\vec{g}$ must be such that $\mathrm{div}(\vec{v}) = 0$ and $\mathrm{div}(\vec{g}) = 0$. So we must have $\delta E = 0$ for all $\vec{g}$ such that $\mathrm{div}(\vec{g}) = 0$. Then by using Lagrange multipliers, we get

$$\frac{\partial}{\partial t}\vec{v} + (\vec{v} \cdot \vec{\nabla})(\vec{v}) + \frac{1}{2}\vec{\nabla}(\| \vec{v} \|^2) = \vec{\nabla}\tilde{p}$$

for some $\tilde{p}$, or, combining the two gradient terms,

$$\frac{\partial}{\partial t}\vec{v} + (\vec{v} \cdot \vec{\nabla})(\vec{v}) = \vec{\nabla}p$$

for some p. This equation is Euler's equation for inviscid, incompressible fluid (p is a pressure and $(\vec{v} \cdot \vec{\nabla})(\vec{v})$ is called the "transport term").

Do solutions to the general EPDiff equation exist and are they unique? It seems likely that *if the degree of differentiability k in the metric is large enough compared to the dimension of the space*, then (a) there is always at least one geodesic joining any two diffeomorphisms; (b) it is unique if the diffeomorphisms are sufficiently close; (c) given a tangent vector at a diffeomorphism ϕ, there is a unique geodesic through ϕ with this tangent vector; and, finally, (d) local geodesics can always be prolonged indefinitely. Trouvé and Younes [215, 217] have proven most of this, although in a slightly different setting. This is a much more technical matter than this book can treat.

Geodesics on $\mathrm{Diff}(\mathcal{D})$ have been applied to the analysis of medical image data by the team led by M. Miller [18, 161, 162]. They use the linearization technique introduced in Section 5.3.3 and have, for example, found mean shapes and modes of variations of the hippocampus for normal, Alzheimer's and schizophrenic patients, as wellas of normal and diseased hearts. An example of their work is shown in Figure 5.15.

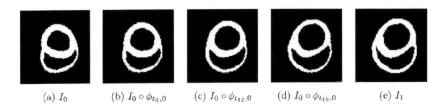

(a) I_0 (b) $I_0 \circ \phi_{t_6,0}$ (c) $I_0 \circ \phi_{t_{12},0}$ (d) $I_0 \circ \phi_{t_{19},0}$ (e) I_1

Figure 5.15. A geodesic in Diff($\mathbb{R}^2$) which joins a slice of a normal canine heart shown in (a) I_0 with the corresponding slice (e) I_1 of a failing canine heart. Images (b)–(d) show intermediate points on this geodesic. (Courtesy of Yan Cao [18].)

5.4.2 The Space of Landmark Points

When any actual computation is done with the group of diffeomorphisms, a finite-dimensional approximation must be made. The standard way this is done is to replace the continuum $\mathbb{R}^n$ by a finite set of points $\{P_1, \cdots, P_N\}$ and to represent a diffeomorphism ϕ by the discrete map $\{P_1, \cdots, P_N\} \mapsto \{\phi(P_1), \cdots, \phi(P_N)\}$. Define the space of N landmark points in $\mathbb{R}^n$ as

$$\mathcal{L}^N(\mathbb{R}^n) = \left\{ \{P_1, \cdots, P_N\} \;\middle|\; \begin{array}{l} \text{if } n = 1,\, P_1 < P_2 < \cdots < P_N \\ \text{if } n > 1,\, \forall i < j,\, P_i \neq P_j \end{array} \right\}.$$

The mathematical study of this space and its quotients by translations and orthogonal transformations of $\mathbb{R}^n$ was initiated by D. Kendall in 1977. There is now an extensive literature about this space in the statistics literature (see, e.g., [27, 63]). Landmark point theory has been developed in these books by using either (a) the sum-of-squares Euclidean distances between corresponding landmarks—which ignores the stretches and squeezes created by moving very close landmarks apart or moving distant landmarks close together—or (b) distances constructed with splines that did not force diffeomorphic warpings.

Instead, we describe here the Riemannian metric on landmark point space induced by the metric on Diff($\mathcal{D}$), which we believe is more natural and better adapted to most applications. This approach is based on the fact that the group Diff($\mathbb{R}^n$) of diffeomorphisms of $\mathbb{R}^n$ acts on $\mathcal{L}^N$. This action is transitive: any N-tuple can be taken to any other by some diffeomorphism. If we choose a basepoint $\{P_1^{(0)}, \cdots, P_N^{(0)}\} \in \mathcal{L}^N$, then mapping ϕ to $\{\phi(P_k^{(0)})\}$ gives a surjective map Diff($\mathbb{R}^n$) $\to \mathcal{L}^N$ and this leads to a metric on landmark point space.

More generally, when a Lie group G acts transitively on a manifold M, then every group element $g \in G$ defines a map $x \mapsto g(x)$ from M to M and every element $X \in \mathfrak{g}$, its Lie algebra, defines an infinitesimal map from M to M, that is, a vector field $\overline{X}$ on M. If we have an inner product on the Lie algebra $\mathfrak{g}$, then

5.4. Modeling Geometric Variations by Metrics on Diff

we get a "quotient" metric on M, defined by

$$\forall v \in T_x M, \; \| v \| = \inf_{X \in \mathfrak{g}} \left\{ \| X \| \; \big| \; \overline{X}(x) = v \right\}.$$

We can also think of this as an example of a *Riemannian submersion*. In general, a smooth mapping between Riemannian manifolds $f : M \to N$ is called a submersion if, for all $x \in M, y = f(x) \in N$, $T_x M$ splits orthogonally:

$$T_x M = T_x M^{\text{vert}} \oplus T_x M^{\text{hor}},$$

where $D_x f$ is zero on the vertical part and is an isometry between the horizontal part and $T_y N$. In our case, we choose a basepoint on $p^{(0)} \in M$ and define $f : G \to M$ by $f(g) = g(p^{(0)})$. The inner product on $\mathfrak{g}$ extends to a right-invariant metric on G, and, by using the quotient metric on M, f is a Riemannian submersion.

Going back to the group of diffeomorphisms and the space of landmark points, if k is large enough, the Sobolev metrics $\| v \|_k$ that we have defined on the group of diffeomorphisms define quotient metrics on the finite-dimensional manifolds $\mathcal{L}^N$. To describe these, we need the idea of *Green's functions*. A good introduction using one of the simplest cases $L = \triangle, n = 3$, can be found in [207], Chapter 7, especially Theorem 2, Section 7.2; or in [204], Chapters 1 and 2. Suppose L is a positive definite self-adjoint differential operator on some space M. Then the Green's function $G(x, y)$ is defined as a function on $M \times M$ with the property: for all functions h on M, the convolution $f = G * h$ is the solution of $Lf = h$. In other words, the differential operator L and convolution operator $G*$ are inverse to each other. Because L is self-adjoint, so is $G*$, which means $G(x, y)$ is symmetric in x and y. The function G is a nice smooth function except on the diagonal $x = y$, where it has some sort of singularity. If, as above, L is used to define a symmetric inner product $\langle f, g \rangle_L = \int_M Lf(x) \cdot g(x) dx$ on functions, then we get

$$\langle f, G(\cdot, P) \rangle_L = \int_M Lf(x) G(x, P) dx = (G * Lf)(P) = f(P),$$

which is called the *reproducing kernel property* of G. This property is also described by saying:

$$LG(\cdot, P) = \delta_P, \quad \text{the "delta function" at } P.$$

Delta functions are the simplest examples of *Schwartz distributions*. We will review the basic ideas behind distributions in Chapter 6, Section 6.5.2. If we smooth the Green's function $G(x, P)$ near $x = P$ and compute LG, or if we calculate LG by finite differences, we will get a result that is zero (or near 0) away from P but with a very large positive value very near P whose integral is approximately 1. See Exercise 6 at the end of this chapter for some simple explicit Green's functions and their singularities.

We apply this to the differential operator $L_k = (I - \triangle)^k$ on $\mathbb{R}^n$. Then the corresponding G_k is easy to determine by using Fourier transforms. First, because L_k is translation invariant, so is G_k; hence, $G_k(x, y) = G_k(x - y, 0)$ and we can just think of G_k as a function of one variable x. Second, by the way derivatives behave under Fourier transform, we have $\widehat{L_k f} = \sum_i (1 + |\xi_1|^2 + \cdots + |\xi_n|^2)^k \cdot \hat{f}$. If $\widehat{G_k}$ is the Fourier transform of G_k, the equation $f = G_k * L_k f$ is equivalent to $\hat{f} = \widehat{G_k} \cdot (1 + \|\xi\|^2)^k \cdot \hat{f}$, or

$$G_k = \left((1+ \|\xi\|^2)^{-k}\right)\widehat{},$$

where the "hat" at the end means to take the inverse Fourier transform of the preceding term. In fact, this is a 19th century special function, essentially a Bessel K-function, which means, for instance, it is available in MATLAB. But, for us, the main usefulness of this formula is to deduce that G_k has $2k - n - 1$ continuous derivatives. This follows immediately by applying the elementary fact that the Fourier transform of an L^1 function is continuous. (In fact, for $2k - n < 0$, G_k has a pole $\|\vec{x}\|^{2k-n}$ at $\vec{x} = 0$ and, for $2k - n = 0$, it looks like $\log(\|\vec{x}\|)$ near $\vec{x} = 0$.)

We are now in a position to work out the quotient metric on $\mathcal{L}^N$, giving us a finite-dimensional approach to diffeomorphism metrics. We assume $k \geq (n + 1)/2$, so G_k is everywhere continuous. Consider the mapping taking the Lie algebra of Diff, that is, the vector space of vector fields $\vec{v}$ on $\mathbb{R}^n$, to the tangent space $T_\mathcal{P} \mathcal{L}^N$, where $\mathcal{P} \in \mathcal{L}^N$, by

$$\vec{v} \mapsto \vec{v}(P_1), \cdots, \vec{v}(P_n).$$

The vertical subspace is the set of $\vec{v}$ such that $\vec{v}(P_i) = 0$ for all i. Let $\vec{e}_p$ be the pth unit vector in $\mathbb{R}^n$. The main property of the Green's function is that $\langle \vec{v}, G(\cdot - P_i)\vec{e}_p \rangle_k = \vec{v}_p(P_i)$; hence, the $N \cdot n$ functions $G(x - P_i)\vec{e}_p$ are all in the horizontal subspace. Moreover, the functions $\{G(x - P_i)\}$ are linearly independent because

$$\sum a_i G(x - P_i) \equiv 0 \Rightarrow \forall \vec{v}, \sum_i a_i \vec{v}_p(P_i) = \langle \vec{v}, \sum_i a_i G(\cdot - P_i)\vec{e}_p \rangle = 0.$$

Since $N \cdot n$ is the dimension of $\mathcal{L}^N$, these must span the horizontal subspace. Note that, by the main property,

$$\langle G(\cdot - P_i), G(\cdot - P_j) \rangle_L = G(P_i - P_j).$$

5.5. Comparing Elastic and Riemannian Energies

Because the functions $\{G(x-P_i)\}$ are linearly independent, the matrix $G(\mathcal{P})_{ij} = G(P_i - P_j)$ is a positive definite symmetric matrix. Let $G(\mathcal{P})_{i,j}^{-1}$ be its inverse. Then, we get Proposition 5.6:

Proposition 5.6. *The horizontal lift of the tangent vector* $(\vec{t}_1, \cdots, \vec{t}_N) \in T_\mathcal{P} \mathcal{L}^N$ *is the global vector field*

$$\vec{v}(x) = \sum_i G(x - P_i)\vec{u}_i, \quad \text{where } \vec{u}_i = \sum_j G(\mathcal{P})_{ij}^{-1}\vec{t}_j,$$

and its length is

$$\| (\vec{t}_1, \cdots, \vec{t}_N) \|_\mathcal{P}^2 = \sum_{i,j} G(\mathcal{P})_{i,j}^{-1} \langle \vec{t}_i, \vec{t}_j \rangle.$$

Proof: With $\vec{u}_i$ as above, we note that

$$\vec{v}(P_k) = \sum_i G(x - P_i)\vec{u}_i \Big|_{x=P_k} = \sum_{i,j} G(\mathcal{P})_{ki} G(\mathcal{P})_{i,j}^{-1} \vec{t}_j = \vec{t}_k.$$

Hence, $\vec{u}$ gives the horizontal lift of $(\vec{t}_1, \cdots, \vec{t}_N)$, and its length is thus

$$\left\langle \sum_i G(x - P_i)\vec{u}_i, \sum_j G(x - P_j)\vec{u}_j \right\rangle_k = \sum_{i,j} G(\mathcal{P})_{ij} \langle \vec{u}_i, \vec{u}_j \rangle$$

$$= \sum G(\mathcal{P})_{ij} G(\mathcal{P})_{ik}^{-1} G(\mathcal{P})_{jl}^{-1} \langle \vec{t}_k, \vec{t}_l \rangle$$

$$= \sum G(\mathcal{P})_{kl}^{-1} \langle \vec{t}_k, \vec{t}_l \rangle. \qquad \square$$

This gives a quite simple description of the quotient metric using only the values of the Green's function G_k. The geodesic equation on $\mathcal{L}^N$ is worked out in the appendix to this chapter (Section 5.8), and we simply state the result here:

$$\frac{dP_i}{dt} = \sum_j G(P_i - P_j)\vec{u}_j, \quad \frac{d\vec{u}_i}{dt} = -\sum_j \nabla G(P_i - P_j)(\vec{u}_i, \vec{u}_j).$$

5.5 Comparing Elastic and Riemannian Energies

At this point, we have two quite different ways to assign "energies" to diffeomorphisms, ways to measure how far they differ from the identity:

$$E_{\text{elast}}(\phi) = \int_\mathcal{D} e_s(\nabla \phi(\vec{x})) d\vec{x},$$

$$E_{\text{Riem}}(\phi) = \inf_{\{\Phi(\vec{x},t) \mid \Phi(\cdot,0) = \text{Id}, \Phi(\cdot,1) = \phi\}} \int_0^1 \int_\mathcal{D} \sum_{|\alpha| \le k} \| D^\alpha(\Phi_t \circ \Phi^{-1})(\vec{y}) \|^2 d\vec{y} dt.$$

These are, roughly speaking, solid and liquid approaches to deformation, respectively. The *principle of least action* gives a common framework in which the two approaches can be compared. Newton's law $F = ma$ leads to differential equations for predicting the evolution of physical systems, but an equivalent formulation is that the evolution from time $t = 0$ to time $t = 1$ is that path through state space that minimizes the integral of kinetic energy minus potential energy. In the case of continuum mechanics, this means choosing a path Φ in the diffeomorphism group that minimizes

$$\text{action} = \int_0^1 \int_{\mathcal{D}} \left(\rho \parallel \Phi_t(\vec{x}, t) \parallel^2 - e_s(\nabla \Phi(\vec{x}, t)) \right) d\vec{x} dt,$$

where ρ is the density in material space and all integrals are over material space.

The elastic approach ignores the kinetic energy and just looks at the difference of the potential energy in the deformed and undeformed configurations. In contrast, the liquid approach ignores the potential energy and considers analogs of the kinetic energy. The relationship is simplest for incompressible fluids: in this case, ρ is constant and e_s is taken to be 0 for volume-preserving diffeomorphisms and infinity otherwise. Thus, we replace potential energy by the hard constraint that the Jacobian of Φ is identically equal to 1. Then the action becomes the same as the Riemannian path energy for $k = 0$ in the Sobolev norm. In other words, the least action paths are just the geodesics on the group of volume-preserving diffeomorphisms for the L^2 metric. This was discovered by V. Arnold in his groundbreaking paper [9].

The Riemannian approach takes as its starting point Arnold's discovery that physical motion in fluids are like geodesics, but generalizes it to compressible fluids. We start with the fact that, in fluids, particles "forget where they came from." Thus, if $d(\phi_1, \phi_2)$ represents how far ϕ_1 is from ϕ_2, the fact that the fluid forgets its past means that we should use right-invariant distance measures,

$$d(\phi_1, \phi_2) = d(\phi_1 \circ \psi, \phi_2 \circ \psi).$$

The function E_{Riem} is the square of the Riemannian metric distance, and this has two consequences:

$$E_{\text{Riem}}(\phi) = d(\phi, \text{Id})^2 = d(\text{Id}, \phi^{-1})^2 = E_{\text{Riem}}(\phi^{-1}),$$

and the energy triangle inequality

$$E_{\text{Riem}}(\psi \circ \phi)^{1/2} = d(\psi \circ \phi, \text{Id}) \leq d(\psi \circ \phi, \phi) + d(\phi, \text{Id})$$
$$= d(\psi, \text{Id}) + d(\phi, \text{Id}) = E_{\text{Riem}}(\phi)^{1/2} + E_{\text{Riem}}(\psi)^{1/2}$$

What are the essential differences between these two energy measures and how do we choose one or the other? The differences seem to be

5.5. Comparing Elastic and Riemannian Energies

1. E_{elast} involves only the first spatial derivative $\nabla\phi$, whereas E_{Riem} involves derivatives of any order $D^\alpha\phi$—and, in fact, if we want this energy to go to infinity when ϕ becomes singular (i.e., its derivative $D\phi$ or its inverse $D\phi^{-1}$ diverges at some point), then it *must* use higher derivatives.

2. E_{Riem} satisfies the energy triangle inequality, whereas, if we want this energy to go to infinity when ϕ collapses, E_{elast} never does. In fact, it must have faster growth when the Jacobian of ϕ goes to 0 than the triangle inequality allows.

We discuss these two points in the next two subsections.

5.5.1 Distance from the Identity to the Boundary of $\mathcal{G}$

It is instructive to look at a very simple example to see how the presence of higher derivatives keeps the boundary of $\mathcal{G}$ far away. Take the simplest case, $\mathcal{D} = [0,1] \subset \mathbb{R}$, and look at a specific path from the identity to a boundary point of $\mathcal{G}$, that is, to a map from $[0,1]$ to itself, which is not a diffeomorphism. For all $a \in]0,1[$, define the function f_a from $[0,1]$ to itself by $f_a(x) = x/a$ if $x \in [0,a]$, and $f_a(x) = 1$ for $x \geq a$. Then, as diagrammed in Figure 5.16, define a path Φ from e to the non-homeomorphism f_a by

$$\Phi(x,t) = \begin{cases} x(1 + \frac{t}{a} - t) & \text{if } 0 \leq x \leq a, \\ x + t - xt & \text{if } a \leq x \leq 1. \end{cases}$$

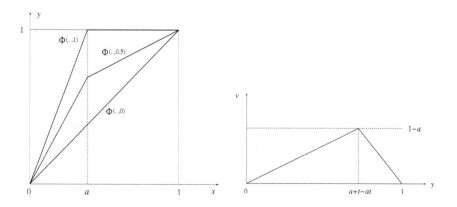

Figure 5.16. An example of a path from the identity to a non-homeomorphism (left), and the corresponding velocity (right).

The velocity of the motion is then

$$v(y,t) = \frac{\partial \Phi}{\partial t}(\Phi^{-1}(y,t),t) = \begin{cases} \frac{y(1-a)}{a+t-at} & \text{if } 0 \leq y \leq a+t-at, \\ \frac{1-y}{1-t} & \text{if } a+t-at \leq y \leq 1. \end{cases}$$

We suppose first that the norm on v is defined using only the function itself (i.e., $\|v\|_S^2 = \int_{\mathcal{D}} v(y)^2 dy$). Then the length of the path Φ is

$$\ell(\Phi) = \int_0^1 \left(\int \left\| \frac{\partial \Phi}{\partial t}(\Phi^{-1}(\vec{y},t),t) \right\|_S^2 d\vec{y} \right)^{1/2} dt = \frac{1}{\sqrt{3}}(1-a).$$

As a tends to 1, this shows that the distance from the identity to the boundary of $\mathcal{G}$ can be made arbitrarily small. The conclusion is that we did not use a good metric. In fact, in this metric the infimum of path lengths between any two diffeomorphisms is zero; see Exercise 5 at the end of this chapter.

Let's see what happens if instead we take a Sobolev norm having more derivatives in it; for example,

$$\|v\|_S^2 = \int_{\mathcal{D}} (v(y)^2 + v'(y)^2) dy \quad \text{or} \quad \int_{\mathcal{D}} (v(y)^2 + v'(y)^2 + v''(y)^2) dy.$$

In order to use the same example, we have to smooth the functions a bit (see Figure 5.17). If we calculate the Sobolev norm of the velocity, we obtain for the

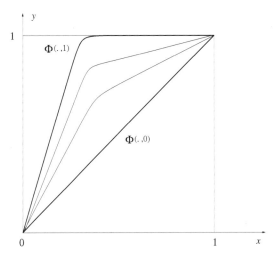

Figure 5.17. Example of a smooth path from the identity to a non-homeomorphism.

5.5. Comparing Elastic and Riemannian Energies

first derivative of the velocity

$$\left(\int_0^1 v'(y)^2 dy\right)^{1/2} \simeq \sqrt{\frac{1-a}{(1-t)(a+t-at)}}.$$

(The approximation comes from the fact that the vector is smoothed near a but the error can be made as small as we want.) Then if we integrate with respect to t, since $\int_0^1 \frac{dt}{\sqrt{1-t}}$ is convergent (with value 2), we will still have a finite distance from the identity to f_a and, for a near 1, it is approximately $2\sqrt{1-a}$. Again, the distance from the identity to boundary of $\mathcal{G}$ can be made arbitrarily small. However, taking infimum over path lengths, L. Younes discovered that this Riemannian metric leads to an elegant global metric on $\mathcal{G}$. It is given in terms of the quite nonphysical elasticity energy $e_s(w) = \sqrt{w}$, $E_s(\phi) = \int_0^1 \sqrt{\phi'(x)} dx$; see Exercise 3 at the end of this chapter.

But now if we look at the second derivative of the velocity, we see that something new happens. Let $\Phi(x,t)$ be any smooth path from Id to f_a. The velocity of this path is $v(y,t) = \frac{\partial \Phi}{\partial t}(\Phi^{-1}(y,t),t)$. The first derivative of the velocity is then

$$v'(y,t) = \frac{\partial v}{\partial y}(y,t) = \frac{\partial^2 \Phi}{\partial x \partial t}(\Phi^{-1}(y,t),t) \times \frac{1}{\frac{\partial \Phi}{\partial x}(\Phi^{-1}(y,t),t)}$$

$$= \frac{\partial}{\partial t} \log\left(\frac{\partial \Phi}{\partial x}\right)(\Phi^{-1}(y,t),t).$$

Now, if we look at the length of this path, we can use Cauchy's inequality to get a lower bound:

$$\ell(\Phi) \geq \int_0^1 \left(\int_0^1 v''(y,t)^2 \, dy\right)^{1/2} dt \geq \int_0^1 \left|\int_0^1 v''(y,t) dy\right| dt$$

$$= \int_0^1 |v'(1,t) - v'(0,t)| dt \geq \left|\int_0^1 (v'(1,t) - v'(0,t)) dt\right|$$

$$= \left|\log(\frac{\partial \Phi}{\partial x}(1,1)) - \log(\frac{\partial \Phi}{\partial x}(1,0)) - \log(\frac{\partial \Phi}{\partial x}(0,1)) + \log(\frac{\partial \Phi}{\partial x}(0,0))\right|.$$

Since $\Phi(x,0) = x$, $\Phi(x,1) = f_a(x)$, these four terms are $\log(0)$, $\log(1)$, $\log(1/a)$, $\log(1)$, respectively, so the value is infinite.

These examples illustrate the fact that Riemannian metrics need a certain number of derivatives before the distance to the boundary is infinite. In fact, the general result is that if we are considering diffeomorphisms that are the identity outside a fixed ball in $\mathbb{R}^n$, this group has finite diameter[3] in all Sobolev metrics with less than $n/2+1$ derivatives. In contrast, if we use more $n/2+1$ derivatives, the distance to the boundary is infinite.

5.5.2 Elasticity and the Triangle Inequality

No power E_s^α of elasticity energies can satisfy the energy triangle inequality. This conclusion follows using the same example as in Section 5.5.1, the path Φ from the identity to the non-homeomorphism f_a. We give a proof by contradiction.

Proof: Let $b = 1 - a$ to simplify notation. Along the path Φ, $\Phi(a,t)$ moves linearly from a to 1. Let t_n be the increasing sequence of times when $\Phi(a,t_n) = 1 - b^n$. Then define maps g_n—which carry $1 - b^n$ to $1 - b^{n+1}$ and are linear in between—by

$$\Phi(\cdot, t_{n+1}) = g_n \circ \Phi(\cdot, t_n)$$

$$g_n(x) = \begin{cases} x \cdot \dfrac{1-b^{n+1}}{1-b^n} & \text{if } 0 \leq x \leq 1 - b^n, \\ (1 - b^{n+1}) + b \cdot (x - 1 + b^n) & \text{if } 1 - b^n \leq x \leq 1. \end{cases}$$

The triangle inequality for E_s^α would imply

$$E_s^\alpha(\Phi(\cdot, t_{n+1})) \leq E_s^\alpha(\Phi(\cdot, t_n)) + E_s^\alpha(g_n);$$

hence, by induction,

$$E_s^\alpha(\Phi(\cdot, t_n)) \leq \sum_{k=1}^{n-1} E_s^\alpha(g_k).$$

But

$$E_s(\Phi(\cdot, t_n)) = \int_0^1 e_s(\Phi_x(x,t))dx = a \cdot e_s\left(\frac{1-b^n}{a}\right) + b \cdot e_s(b^{n-1}),$$

$$E_s(g_n) = \int_0^1 e_s(g_n'(x))dx = (1 - b^n) \cdot e_s\left(\frac{1-b^{n+1}}{1-b^n}\right) + b^n \cdot e_s(b).$$

As $n \to \infty$, $b^n \to 0$; hence, $e_s(b^n) \to \infty$, so by the above, $E_s(\Phi(\cdot, t_n)) \to \infty$.

$$E_s(g_n) \leq (C + e_s(b)) \cdot b^n; \quad \text{hence } \sum_{k=1}^\infty E_s^\alpha(g_k) < \infty.$$

This is a contradiction, so elastic energies cannot satisfy the triangle inequality. □

In fact, the same argument also applies not only to powers but to energies of the form $F(\int e_s(J))$ for any Hölder continuous increasing F. Another way to see

---

[3]There is a constant C, depending on the ball and the metric, such that any two diffeomorphisms can be joined by a path of length at most C.

what is happening is the following: in terms of the energy density e_s, if $\alpha \leq 1$, a slight strengthening of the energy triangle inequality for E_s^α is given by the local inequality:

$$e_s^\alpha(X \cdot Y) \leq e_s^\alpha(X) \cdot \det(Y)^\alpha + e_s^\alpha(Y). \tag{5.4}$$

In fact, let $p = 1/\alpha$ and let $\| f \|_p$ be the L^p norm. Using this local inequality, we get back the triangle inequality,

$$E_s^\alpha(\phi \circ \psi) = \| e_s^\alpha((D\phi \circ \psi) \cdot D\psi) \|_p \leq \| e_s^\alpha(D\phi \circ \psi) \cdot \det(D\psi)^\alpha \|_p$$
$$+ \| e_s^\alpha(D\psi) \|_p$$
$$= \| e_s^\alpha(D\phi) \|_p + \| e_s^\alpha(D\psi) \|_p = E_s^\alpha(\phi) + E_s^\alpha(\psi)$$

(by using the change of variables $\vec{y} = \psi(\vec{x})$ in the first term). But the inequality (5.4) is much too strong an inequality on e_s and prevents it from growing as it must when its matrix argument goes to the boundary of $GL(n)$.

5.6 Empirical Data on Deformations of Faces

What deformations do faces actually undergo? We can approximate the diffeomorphism ϕ by using landmark points $\{P_\alpha\}$ and analyze this from the data. Ideally, this should be done by using the natural metric on landmark space described above, but this has not been done. The following two groups, however, have done analyses using the naive linearization of the space of diffeomorphisms given by the coordinates $\phi(P_\alpha)$.

1. Cootes and Taylor [66] took a data base of a large number of faces, and then labeled by hand 169 landmark points p_k on each (see Figure 5.18). If the diffeomorphism ϕ is

   $$\phi : \mathcal{D}_0 \longrightarrow \mathcal{D}_1,$$

   where $\mathcal{D}_0$ is the standard face and $\mathcal{D}_1$ is the observed face, then they replaced ϕ by the vector of coordinates $\{\phi(p_k)\} \in \mathbb{R}^{2 \times 169}$. As above, they got a finite-dimensional approximation to the infinite-dimensional group $\mathcal{G}$:

   $$\mathcal{G} \longrightarrow \mathbb{R}^{2 \times 169}$$

   $$\phi \longrightarrow \{\phi(p_k)\}.$$

   Conversely, having the set of $\phi(p_k)$, we can define a particular ϕ with these values by triangulating the full face with the vertices p_k and using bilinear interpolation.

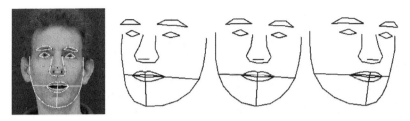

Figure 5.18. Figures from Cootes-Taylor [66] showing the contours along which landmark points were chosen and the diffeomorphism resulting from rotation of the head. (Courtesy of T. Cootes.)

2. Hallinan [97, 99] took a *dense grid* of landmark points and approximated ϕ by the set $\{\phi(p_{ij})\}_{1 \leq i,j \leq N}$, again replacing the diffeomorphism ϕ by a finite-dimensional vector $\vec{\phi} \in \mathbb{R}^{2N^2}$. He then solved for ϕ on large data sets by using methods of modified gradient descent (described below) with one of the elasticity priors E described in Section 5.2. This then gave him empirical data for the deformations that actually occur and so allowed him to bootstrap the process and get a refined prior for deformations, one based on the data.

The method is to start with a lot of faces α, to form for each a discretized estimate $\phi^{(\alpha)} \in \mathbb{R}^{2N^2}$ of ϕ, and then apply to principal components analysis to the $\phi^{(\alpha)}$ in the naive linear coordinates $\mathbb{R}^{2N^2}$. We let $\overline{\vec{\phi}}$ be the mean of the data $\phi^{(\alpha)}$, and ψ_i be the ordered eigenvectors of the covariance matrix of the data $\phi^{(\alpha)}$; ψ_i is also called the ith eigenwarp. Then we can write for an observed face $\vec{\phi}$:

$$\vec{\phi} = \overline{\vec{\phi}} + \sum_{i=1}^{2N^2} \lambda_i \psi_i.$$

This is the Gaussian approximation of the distribution of ϕ in the vector space of landmark point coordinates. The results of Hallinan's calculation of the eigenwarps are illustrated in Figure 5.19, which shows, for each of the first few ψ, the four maps: $\vec{x}+3\psi(\vec{x}), \vec{x}+\psi(\vec{x}), \vec{x}-\psi(\vec{x})$, and $\vec{x}-3\psi(\vec{x})$.

We can also use the ψ_i to create a customized strain energy density e_s to reproduce the specific variations of faces. If we triangulate the domain $\mathcal{D}$, then $\int e_s(D\phi(\vec{x}), \vec{x})d\vec{x}$ may be replaced by the following finite approximation:

$$\sum_{\sigma = \langle P,Q,R \rangle} e_\sigma(\phi(Q) - \phi(P), \phi(R) - \phi(P)),$$

where the sum is over the triangles in the triangulation. With this approach, we can model relatively rigid pairs of parts and specific relative flexibility. What

5.6. Empirical Data on Deformations of Faces

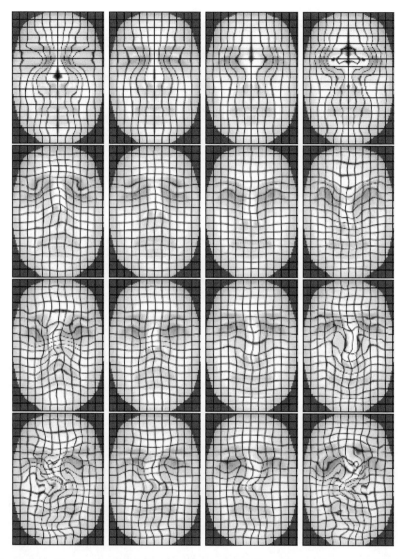

Figure 5.19. The first four eigenwarps from the work of Hallinan. Each row shows one warp, by moving a grid a little or a lot forward or backward under the warp. The top row is a warp that moves the eyes closer together or farther apart. The second row is a warp in which the nose is moved up or down relative to the eyes. The third row shows a warp in which the nose is shortened or lengthened. The bottom row shows the nose twisting left or right. (From [97, Figure 4.18], [99].)

we cannot model is the requirement or tendency to bilateral symmetry (e.g., the fact that the two eyes move symmetrically toward or away from the midline in different faces).

5.7 The Full Face Model

5.7.1 The Model

The complete model for faces following Hallinan combines (a) intensity variations given by eigenfaces reflecting varying illumination, and (b) geometric variations given by a warping and a noise residual. The random variables of such a model are:

- the observed face I on a domain $\mathcal{D}$;
- the illumination coefficient vector $\vec{c}$;
- the warping $\phi : \mathcal{D} \longrightarrow \mathcal{D}$.

Finally, putting all this together, we get an energy function

$$E(I, \vec{c}, \phi) = a_1 \iint \left[I(\phi(x,y)) - \bar{I}(x,y) - \sum_\alpha c_\alpha F^{(\alpha)}(x,y) \right]^2 dx dy$$
$$+ a_2 \iint \operatorname{tr}(D\phi^t \cdot D\phi)(1 + (\det D\phi)^{-2}) \, dx dy$$
$$+ a_3 \Big(\text{quadratic function of } \vec{c}\Big).$$

Notice that this model uses strain energy density. In finite dimensions, we can get a probability distribution setting

$$P(I, \vec{c}, \phi) = \frac{1}{Z} e^{-E(I,\vec{c},\phi)}.$$

Of course, the second term of the energy function could use Riemannian metrics on the space of diffeomorphisms also. We mention in passing yet another approach, which combines metrics on Diff with compensation for image noise: the *metamorphosis* model of Younes [216]. He takes the product of the space of images and the space of diffeomorphisms, puts a product Riemannian metric on them, and solves for geodesics between two images for the quotient metric on the space of images. This means we seek a path of shortest length in which, at every step, we morph a little bit and add a little bit of "noise." This allows a child's face to form wrinkles or—in a typical medical application—allows a tumor to be formed while deforming a template body map to a patient's scan.

The basic calculation required to use this class of models is: given the observed face I, find or approximate the hidden variables $\vec{c}$ and ϕ that minimize E. In our case, E is quadratic in the c_α and so minimizing E in $\vec{c}$ given I and ϕ is just a simple problem of linear algebra. To minimize E in ϕ, one can use general

5.7. The Full Face Model

techniques from the theory of numerical nonlinear optimization. These all start from the gradient descent method. The idea is to generate a sequence ϕ_n such that $E(\phi_n)$ is decreasing (where $\vec{c}$ is fixed) by setting

$$\phi_{n+1} = \phi_n - \varepsilon \frac{\delta E}{\delta \phi}.$$

But this is *very* slow! We need to analyze this problem to improve this algorithm.

5.7.2 Algorithm VII: Conjugate Gradient

The general problem is to compute numerically the minimum of a real-valued function f. One possibility, as described above, is to make a gradient descent with a fixed step size ε. Another possibility is to choose this step in an "optimal" way: this is the aim of enhancing the gradient descent algorithm with line searches (also called the steepest gradient descent).

Algorithm (Gradient Descent with Line Search).

1. Start at an arbitrary point x_0.

2. Assume x_k is constructed, and let $g_k = \nabla f(x_k)$. Then construct, by a one-dimensional search, the next x_{k+1} as the point achieving the minimum of f on the half-line starting from x_k in the direction of $-g_k$.

3. Iterate until one or both of $\| g_k \|$ and $|f(x_k) - f(x_{k-1})|$ become sufficiently small.

But recall our discussion of gradients in Section 5.3.4. A function on $\mathbb{R}^n$ has a differential, its linear approximation. To convert a differential to a vector requires choosing an inner product. The usual ∇f is the vector obtained by the usual sum of squares inner product. But this inner product has no natural connection to the problem of minimizing f.

We now analyze the case of a quadratic function f defined on $\mathbb{R}^n$ by

$$f(x) = c + \langle b, x \rangle + \frac{1}{2} x^t A x, \tag{5.5}$$

where A is a symmetric positive-definite matrix, $b \in \mathbb{R}^n$ and $c \in \mathbb{R}$. At each point x, the gradient of f is given by $\nabla f(x) = Ax + b$ and the Hessian matrix is $D^2 f(x) = A$. Function f has a global minimum at $x = A^{-1} \cdot b$ where the

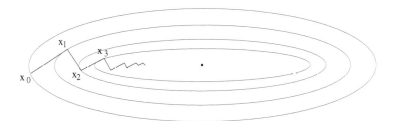

Figure 5.20. Example of a steepest gradient descent on a quadratic function f. At each step, the next search direction is orthogonal to the previous one. The algorithm will need a large number of steps to converge to the minimum.

gradient vanishes. The minimum of f on the line starting from x_k in the direction of $-g_k = -\nabla f(x_k)$ is achieved at

$$x_{k+1} = x_k - \alpha_k g_k \text{ with } \alpha_k = \frac{\langle g_k, g_k \rangle}{\langle g_k, A g_k \rangle}.$$

And the next direction will be $g_{k+1} = \nabla f(x_{k+1})$, which is orthogonal to g_k.

Thus, the main problem with the steepest gradient descent is that at each step we go in a direction that is orthogonal to the previous one (as illustrated by the "zigzag" in Figure 5.20), and moreover it may takes a large number of steps to get near the minimum. The algorithm is efficient when the minimum is "spherical" (i.e., when all the eigenvalues of A are quite the same), but it becomes very inefficient when it is in a "narrow valley" (i.e., when A has both very large and very small eigenvalues). The root of the problem is that we should not use the sum-of-squares inner product to define the gradient, but rather the inner product natural to the problem, namely, that defined by the Hessian of f: $x^t A x$. The modified gradient associated with this inner product is the vector $\tilde{g}$ such that

$$\forall y, \, (df, y) = y^t A \tilde{g},$$

where $(df, y) = $ directional derivative of f in the direction y. This means, simply, that $y^t \cdot (b + Ax) = y^t A \tilde{g}$, or

$$\tilde{g} = A^{-1} \cdot b + x = A^{-1} \cdot \nabla f(x).$$

Then, the global minimum is attained in a single step from x to $x - \tilde{g}$. Applying this to general nonlinear functions f is *Newton's method*:

At each x_k, define $g_k = D^2 f(x_k)^{-1} \cdot \nabla f(x_k)$.

This is great near the minimum for f but needs to be modified elsewhere (e.g., $D^2 f$ might not be invertible). This leads to the class of so-called *quasi-Newton methods*. See Chapter 10 in the excellent text [181] for details.

5.7. The Full Face Model

Conjugate gradients provide a remarkably simple way to imitate Newton's method without having to compute the Hessian $D^2 f$ at each step. In very high-dimensional situations, the Hessian can be difficult to store, let alone compute. As in Newton's method, the idea is that instead of looking for the minimum on the line orthogonal to the previous direction for the usual inner product of $\mathbb{R}^n$, we take the direction orthogonal for the scalar product defined by the Hessian matrix. Thus, the method looks like a Gram-Schmidt orthogonalization procedure.

Algorithm (Conjugate Gradient Algorithm).

1. Start at an arbitrary point x_0, and take for initial direction $h_0 = \nabla f(x_0) = g_0$.

2. Assume that x_k and h_k are constructed. Then define x_{k+1} as the point that achieves the minimum of f on the line starting from x_k in the direction of h_k. Let $g_{k+1} = \nabla f(x_{k+1})$, so that g_{k+1} and h_k are perpendicular, and let the next direction be the combination of these:

$$h_{k+1} = g_{k+1} + \lambda_k h_k, \quad \text{where} \quad \lambda_k = \frac{\langle g_{k+1} - g_k, g_{k+1} \rangle}{\langle g_k, g_k \rangle}.$$

3. Iterate until some criterion is met.

One can view adding a multiple of h_k to the gradient at x_{k+1} to get the new direction h_{k+1} as giving the gradient descent a sort of momentum. But where does this specific coefficient λ_k come from? To understand the behavior of this algorithm, we consider again the particular case of the quadratic function f given by Equation (5.5). In that case, $\nabla f(x) = b + A \cdot x$, hence $g_{k+1} - g_k = A \cdot (x_{k+1} - x_k)$, which is a multiple of $A \cdot h_k$. To find the multiple, we take the dot product with h_k and find

$$g_{k+1} = g_k - \mu_k A h_k, \quad \text{where} \quad \mu_k = \frac{\langle g_k, g_k \rangle}{\langle h_k, A h_k \rangle}.$$

We can prove in this case by simple induction, that the following conditions of orthogonality and conjugacy are satisfied:

1. $\text{span}(g_0, \cdots, g_k) = \text{span}(h_0, \cdots, h_k) = \text{span}(h_0, A h_0, \cdots, A^k h_0)$

2. $\{h_0, \cdots, h_k\}$ are orthogonal with respect to A.

It follows that after at most n steps, $g_k = 0$, and we have reached the global minimum of f. If, for example, the function f is a quadratic form on $\mathbb{R}^2$, the conjugate gradient algorithm will converge in at most two steps, even in the case of a "narrow valley" (see Figure 5.21).

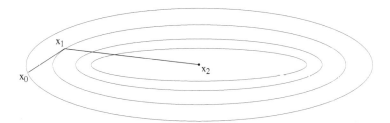

Figure 5.21. When f is a quadratic function on $\mathbb{R}^2$, the conjugate gradients algorithm converges in two steps.

5.7.3 Other Computational Issues

Gradient descent algorithms seem like simple straightforward methods, but there are several issues that they raise:

1. This sequence may converge only to a local minimum and not to the global one. Often we are not even aware of the many local minima lurking in a seemingly innocent problem. Hallinan [97] performed an experiment in which he took a set of random images of scenes J from real life, changed them to $I(x,y) = J(T_0(x,y))$ by an affine transformation T_0, and attempted to solve for this affine transformation by doing gradient descent with respect to T on

$$\iint (I(x,y) - J(T(x,y)))^2 dx dy.$$

Although this usually worked, it sometimes got stuck in surprising ways. Figure 5.22 shows two examples. Note that in the first, the white shirts of the girls are matching the white sand incorrectly, and in the second, the columns have locked in the wrong way.

2. As noted, to minimize any function f on a manifold $\mathcal{M}$, we need an inner product on the tangent space $T\mathcal{M}$ to convert df, which is a co-vector to ∇f, which is a vector (see also Subsection 5.3.4). If we are on some manifold other than Euclidean space, some inner product must be chosen, even for the conjugate gradient algorithm. If this is done badly, it can make gradient descent fail. We need to give the correct variables a chance to control the descent at the right time.

3. Discretization can cause artifacts; for example, it may create a lot of artificial local minima if we aren't careful (see Figure 5.23).

5.7. The Full Face Model

Figure 5.22. Two examples of gradient descent for fitting a warp falling into a local strain energy minimum: see text. In both cases, an image has been translated and rotated to see if matching this with the original undid the translation and rotation. The figure shows where realigning the modified image got stuck. But on the left, the girls shirts on the bottom have been incorrectly aligned with the sand between their skirts on the top. On the right, three of the columns have been matched to three wrong ones and fourth is matched to two columns. (From [97, Figure 4.5], [99].)

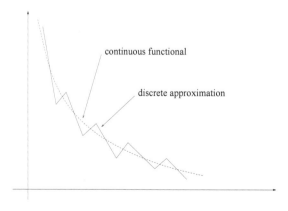

Figure 5.23. Discretization may create artificial local minima.

Figure 5.24. The present location is too far away from the sought-for minimum of the functional.

4. If we replace $\int_D (I_1(\phi(\vec{x})) - I_2(\vec{x}))^2 \, d\vec{x}$ by the approximation

$$\sum_k (I_1(\phi(p_k)) - I_2(p_k))^2 ,$$

then if $D\phi$ gets large we can miss parts of the domain of I_1 and get what looks like a perfect match, even though major features of I_1 have been omitted.

5. A major problem is that a decrease in E may be a little too far away to make its presence felt, that is, to affect the gradient (see Figure 5.24). This occurs, for instance, in E if we are trying to deform one face to another, and the eyes don't even overlap. Once they overlap, moving them closer will make the images more similar. Addressing this problem leads to a very central idea in applying gradient descen—the hierarchical approach. Here we describe two variants:

- *Method 1.* (used by Grenander and Matejic [153]) We factor ϕ into $\phi_1 \circ \phi_2 \circ \phi_3$ with the ϕ_k belonging to a sequence of subgroups. For example, we can have first a translation, then we might have a scaling-rotation, and finally a general diffeomorphism that is constrained to be small. This approach is shown in Figure 5.25.
- *Method 2.* (using multiple scales) The method here is to replace $\int_D (I_1(\phi(\vec{x})) - I_2(\vec{x}))^2 \, d\vec{x}$ by

$$\int_D ((G_\sigma * I_1)(\phi(\vec{x})) - (G_\sigma * I_2)(\vec{x}))^2 \, d\vec{x},$$

which means that we smooth the faces and so we lose small details. The key point of this method is that if we had disjoint eyes, say, then after having blurred our image, the eyes are no longer disjoint and then we can match them roughly. Once matched roughly, we decrease σ, making fine details such as the irises visible, and we can match them more precisely.

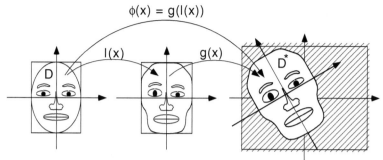

Figure 5.25. Factoring a general warp into a local small diffeomorphism and a global affine map. From [97, Figure 4.1].

5.8 Appendix: Geodesics in Diff and Landmark Space

5.8.1 Momentum and Arnold's General Description of Geodesics

In his paper, V. Arnold [9], gives a very general description of geodesics on a Lie group. He characterizes them as curves along which momentum is preserved. Let us describe this more precisely, using the notations and definitions introduced in Section 5.3. We will then see what this momentum becomes in the case of the special orthogonal group $SO(n)$ and in the case of the group of diffeomorphisms $\mathcal{G}$.

Let G be a Lie group and let $\mathfrak{g}$ denote its Lie algebra. Let $\langle u, v \rangle$ be a positive definite inner product on $\mathfrak{g}$, and use it to define a left invariant metric on G. Let $\beta : \mathfrak{g} \times \mathfrak{g} \to \mathfrak{g}$ be the operator such that

$$\forall u, v, w \in \mathfrak{g}, \quad \langle u, [v, w] \rangle = \langle \beta(u, v), w \rangle.$$

In Section 5.3, we defined the adjoint representation Ad, which associates with each $g \in G$ the derivative at e of conjugation by g, a map $\mathrm{Ad}_g : \mathfrak{g} \to \mathfrak{g}$. Loosely speaking, $\mathrm{Ad}_g(u) = gug^{-1}$. Its transpose $\mathrm{Ad}_g^\dagger : \mathfrak{g} \to \mathfrak{g}$ is defined by

$$\forall \xi \in \mathfrak{g}^\dagger, \quad \forall \eta \in \mathfrak{g}, \quad \langle \mathrm{Ad}_g^\dagger \xi, \eta \rangle = \langle \xi, \mathrm{Ad}_g \eta \rangle.$$

Arnold proved in [9], Theorem 5.7, characterizing geodesics on Lie groups in left[4] invariant metrics, together with a law of conservation of momentum:

---

[4] With respective changes made, Theorem 5.7 holds for right invariant metrics

Theorem 5.7. *Let $t \mapsto g(t)$ be a geodesic on G and, using the map $D_g L_{g^{-1}} : T_g G \to T_e G = \mathfrak{g}$, let $u(t) = D_g L_{g^{-1}} \dot{g}(t)$. Then*

$$\forall t, \quad \dot{u}(t) = \beta(u(t), u(t)) \tag{5.6}$$

$$\mathrm{Ad}^\dagger_{g(t)^{-1}} u(t) = \text{constant}. \tag{5.7}$$

Equation (5.7) is a sort of conservation of momentum for a Riemannian metric on a group, and it is an integrated form of the geodesic equation (5.6).

Instead of giving an abstract version of the proof, we present here a computational version using the approximating polygons $g_i = g(i\Delta t)$. We can approximate $u(t)$ by $\{u_i\}$ satisfying $g_i^{-1} \circ g_{i+1} = \exp(u_i \Delta t)$. Let $\widetilde{g}_i = g_i \circ \exp(\eta f_i)$ be a small perturbation of the path, 0 at the endpoints. What is the corresponding perturbation of u_i? It is

$$\exp(\widetilde{u}_i \Delta t) = \widetilde{g}_i^{-1} \circ \widetilde{g}_{i+1} = \exp(-\eta f_i) \exp(u_i \Delta t) \exp(\eta f_{i+1})$$
$$\approx \exp(u_i \Delta t) \circ [\exp(-u_i \Delta t), \exp(-\eta f_i)] \circ \exp(\eta(f_{i+1} - f_i),)$$

where the middle term stands for the group commutator $aba^{-1}b^{-1}$. This commutator is approximately the exponential of the Lie algebra bracket; hence,

$$\exp(\widetilde{u}_i \Delta t) \approx \exp\left(u_i \Delta t + \eta \Delta t. \left([u_i, f_i] + \frac{f_{i+1} - f_i}{\Delta t}\right)\right)$$

$$\widetilde{u}_i \approx u_i + \eta \left([u_i, f_i] + \frac{f_{i+1} - f_i}{\Delta t}\right).$$

Plugging this into the formula $\sum_i \langle u_i, u_i \rangle$ for the energy of the polygonal path, we find that the perturbation in the energy is:

$$2\eta \sum_i \left\langle u_i, \left([u_i, f_i] + \frac{f_{i+1} - f_i}{\Delta t}\right) \right\rangle_e = 2\eta \sum_i \left\langle \beta(u_i, u_i) - \frac{u_i - u_{i-1}}{\Delta t}, f_i \right\rangle_e.$$

Since this must vanish for all f_i, the first term in the bracket must vanish and we have a discrete form of Arnold's differential equation for $u(t)$. It is straightforward that the derivative of the conservation law is this differential equation, by using the fact that the derivative of $\mathrm{Ad}_{g(t)^{-1}}$ with respect to t is $\mathrm{ad}_{u(t)} \circ \mathrm{Ad}_{g(t)^{-1}}$ and taking transposes.

The momentum point of view is based on introducing the dual space $\mathfrak{g}^*$, the space of linear functionals from $\mathfrak{g}$ to $\mathbb{R}$. If (l, u) is the natural pairing of $l \in \mathfrak{g}^*, u \in \mathfrak{g}$, then inner products $\langle u, v \rangle$ on $\mathfrak{g}$ are the same as self-adjoint maps $L : \mathfrak{g} \to \mathfrak{g}^*$ via the correspondence $\langle u, v \rangle = (Lu, v)$. Instead of the transpose $\mathrm{Ad}^\dagger : \mathfrak{g} \to \mathfrak{g}$ of Ad, we now use the dual $\mathrm{Ad}^* : \mathfrak{g}^* \to \mathfrak{g}^*$. Since $L\mathrm{Ad}^\dagger = \mathrm{Ad}^* L$,

5.8. Appendix: Geodesics in Diff and Landmark Space

the conservation law is now $\mathrm{Ad}^*_{g(t)^{-1}} Lu(t) = $ constant. This can be reinterpreted as: any $g \in G$, the metric on the tangent space to G at g defines a map $L: T_g G \to T_g^* G$, and left translation $D_e L_g : \mathfrak{g} \to T_g G$ has a dual $(D_e L_g)^* : T_g^* G \to \mathfrak{g}^*$. Then we can check that $\mathrm{Ad}^*_{g(t)^{-1}} Lu(t) = (D_e T_{g(t)})^*(L\dot{g}(t))$; hence, Arnold's conservation law becomes

$$DT^*_{g(t)}(L\dot{g}(t)) = \text{constant},$$

and $L\dot{g}(t)$ is then called the momentum of the path, and the $(D_e T_{g(t)})^*(L\dot{g}(t))$ are the momenta along the path all carried back to the identity e.

Example (Geodesics on the Group $SO(n)$). Let $G = SO(n)$ be the group of $n \times n$ orthogonal matrices with determinant 1. We already saw in Section 5.3 that G is a Lie group and that its Lie algebra $\mathfrak{g}$ is the space of skew symmetric matrices. To define the length of a path on G, we first need an inner product on $\mathfrak{g}$. Let S be a positive diagonal matrix $\{s_i \delta_{ij}\}$. We then define

$$\forall A, B \in \mathfrak{g}, \quad \langle A, B \rangle_\mathfrak{g} = \mathrm{tr}(A^t SB) = -\mathrm{tr}(ASB) = -\frac{1}{2}\mathrm{tr}(A(SB + BS)).$$

The corresponding operator is thus $L : \mathfrak{g} \to \mathfrak{g}$, with $LB = SB + BS$, and $(LB)_{ij} = (s_i + s_j) \cdot B_{ij}$. It is an isomorphism, and we denote its inverse by L^{-1}. We then need to compute the operator $\beta : \mathfrak{g} \times \mathfrak{g} \to \mathfrak{g}$. According to its definition, it has to be such that for all A, B, and C skew symmetric,

$$\mathrm{tr}(\beta(A, B)SC) = \mathrm{tr}(AS(BC - CB)) = \mathrm{tr}(ASBC) - \mathrm{tr}(BASC).$$

Since this holds for every skew symmetric C, it implies that the skew symmetric part of $\beta(A, B)S - ASB + BAS$ has to be 0. Writing this out, we get

$$\beta(A, B) = L^{-1}(ASB - BSA - BAS + SAB).$$

In particular, we get $L\beta(A, A) = -A^2 S + SA^2 = [S, A^2]$. Let $t \mapsto g(t)$ be a path on G. Let $\Omega(t) = g^{-1}(t)\dot{g}(t)$ be the so-called angular velocity. The energy of this path is then

$$\int \langle g(t)^{-1}\dot{g}(t), g(t)^{-1}\dot{g}(t) \rangle_\mathfrak{g} dt = -\int \mathrm{tr}(\Omega(t) S \Omega(t)) dt.$$

The equations for geodesics are thus, in this, case

$$L\dot{\Omega}(t) = L\beta(\Omega, \Omega) = [S, \Omega^2]. \tag{5.8}$$

The momentum which is preserved is

$$M(t) = g(t)(\Omega(t)S + S\Omega(t))g(t)^{-1} = \text{constant}.$$

When the dimension is $n = 3$, it is customary to write

$$S = \begin{pmatrix} I_2 + I_3 & 0 & 0 \\ 0 & I_3 + I_1 & 0 \\ 0 & 0 & I_1 + I_2 \end{pmatrix}$$

and

$$\Omega = \begin{pmatrix} 0 & -\Omega_3 & \Omega_2 \\ \Omega_3 & 0 & -\Omega_1 \\ -\Omega_2 & \Omega_1 & 0 \end{pmatrix}.$$

Then the equation of geodesics becomes

$$\begin{cases} I_1 \dot{\Omega}_1 - \Omega_2 \Omega_3 (I_2 - I_3) = 0, \\ I_2 \dot{\Omega}_2 - \Omega_3 \Omega_1 (I_3 - I_1) = 0, \\ I_3 \dot{\Omega}_3 - \Omega_1 \Omega_2 (I_1 - I_2) = 0. \end{cases}$$

These are exactly the usual Euler's equations in mechanics. Here S is the inertia matrix and $(\Omega_1, \Omega_2, \Omega_3)$ are the angular velocities.

Example (Geodesics on the Group of Diffeomorphisms $\mathcal{G}$). Consider the group $\mathcal{G} = \mathrm{Diff}(\mathbb{R}^n)$ of all smooth diffeomorphisms, which are the identity outside a compact subset of $\mathbb{R}^n$. Its Lie algebra $\mathfrak{g}$ is the space of smooth vector fields with compact support. As we already saw, we can equip $\mathrm{Diff}(\mathbb{R}^n)$ with a *right* invariant metric (not left invariant as is customary for finite-dimensional Lie groups) given by an inner product on $\mathfrak{g}$:

$$\langle X, Y \rangle_{\mathfrak{g}} = \int_{\mathbb{R}^n} \langle LX(x), Y(x) \rangle \, dx = \int_{\mathbb{R}^n} \langle X(x), LY(x) \rangle \, dx,$$

where L is a self-adjoint operator (we can, for example, take a Sobolev inner product: $L = (1 - \Delta)^k$) and $\langle A, B \rangle = \sum_i A_i B_i$.

What is the form of the conservation of momentum for a geodesic $\Phi(t)$ in this metric—that is, a flow $x \mapsto \Phi(x, t)$ on $\mathbb{R}^n$? For a diffeomorphism $\Phi \in \mathcal{G}$, we need to work out $\mathrm{Ad}_\Phi^\dagger$ first. Using the definition, we see

$$\int_{\mathbb{R}^n} \langle X, L\mathrm{Ad}_\Phi^\dagger(Y) \rangle = \langle X, \mathrm{Ad}_\Phi^\dagger Y \rangle = \langle \mathrm{Ad}_\Phi X, Y \rangle$$

$$= \int_{\mathbb{R}^n} \langle (D\Phi.X) \circ \Phi^{-1}, LY \rangle$$

$$= \int_{\mathbb{R}^n} \det(D\Phi) \langle D\Phi.X, LY \circ \Phi \rangle$$

$$= \int_{\mathbb{R}^n} \langle X, \det(D\Phi).D\Phi^t.(LY \circ \Phi) \rangle.$$

5.8. Appendix: Geodesics in Diff and Landmark Space

Hence,
$$\mathrm{Ad}_\Phi^\dagger(Y) = L^{-1}\left(\det(D\Phi) \cdot D\Phi^t \cdot (LY \circ \Phi)\right).$$

Let $v(t) = \frac{\partial \Phi}{\partial t} \circ \Phi^{-1}$ be the velocity of the flow Φ. Theorem 5.7 about the conservation of momentum for geodesics $\Phi(t)$ says that

$$L^{-1}\left(\det(D\Phi)(t) \cdot D\Phi(t)^t \cdot \left(L(\frac{\partial \Phi}{\partial t} \circ \Phi^{-1}) \circ \Phi\right)\right) \quad \text{is independent of } t.$$

This can be put in a much more transparent form. First, L doesn't depend on t, so we cross out the outer L^{-1}. Now let $u(t) = Lv(t)$, so that:

$$\det(D\Phi)(t) \cdot D\Phi(t)^t \cdot (u(t) \circ \Phi(t)) \quad \text{is independent of } t.$$

We should *not* think of $u(t)$ as a vector field on $\mathbb{R}^n$ because we want $\int_{\mathbb{R}^n} u \cdot v$ to make coordinate invariant sense. This means we should think of u as expanding to the measure-valued differential 1-form:

$$\omega(t) = (u_1 \cdot dx^1 + u_2 \cdot dx^2 + \ldots + u_n \cdot dx^n) \otimes \mu,$$

where $\mu = dx^1 \wedge dx^2 \wedge \ldots \wedge dx^n$ is the volume form on $\mathbb{R}^n$. The 1-form pairs naturally with a vector, and the measure allows us to integrate over $\mathbb{R}^n$ without any further choices. But then the pullback of the differential form $\omega(t)$ is

$$\Phi(t)^*(\omega(t)) = \det(D\Phi)(t) \cdot \langle D\Phi^t \cdot (u \circ \Phi(t)), dx\rangle \otimes \mu,$$

so conservation of momentum says simply:

$$\Phi(t)^*\omega(t) \quad \text{is independent of } t.$$

This extremely simple law is the integrated form of the geodesic equation. If we differentiate it with respect to t, it is not difficult to check that we get again EPDiff, the differential equation for the geodesics on the group of diffeomorphisms.

Example (Geodesics on Landmark Space $\mathcal{L}^N$).

Recall that, by choosing a basepoint in $\mathcal{L}^N(\mathbb{R}^n)$, we get a submersion $\mathrm{Diff}(\mathbb{R}^n) \to \mathcal{L}^N(\mathbb{R}^n)$. For any submersion, it is a theorem that geodesics in the quotient space lift precisely to geodesics in the top space, which are horizontal at one point and hence at all their points. Recall that in the landmark point case, horizontal vectors are given using the Green's function of L. Since this must hold at every point on the geodesic, we see that geodesics on $\mathcal{L}^N(\mathbb{R}^n)$ must be given by the *ansatz*:

$$\vec{v}(x,t) = \sum_i G(x - P_i(t))\vec{u}_i(t),$$

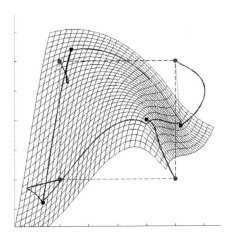

Figure 5.26. A geodesic in $\mathcal{L}^4(\mathbb{R}^2)$ between the four points forming the corners of the dotted square and the remaining four points. The figure shows (a) the paths that the four points follow and (b) the final diffeomorphism of the plane when the geodesic is lifted to a horizontal geodesic in $\text{Diff}(\mathbb{R}^2)$. (Courtesy of Mario Miceli.)

where

$$\dot{P}_i(t) = \vec{v}(P_i(t), t) = \sum_j G(P_i - P_j) u_j(t),$$

$$\vec{u}(x, t) = (L\vec{u})(x, t) = \sum_i \delta(x - P_i(t)) \dot{u}_i(t).$$

We need to work out when these functions satisfy EPDiff. Now, $L\vec{u}$ is not a real function, so the above expressions cannot be a solution in the usual sense. Rather, they are a so-called *weak solution*. A weak solution $\vec{u}$ of a partial differential equation means that (a) we multiply the PDE by an arbitrary smooth test function $\vec{\phi}$ with compact support (here it is vector-valued since we have a system of PDEs) and then (b) integrate by parts to move the derivatives from $\vec{u}$ to $\vec{\phi}$ until the remaining derivatives of $\vec{u}$ (if any) exist and the equation makes sense. Then we ask that it holds for every $\vec{\phi}$. With EPDiff, this means that the following are zero (where we use $i, j, \cdots$ as indices for landmark points as before and $p, q, \cdots$ for coordinates in $\mathbb{R}^n$):

$$\sum_p \int \int \phi_p(x, t) \cdot \bigg((Lv_p)_t + \sum_q (v_q \cdot (Lv_p)_{x_q}$$

$$+ (v_q)_{x_q}(Lv_p) + (Lv_q)(v_q)_{x_p} \bigg) d\vec{x} dt$$

$$= \sum_{p,i} \int\int \delta(x-P_i)\cdot\left(-\phi_{p,t}+\sum_q -(\phi_p\cdot v_q)_{x_q}+(v_q)_{x_q}\phi_p\right)u_{i,p}$$
$$+\phi_p\sum_q u_{i,q}.v_{q,x_p}d\vec{x}dt$$

$$= \sum_{p,i}\int\left(\left(-\phi_{p,t}(P_i)-\sum_q v_q(P_i)\phi_{p,x_q}(P_i)\right)u_{i,p}\right.$$
$$\left.+\phi_p(P_i)\sum_q u_{i,q}v_{q,x_p}(P_i)\right)dt$$

$$= \sum_{p,i}\int\left(-\frac{d}{dt}\phi_p(P_i(t),t).u_{i,p}+\phi_p(P_i)\sum_j G_{x_p}(P_i-P_j)(\vec{u}_i,\vec{u}_j)\right)dt$$

$$= \sum_{p,i}\int \phi_p(P_i)\left(\frac{d}{dt}u_{i,p}+\sum_j G_{x_p}(P_i-P_j)(\vec{u}_i,\vec{u}_j)\right)dt.$$

Since this holds for every $\vec{\phi}$, the geodesic on the landmark space satisfies the coupled ODEs:

$$\frac{d}{dt}P_i = \sum_j G(P_i-P_j)\vec{u}_j$$
$$\frac{d}{dt}\vec{u}_i = -\sum_j \nabla G(P_i-P_j)(\vec{u}_i,\vec{u}_j).$$

This is an intriguing dynamical system that expresses the idea that there is a sort of glue binding the landmarks, stronger as they get closer, weaker when farther apart. An example of a geodesic between two sets of four points in the plane is shown in Figure 5.26.

5.9 Exercises

1. A Puzzle

What is the object depicted in the image, shown as a graph, in Figure 5.27? (*Hint*: the level of the background is the lowest, that is, 0 or black.)

2. Deceptive Clustering in Curved Manifolds

As described in Section 5.3.3, if a data set in a curved manifold is linearized by the exponential map, it can be distorted, some clusters being pulled apart and others formed. Here

Figure 5.27. A graph of an image of an instantly recognizable object. Can you guess what it is? [158, p. 241]

we work out an example of this. The example is a simple surface with coordinates (x, y), but on this surface we assume the metric is given by

$$\| (dx, dy) \|^2 = (y^2 + 1) \cdot (dx^2 + dy^2).$$

Using the notations for a Riemannian metric, this means that the inner product of two tangent vectors $u = (u_1, u_2)$ and $v = (v_1, v_2)$ at a point (x, y) on the surface is given by $\langle u, v \rangle_{(x,y)} = (y^2 + 1)(u_1 v_1 + u_2 v_2)$. Being a surface, the curvature is a single number, the so-called *Gaussian curvature*. This example has been carefully contrived so that this curvature is positive if $|y| > 1$, but negative if $|y| < 1$. In fact,

$$\text{curvature} = \frac{y^2 - 1}{(y^2 + 1)^3}.$$

The series of problems presented here analyzes the distortions caused by linearizing this space via the exponential map.

1. The differential equation for geodesics is a standard topic in all books on differential geometry. It is not at all difficult to work out from scratch, and we sketch how to do this in this first problem. Using a local chart, the whole idea is to compute the first variation of the functional

$$E(\gamma) = \int_0^1 \sum_{ij} g_{ij}(\gamma(t)) \frac{d\gamma^i}{dt} \frac{d\gamma^j}{dt} dt$$

with respect to a small perturbation $\gamma \mapsto \gamma + \delta\gamma$ with compact support. Show that, using integration by parts,

$$\delta E = \int_0^1 \left(\sum_{ijk} \left(\frac{\partial g_{ij}}{\partial x^k} \circ \gamma \right) \delta x^k \frac{d\gamma^i}{dt} \frac{d\gamma^j}{dt} - 2 \sum_{ij} \frac{d}{dt} \left((g_{ij} \circ \gamma) \frac{d\gamma^i}{dt} \right) \delta\gamma^j \right) dt.$$

Expanding the outer derivative in the second term, and changing dummy indices so that both $\delta\gamma$ terms have the same index, show that $\delta E = 0$ for all $\delta\gamma$ is equivalent to

$$\forall j, \quad \sum_i g_{ij} \frac{d^2 \gamma^i}{dt^2} = \frac{1}{2} \sum_{ik} \left(\frac{\partial g_{ik}}{\partial x^j} - \frac{\partial g_{ij}}{\partial x^k} - \frac{\partial g_{jk}}{\partial x^i} \right) \frac{d\gamma^i}{dt} \frac{d\gamma^k}{dt}.$$

5.9. Exercises

Denoting the inverse of the metric matrix g_{ij} by g^{ij}, this is usally rewritten as

$$\frac{d^2\gamma^i}{dt^2} = -\sum_{jk} \Gamma^i_{jk} \frac{d\gamma^j}{dt}\frac{d\gamma^k}{dt}, \quad \text{where } \Gamma^i_{jk} = \frac{1}{2}g^{i\ell}\left(\frac{\partial g_{j\ell}}{\partial x^k} + \frac{\partial g_{\ell k}}{\partial x^j} - \frac{\partial g_{jk}}{\partial x^\ell}\right).$$

2. Apply this result to our surface, where $g_{11} = g_{22} = (1+y^2)$, $g_{12} = 0$. Work out the geodesic equations, then eliminate time from them, reducing them to a second-order equation for $y = f(x)$. By solving this equation, show that the geodesics are one of the following three families of curves:

$$y = A\cosh(\tfrac{x-x_0}{B}), \quad B^2 - A^2 = 1$$
$$y = A\sinh(\tfrac{x-x_0}{B}), \quad B^2 + A^2 = 1$$
$$y = e^{\pm(x-x_0)}$$

In particular, geodesics through the origin with direction θ are given by

$$y = \sin(\theta)\sinh(x/\cos(\theta)).$$

3. By integrating ds along these paths, we get path length as a function of x: it is a simple transcendental function (but there is no simple formula for its inverse). Write a simple program to compute the exponential map for this surface, starting at the origin. Graph rays from this point out to a distance of 6 to get an idea of convergence and divergence of geodesics. In particular, if C is the image of the circle of radius 6 under the exponential, then calculate the max and min of its derivative. We get a ratio of about 30:1. Hence, show that uniformly spaced points on C, pulled back by the exponential map, are strongly clustered at $\theta = 0, \pi$. Conversely, a true cluster on C near the y-axis is not so strongly clustered when pulled back by the exponential.

3. Younes' Metric

One quite beautiful case of a metric on diffeomorphisms that is simultaneously elastic and Riemannian was discovered by L. Younes [230, 231]. Of course, the elastic version cannot be physically reasonable! This is the case where $n = 1, \mathcal{D} = [0,1]$ and the Riemannian path length metric on $\mathcal{G} = \text{Diff}(\mathcal{D})$ is:

$$d_{\text{Younes}}(\phi) = \inf_{\text{paths } \Phi, \, \text{Id to } \phi} \int_0^1 \left(\int_0^1 v'(y)^2 dy\right)^{1/2} dt, \quad \text{where } v(y,t) = \frac{\partial \Phi}{\partial t}(\Phi^{-1}(y,t),t).$$

Younes showed that the map $\phi \mapsto \sqrt{\phi'}$ gives an *isometric* map from $\mathcal{G}$ into the positive quadrant of the unit sphere $S_{L^2(\mathcal{D})}$ in $L^2(\mathcal{D})$. Prove this by considering its derivative from $T\mathcal{G} \to TS_{L^2(\mathcal{D})}$ and showing this is an isometry (actually with a factor $\tfrac{1}{2}$). Since great circles are geodesics on any sphere, deduce that:

$$d_{\text{Younes}}(\phi) = \arccos\left(\int_0^1 \sqrt{\phi'(x)}dx\right) = f(E_s(\phi)),$$

where E_s is the strain based on the strain energy density $e_s(J) = (1+J)/2 - \sqrt{J}$ (with a minimum at $J = 1$, increasing as $J \to 0$ or ∞) and f is the monotone increasing function $\arccos(1-x)$.

4. One-Dimensional Elastic Matching Problem

Consider a one-dimensional matching problem, where two real-valued functions f_1 and f_2 are given on an interval $I = [a, b]$ and one seeks a diffeomorphism $\phi : I \to I$ minimizing

$$\int_a^b (f_1(\phi(x)) - f_2(x))^2 dx + E(\phi), \quad E(\phi) = \int_a^b e_s(\phi'(x)) dx.$$

This problem arises, for instance, in the "correspondence problem" for stereoscopic vision: finding matching points in the images of the same scene viewed independently by the left and right eyes. This reduces to a one-dimensional matching problem because of the "epipolar constraint": any three-dimensional point lies in a plane through the centers of the left and right eyes, and this plane defines lines in the left and right eye images between which the match must take place. So we may restrict the images to points on this pair of lines, giving functions of one variable as above.

1. Find the condition on e_s such that $E(\phi) = \int_a^b e_s(\phi'(x)) dx$ satisfies $E(\phi^{-1}) = E(\phi)$. How can we modify the data term to make it symmetric under interchanging f_1 and f_2?

2. Find e_s for which $E(\phi)$ is the length of the graph of ϕ in $I \times I$, and verify the condition just obtained for this e_s.

3. Assuming f_1 and f_2 are defined by data only at a finite set of points,

$$p_k = a + k\frac{b-a}{n}, \quad 0 \le k \le n,$$

formulate a discrete version of the problem of finding an optimum ϕ by replacing the integral by a sum. Consider alternatives for e_s. Consider $f_1 \equiv 0$ and $f_1(p_k) = C\delta_{k,k_0}$ (with $C >> 0$) to compare some of these (some discrete versions are too simple and should fail).

5. The Collapse of L^2 Metrics on Diff

We have stated that if we use the L^2 inner product on vector fields on any manifold $\mathcal{M}$, $\| \vec{v} \|^2 = \int_{\mathcal{M}} \| \vec{v}(x) \|^2 dx$, as the basis for a right-invariant metric on $\text{Diff}(\mathcal{M})$, then the infimum of path lengths joining any two diffeomorphisms is zero. The essential idea is easily understood in dimension 1. We want to model a shock wave moving from left to right at speed 1 along the real line. Individual points will be at rest until the front of the shock wave hits, then they accelerate to a speed slightly less than 1, falling slowly back relative to the wave itself. After some time, they are left behind by the wave. In this way, all points are moved forward by a fixed amount.

1. In formulae, consider the time-varying vector field on the line $v(x,t) = g(x-t)$, where g is supported on a very small interval $[-\varepsilon, \varepsilon]$, is everywhere non-negative, and has a maximum value $1-\varepsilon$ on, say, half this interval, $[-\varepsilon/2, \varepsilon/2]$. In the (x,t)-plane, draw this vector field and check that graphically a typical solution $x(t)$ of the flow $x'(t) = v(x(t), t)$ looks like the paths in Figure 5.28. Prove that the flow is a very short compression wave, moving at speed 1, but within which the particles move very slightly faster than the wave, so they enter the wave, stay there at least

5.9. Exercises

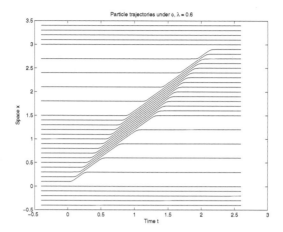

Figure 5.28. A shock wave on a line. Each line shows the path in space-time of a single particle. From [158, p. 241].

one unit of time, and then leave. Prove that, as the wave passes, the whole line is moved forward by a distance

$$\int_{-\infty}^{+\infty} \frac{g(x)}{1-g(x)} dx \geq 1 - \varepsilon.$$

In contrast, show that in a bounded part of the line, the L^2-norm of v goes to zero with ε.

2. Adapt this argument to the circle to show that rotations can be connected to the identity by paths of arbitrarily short length. One way is to (1) start a compression wave such as the one in problem 1, (2) make the wave go around the circle a large number of times, and (3) stop the wave. The effect of (2) is to rotate all points some amount, but the stopping and starting causes some deviation from a perfect rotation, so we need to follow it by a small correction. This becomes messy, but it works. (We can also modify the construction by letting the distance between the leading and trailing edges of the compression wave not be constant, and get general diffeomorphisms of S^1 in this way.)

6. Green's Functions and Delta Functions

For readers unfamiliar with Green's functions, it is helpful to solve the simple equation $\triangle G = \delta$ in $\mathbb{R}^n$, as well as to look at the exercises of Chapter 6, where we introduce distributions (also called generalized functions).

1. If $n = 1$, let $G(x) = \frac{1}{2}|x|$. Show that

$$\forall \text{ smooth, compactly supported } \phi, \int \phi''(x) G(x) dx = \phi(0).$$

This assertion is, by definition, what it means to say $G'' = \delta$.

2. If $n = 2$, let $G(\vec{x}) = \frac{1}{2\pi} \log(\|\vec{x}\|)$. Show that

$$\forall \text{ smooth, compactly supported } \phi, \iint \triangle\phi(\vec{x})G(\vec{x})d\vec{x} = \phi(0),$$

which again means $\triangle G = \delta$. To do this, we convert to polar coordinates, with the appropriate expression for $\triangle$.

3. If $n > 2$, let $G(\vec{x}) = -c_n/\|\vec{x}\|^{n-2}$. Prove the same assertion with a suitable positive constant c_n (depending on the volume of the n-sphere).

7. Collisions of Landmarks

The aim of this exercise is to show the influence of the choice of the norm on the length of a path when two landmark points come together.

1. Consider two landmarks, both on $\mathbb{R}$, moving together:

$$\begin{vmatrix} q_1(t) = f(t) \\ q_2(t) = -f(t) \\ f(0) = 1, \ f(1) = 0, \ f \text{ decreasing} \end{vmatrix}$$

We put the H^1 norm on one-dimensional vector fields v vanishing at ∞:

$$\|v\|_{H^1}^2 = \int (1 - D_x^2)v \cdot v = \int v^2 + \int (D_x v)^2.$$

Choose a simple vector field $v(x, t)$ on $\mathbb{R}$ that carries the two landmark points along:

$$v(q_i(t), t) = \dot{q}_i(t),$$

where the notation $\dot{q}_i$ means, as usual, $\frac{d}{dt}q_i$.

Show that with suitable v, this path has finite energy

$$\int_0^1 \|v(\cdot, t)\|_{H^1}^2 \, dt < \infty.$$

(*Hint:* Make v die off *exponentially* at $\pm\infty$. Also, it is smooth enough to make v piecewise C^1.)

2. Now replace the norm on v by the H^2-Sobolev norm,

$$\|v\|_{H^2}^2 = \int (1 - D_x^2)^2 v \cdot v = \int v^2 + 2\int (D_x v)^2 + \int (D_x^2 v)^2.$$

Assume we have n landmark points. We want to prove that no two landmark points come together on a path with finite energy.

(a) First, find the Green's function $G(x)$ for $(1 - D_x^2)^2$. This is elementary ODE theory: G should be C^2, while G'' has a discontinuity of size 1 at $x = 0$, so that the fourth derivative $G^{(iv)}$ has a δ-function at $x = 0$. Work out the Taylor series approximation to G at 0.

(b) Second, work out the matrix for the metric and its inverse.

5.9. Exercises

(c) Finally, use the Cauchy-Schwarz inequality

$$|<x,u>| \leq \sqrt{x^t Q \cdot x} \cdot \sqrt{u^t Q^{-1} \cdot u}$$

for any positive definite symmetric matrix Q to get a lower bound on the energy when two of the q_i come together.
(*Hint.* Consider $\dot{q}_i - \dot{q}_j$.)

8. Computational Experiments

Eigenimages

It is not difficult to investigate how effectively arbitrary illumination conditions can be reproduced as sums of small numbers of eigenimages. The experiment is this: first choose an object and prepare a room thart can be made totally dark. Put a digital camera on a tripod and take several hundred images of the object while illuminating the object by a spotlight from multiple angles. We get a database of image $\{I^\alpha(p)\}$, where p ranges over $N \times M$ pixels and α over the set of, say, L images. View I as a $NM \times L$ array and find its singular values. *Do not take the eigenvalues of $I \circ I^t$.* This is $NM \times NM$ and is too huge. Take the eigenvalues (and eigenvectors $\vec{c}$ of length L) of $I^t \circ I$. These define the eigenimages $I \cdot \vec{c}$. The interest is whether the eigenvalues decay fast; hence, all the images are well approximated by sums of small numbers of these eigenimages. This is what happened with faces. But if the object is highly specular, such as a glossy vase or a motorcycle helmet, this will not be the case. Also, if the object is highly concave and has cavities, this will happen. It is quite instructive to actually do this and look at what eigenimages result. The ideas in Section 5.1.2 are helpful in understanding what happens.

The initial value problem for geodesics on landmark space

The goal here is to calculate geodesics in the landmark metric starting at a given point with given tangent direction. Assume we have N points in $\mathbb{R}^M$ whose NM position variables are $q_{k,i}, 1 \leq k \leq N, 1 \leq i \leq M$, and NM momenta $p_{k,i}$ that are functions of $\dot{q}_{k,i}$. We deal only with $M = 1$ and $M = 2$. We assume the metric is defined by the differential operator $L = (I - A^2 \triangle)^d$; we deal only with $A = 1, d = 2$. An important fact is that its Green's operator is

$$G(\vec{x}, \vec{y}) = \frac{1}{(2\pi A)^{M/2} \cdot A^d \cdot (d-1)!} \cdot ||\vec{x} - \vec{y}||^{d-M/2} \cdot K_{d-M/2}\left(\frac{||\vec{x} - \vec{y}||}{A}\right),$$

where $K_a(s)$ is the modified Bessel function, given in MATLAB by `besselk(a,s)`. Dealing with this is special function heaven, for which the bible is the book of Abramowitz and Stegun [1]. We can ignore the constant in front, as this just rescales time. The important function for us is $\widetilde{K}_a(s) = |s|^a \cdot K_a(|s|)$, which has the following properties:

1. $\widetilde{K}_a(s)$ is positive, has its max at $s = 0$, and decreases monotonically as $|s| \to \infty$;

2. $\frac{d}{ds}\left(\widetilde{K}_a(s)\right) = -s\widetilde{K}_{a-1}(s)$;

3. $\widetilde{K}_a(s)$ is C^1 with derivative 0 at 0 if $a \geq 1$; it is C^0 if $a > 0$, but it has a log pole at $s = 0$ if $a = 0$;

4. Its value at $s = 0$ is $\Gamma(a).2^{a-1}$. Recall that $\Gamma(n) = (n-1)!$ and $\Gamma(n+1/2) = \frac{(2n-1)!}{2^{2n-1}.(n-1)!}\sqrt{\pi}$.

Now for the problems:

1. Plot $\widetilde{K}_a(s)$ for $a = 0, 1/2, 1, 3/2, 2$.
2. Work out the ODEs for the variables $\{q_{k,i}, p_{k,i}\}$ by using the functions $\widetilde{K}_a(s)$.
3. Code the solutions of these ODEs using MATLAB's function ode45 (known as Runge-Kutta). Coding the derivatives for the functional input to ode45 shouldn't be more than 5 or 10 lines of code.
4. Solve for the following two cases:

    (a) $M = 1, d = 2, A = 1, N = 4$, with initial conditions $q_1 = 2, q_2 = .5, q_3 = 0, q_4 = -.5$ and $p_1 = 0, p_2 = 1, p_3 = 0, p_4 = -.5$. Integrate this forward to $t = 10$ and backward to $t = -5$. Plot the four q_i against time and the four p_i against time.

    (b) $M = 2, d = 2, A = 1, N = 2$, with initial conditions $q_1 = (0, 1), q_2 = (0, -1), p_1 = p_2 = (1, 0)$. Integrate forward and backward with $|t| \le 5$; do this again but with $p_2 = (-1, 0)$.

Hippopotaffe and Giramus.

Find images on the web of a giraffe and a hippopotamus. We have also provided two on our website. Deform one animal into the other. The first step is to load them via imread into MATLAB. On each, using ginput, find 20 or so corresponding points. (Corresponding? It's a basic principle of organismic biology that all vertebrates are diffeomorphic. So use what you presume is the "true" diffeomorphism to pick out corresponding points.) Choose these points so that moving them will move the whole animal as coherently as possible. This gives two sets of landmark points in the plane.

The second step is to compute a geodesic between two sets of landmark points. One set of code for doing this is also on the website. The file matchLandmarks.m defines a function that computes the geodesic (calling the basic routine conjgrad.m). When this is computed, the file flow.m will compute the associated global flow of $\mathbb{R}^2$. Use this to deform the whole image of the giraffe following its landmark points to their position in the hippo. Plot the result. Then do the reverse, deforming the image of the hippo while its landmark points move to their locations in the giraffe. Plot this result too. These are wonderful creatures.

Figure 6.1. On the top, a river scene from Maine in the 1950s in which objects and textures occur at various sizes. Such multiscale "clutter" is a universal property of natural scenes. On the bottom, a portion of Paolo Uccello's painting *The Hunt by Night*. Note that the trees and the dogs all have different sizes relative to the image itself: varying the viewing distance literally shrinks or expands objects, leading to the scaling properties of images. (Top photograph from *Night Train at Wiscasset Station*, by Lew Dietz and Kosti Ruohomaa, Doubleday, 1977; courtesy of Kosti Ruohomaa/Black Star. Bottom image from http://commons.wikimedia.org/wiki/File:Paolo_Uccello_052.jpg.)

– 6 –

Natural Scenes and Multiscale Analysis

THIS CHAPTER IS more ambitious than the previous ones. Our goal is consider the *totality* of natural scenes, images of the world around us—outside, inside, people, buildings, objects, etc.—and ask what statistical regularities this ensemble of images has. It is rather surprising that there are many quite distinctive properties that capture a nontrivial amount of the "look-and-feel" of natural images. If we use the language analogy, the structures in speech are often divided into a hierarchy of levels: phonology, syntax, semantics, and pragmatics. In this analogy, we are studying in this chapter the phonology of images. Syntax was the main subject of Chapters 3 and 4 and semantics refers to specific object classes, such as faces as discussed in Chapter 5. But as we will see, this very basic low level of analysis is not a poor relation. However, we discuss in this chapter only the construction of data structures and probability models incorporating fundamental image statistics, especially *scale invariance* (see bottom image in Figure 6.1). We do not discuss their use in analyzing specific images via Bayes' theorem. In fact, in the present state of the art, finding significant objects at all scales—distinguishing clutter made up of smaller objects from texture that is not (see top image in Figure 6.1), has not been widely addressed (but see [79] for a multiscale segmentation algorithm). It seems important to know the basic statistical nature of generic real-world images first. For example, if hypothesis testing is used to check whether some part of an image is a face, surely the null hypothesis should be a realistic model incorporating these statistics.

Our goal, then, is to build a probability distribution on the space of all images whose samples have the same basic statistical properties as large databases of natural images. By "basic statistical properties" we mean the joint histograms of small sets of simple filters—but we want to keep the meaning of the words "small" and "simple" loose. Thus, we do not want to model the existence of long-range groupings nor the nature of the particular objects with which our images are populated. Moreover, the images in question may be discrete, defined on a finite set of pixels with one byte values, or they may be continuous, defined on a continuum with real values. Sometimes a discrete model is simpler; sometimes a continuous one. We introduce new tools for dealing with the continuous case, which is mathematically more complex than those needed in the discrete one. In particular, we need the concept of Schwartz distributions, also called "generalized functions," and bits of the theory of probability measures on function spaces.

In Sections 6.1, 6.2, and 6.4, we discuss the three basic low-level statistical properties of images: high kurtosis filter responses, scale invariance, and the existence of characteristic "local" primitive shapes. In Section 6.3, we introduce the *image pyramid*, a three-dimensional data structure that describes the image in "scale-space," that is, at all scales at once. Combining scale invariance and the primitive local shapes leads us to the idea of *frames* and *wavelets*, which we describe briefly in Section 6.4. Many books are available that study them in more detail. We then turn to the construction of image models that incorporate all three basic properties. In Section 6.5, we note that distributions are essential if we want any scale- and translation-invariant measure on images, and we introduce them in Section 6.5.2. In Section 6.6, we extend the construction of Brownian motion in Chapter 3 and discuss more of the basics on constructing measures on function spaces. This leads to the definition in Section 6.7 of the unique scale-, rotation- and translation-invariant Gaussian distribution on images. This has all marginals with kurtosis 3 and allows no shape primitives either. In Section 6.8, we introduce infinitely divisible distributions and use them to construct *random wavelet* models, which do carry the three basic properties. But these models are not wholly satisfactory either, and we discuss in Section 6.9 a further class of models, called *dead leaves* models by the school of Matheron and Serra (see [192]).

6.1 High Kurtosis in the Image Domain

We saw in Chapter 2 that signals from the world typically have discontinuities, or places where they change very suddenly. For images, such discontinuities are found in the differences of adjacent pixel values as well as in the statistics derived from them. We also saw in Chapter 4 that when we study a textured image I, the occasional large values in $F*I$, for some filter F, are absolutely crucial to texture perception. As in Section 2.4.1, these large values are outliers in the histogram of the filter output and cause this histogram to have high kurtosis. This is the first universal property of image statistics that needs to be incorporated into our model.

Let's look at some data from images. We suppose that F is the filter such that $F(0,0) = 1$, $F(1,0) = -1$, and F is 0 elsewhere, so that

$$(F*I)(i,j) = I_x(i,j) = I(i,j) - I(i-1,j);$$

6.1. High Kurtosis in the Image Domain

that is, it is the first difference in the x direction. Thus $F * I$ will have large values where there is a roughly vertical sharp edge with high contrast. If I is an image of the real world, then typical values that we find for the kurtosis κ of this distribution are

$$\kappa(\text{distrib. of } I_x) \in [10, 20],$$

which is much bigger than the $\kappa = 3$ of a normal distribution. If we look at the tail of the distribution of I_x, we find that

$$\mathbb{P}(|I_x| > 4 \times SD) \simeq 1\%,$$

where SD denotes standard deviation, which is a huge tail compared to the one of the normal distribution. In fact, if $\mathcal{X}$ is a random variable with Gaussian distribution with mean 0 and standard deviation σ, then

$$\mathbb{P}(|\mathcal{X}| \geq 4\sigma) = \frac{2}{\sigma\sqrt{2\pi}} \int_{4\sigma}^{+\infty} e^{-x^2/2\sigma^2} dx = \frac{2}{\sqrt{2\pi}} \int_{4}^{+\infty} e^{-x^2/2} dx \simeq 0.006\%,$$

which is about 150 times smaller than the tail of the distribution of I_x.

Typically, the first differences I_x and I_y have big responses at the edges, and their distribution also has a peak at 0 due to the numerous blank parts of an image. The distribution generally looks like the graph in Figure 6.2(a). Note that, as in Chapter 2, we have plotted *logs* of the histograms, not the actual histograms. These give a much clearer picture of the tails. Figure 6.2(b) contains

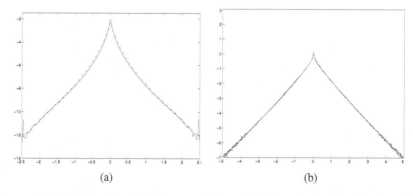

Figure 6.2. (a) The *logarithm* of the histogram of values of derivatives from two large data bases: the van Hateren Natural Stimuli Collection (dashed line) and the British Aerospace Sowerby Collection (solid line). The data bases are normalized and a log is taken, so derivatives are logs of ratios of true intensity values at adjacent pixels, which are dimension-free numbers. The horizontal axis shows these values, and this gives a standard deviation of about 0.25, and a kurtosis of about 15. (b) The log histogram of 8 × 8 mean 0 *random* filters applied to the van Hateren data base. Here, the x-axis shows multiples of the standard deviation, and the kurtosis is between 7 and 9. (From [110, Figure 2.47].)

sample histograms from random zero-mean 8×8 filters, that is, convolutions with $F_{i,j}, -4 < i, j \leq 4$, where the values $F_{i,j}$ are independent normal random variables with $\sum_{i,j} F_{i,j} = 0$. Although not as pronounced, again we see large tails—larger than Gaussians would have and with kurtosis around 8. All this shows that we have to abandon Gaussian models for images of the real world because outliers are very important.

To get accurate image statistics, it is important to use calibrated image data, that is, to use images in which the pixel values represent measurements of the actual light energy incident on the sensor. Typical images are not calibrated because:

- They have "gamma" correction, which means that the measured image I is taken to be $I \sim$ (sensor energy)$^\gamma$ for some γ such as 0.5. This is done to compress the range of contrast to one the system can handle. Typical natural scenes without gamma correction have contrasts of 1000:1, and such a range is not reproducible on screens, film, or paper.

- They have saturation of extreme blacks and whites, in spite of the gamma correction.

- Someone may have played with them in exotic ways in Adobe Photoshop to make them prettier.

Such processing is almost never documented, and much of it is irreversible. Fortunately van Hateren produced a very large database of calibrated images, publically available at http://hlab.phys.rug.nl/imlib/index.html. If I_0 is a calibrated image measured in units of energy, then it is natural to consider $I = \log I_0$ because it is, up to an additive constant, dimensionless. More precisely, the first difference $I(\alpha) - I(\beta)$ for α and β neighbors (denoted by $\alpha \sim \beta$) is equal to $\log(I_0(\alpha)/I_0(\beta))$, which is a dimensionless number, and the density of its distribution, call it $p_{\text{diff}}(x)$, is an invariant of natural scenes.

If we plot the log of the observed probability density $\widehat{p_{\text{diff}}}(x)$ for different kinds of images of the real world, we find results similar to those in Figure 6.3. These histograms are based on a data base of British Aerospace of scenes around Bristol, England. British Aerospace's Sowerby lab hand-segmented the pixels into different categories, which we have simplified into four types: sky, vegetation, road, and man-made objects. It seems that there are two universal characteristics, regardless of scene type:

1. a peak of p_{diff} at $x = 0$, with a singularity;

2. fat tails compared to those of Gaussians (see, e.g., Figure 2.9).

6.1. High Kurtosis in the Image Domain

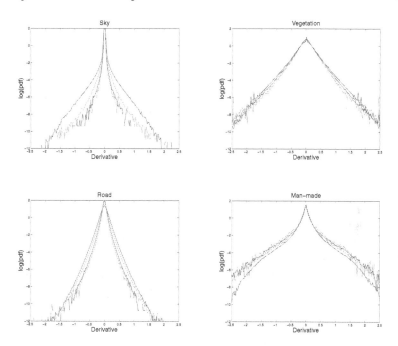

Figure 6.3. Log histograms of derivatives from four categories of pixels in the British Aerospace data base: sky, vegetation, roads, and man-made objects. Each plot shows the histograms of derivatives for the original image and for coarser versions in which blocks of pixels are averaged (see text). The four colors red, green, blue, and yellow are data for $n \times n$ block averaged images, $n = 1, 2, 4, 8$, respectively. (From [110, Figure 2.10]. See Color Plate VIII.)

The peak at $x = 0$ is produced by the large regions of the image I with no contrast variation: we call this the "blue sky effect" (indeed, it is most prominent in the sky regions). There are always many outliers in the distribution (i.e., the fat tails) which are produced by the contrast edges of the image I. The standard deviation of p_{diff} for large mixed data bases seems to be about 0.25 (note that this number is dimension-free). For sky, it is naturally much less, at only 0.07 whereas for vegetation it is more, 0.35. Similarly, the kurtosis has a much higher range: we found it ranging between 7 and 60! Some of the variability in these distributions can be described by a parameter that may be called "clutter." For example, the vegetation scenes which are very cluttered, tend to have derivatives with a probability distribution like $e^{-|x|}$, and man-made scenes, which are less cluttered, tend to have a quite particular sort of distribution: (a) a very sharp peak at 0, (b) very slow fall-off on the shoulders, but (c) a rapid decrease when $|x| > 2.5$. This seems to reflect a human preference for plain unmarked walls, with sharp high-contrast edges but an avoidance of the extremes of light and dark that are encountered in nature (and cause humans to wear sunglasses).

As an aside on data analysis, when looking at summary statistics, kurtosis is very unreliable because when we have fat tails, the empirical fourth moment is very sensitive to how many outliers we happen to have collected. In one experiment, we found a kurtosis approximately between 13 and 16, but the full probability distribution was well fit by $e^{-\sqrt{|x|}}$ whose kurtosis is 25.2. In fact, if we cut off this distribution at 9 standard deviations, then we underestimate the kurtosis by 50 percent. A more reliable summary statistic is $\mathbb{P}(|\nabla I| \geq 4 \times SD)$. As previously mentioned, this was about 1 percent in our data, which is about 150 times more than for the normal distribution.

Are there explicit models for one-dimensional probability densities that fit the filter response data? One reasonable model, suggested by people working on wavelets, is a Laplace distribution, $ce^{-|x|^\alpha}$, where α is in the range $(0.5, 1.0)$ and c is a normalizing constant. Another model is given by Bessel functions, $c|x|^s K_s(|x|)$, where K_s is the modified Bessel function (besselk in MATLAB), $0 \leq s \leq 0.5$. Without going into the specialized theory of Bessel functions, this distribution is best understood as the Fourier transform of $1/(1 + \xi^2)^{s+.5}$. On first glance, both distributions look alike and both generalize the double exponential $\frac{1}{2}e^{-|x|}$ (for the Laplace take $\alpha = 1$, for the Bessel take $s = .5$). But the Laplace distribution has larger tails when $\alpha < 1$ because the Bessel distributions are all asymptotically like $e^{-|x|}$. Which one really fits the data better is not clear.

A strong argument, however, for the Bessel models (proposed originally by Grenander [92, 203]) is that these arise as the distributions of products $x \cdot y$, where x is Gaussian and y is the positive square root of a variable with a gamma distribution. It is tempting to assume that this explains the kurtotic distribution of random mean 0 filter responses (see Figure 6.2). This could happen if x was a *local* Gaussian factor in the image and y was a *slowly varying* contrast factor, coming from illumination or other scene variables.

6.2 Scale Invariance in the Discrete and Continuous Setting

6.2.1 Definitions

In this section, we are interested in a second remarkable empirical fact—the statistics of images seem to be very nearly scale-invariant. Roughly speaking, this means that

$$I(x,y) \text{ is equiprobable with } I(\sigma x, \sigma y)$$

6.2. Scale Invariance in the Discrete and Continuous Setting

for all positive real scaling factors σ. Consider the painting of the hunting scene in Figure 6.1; the trees and dogs appear in the painting at many different distances, and their representation on the image plane shrinks or expands accordingly. More generally, any object in the world may be imaged from near or far so it is represented by larger or smaller sets of pixels in the image. We can say that an image results from a world scene and a position of a camera in this scene. To some extent, these are independent random variables.[1] Note that this makes visual signals completely different from auditory and tactile signals. There is nothing the observer can do to speed up or slow down the sound being heard, nor to make their fingers cover less or more of a surface. In temporal signals and in sensations arising from spatial contact, the scale of the signal is fixed. But in vision, people can move closer to get a better look or move away to see more of a scene. Because of perspective corrections, this is not exactly the same as replacing $I(x, y)$ by $I(\sigma x, \sigma y)$. When you approach an object, its nearer parts get bigger faster than the more distant parts. Nonetheless, it is very nearly true that images are changed by scaling over large ranges of viewing distance, and this produces a very real and important effect on image statistics.

When we make the concept of scale-invariance precise, we have to distinguish the case of images defined on discrete or on continuous domains. For continuous images $I(x, y)$, we can simply substitute σx and σy for x and y, but we cannot do this for discrete ones. For discrete images $I(i, j)$, the simplest way to study scaling is via the *image pyramid*. The simplest version of the idea, which goes back Leonard Uhr [219] and Steven Tanimoto and Theo Pavlidis [210] in the 1970s , as well as to physicists studying scaling behavior, is to look at all the derived images obtained from the original I by multiple 2×2 block averaging. The first step is

$$I^{(1)}(i,j) = \frac{1}{4}\left(I(2i-1, 2j-1) + I(2i, 2j-1) + I(2i-1, 2j) + I(2i, 2j)\right),$$

and this operation is then repeated. Say, for simplicity, that the original image has size $2^n \times 2^n$. Then for each k, $0 \leq k \leq n$, we can define the k-fold "blown-down" image, of size $2^{n-k} \times 2^{n-k}$, by

$$I^{(k)}(i,j) = \frac{1}{4^k} \cdot \sum_{r=0}^{r=2^k-1} \sum_{s=0}^{s=2^k-1} I(2^k i - r, 2^k j - s).$$

Here, the original I is $I^{(0)}$ and the final $I^{(n)}$ is just a constant equal to the mean pixel value. These are conceived of as layers of a pyramid, as shown in Figure 6.4.

---

[1]The photographer is, of course, biased in his/her choice of viewpoint, seeking to produce beautiful or informative images, and this creates significant but subtle dependencies.

Figure 6.4. The idea of the image pyramid (left): a 32 × 32 image of a handwritten 5 is at the base; above it are five block-averaged images. The dotted gray lines connect corresponding points. A 128×128 image (top right). Two 64×64 images (bottom right) obtained by extracting the central portion and averaging 2×2 blocks in the bigger image. (From [110, Figure 1.5].)

We can consider the smaller images $I^{(k)}$ as the same scene taken with a camera with lower resolution. By using this averaging, we can define scale-invariance in a completely discrete setting. We assume we have a probability measure on the space $\mathbb{R}^{4N^2}$ of $2N \times 2N$ real-valued images $I : (2N \times 2N \text{ grid}) \to \mathbb{R}$. Then we can form *marginals* of this measure by certain linear projections $r : \mathbb{R}^{4N^2} \to \mathbb{R}^{N^2}$. Either we can extract an $N \times N$ subimage of the bigger image by *restricting* I to a subgrid of size $N \times N$ in any position (in the figure, the central subgrid is chosen), or we can *average* I over 2×2 blocks. In formulas, the maps are

$$r_{a,b}(I)(i,j) = I(i+a, j+b) \quad \text{or}$$
$$r_{\text{block}}(I)(i,j) = \frac{1}{4}\left(I(2i-1, 2j-1) + I(2i, 2j-1) + I(2i-1, 2j)\right.$$
$$\left. + I(2i, 2j)\right).$$

We define the measure to be a *renormalization fixed point* if its marginals under all these linear maps r are identical. Even if we replace the pixel values by single bytes, we can approximate this property by rounding the block average randomly back to one byte. As we discuss later, *random natural images appear to follow a distribution with this property*.

More generally, assume we have fixed measures on the spaces of $N_1 \times N_2$ images for all N_1, N_2. Then scale invariance should mean that taking its marginals

6.2. Scale Invariance in the Discrete and Continuous Setting

by either taking subimages in any position or by taking 2×2 block averages always leads to a measure in the same family. Thus, in forming the pyramid of an image, the probability of the image $I^{(k)}$ on each layer should be the sum of the probabilities of all high-resolution images $I^{(0)}$ on the bottom giving this averaged image at layer k.

In contrast, it is in many ways more natural to pass to the limit and treat discrete images as sampled versions of a more complete underlying continuous image. To do this, we need probability measures on function spaces. We give more background on these later, but let's just work with the idea here. One way to formulate that $I(x, y)$ and $S_\sigma I(x, y) = I(\sigma x, \sigma y)$ are equiprobable is to go back to what it means that some object is present in an image. This can be defined by asking that a collection of sensors Φ_k satisfy inequalities such as

$$|\langle I, \Phi_k \rangle - a_k| \leq \varepsilon, \quad \text{where } \langle I, \Phi_k \rangle = \iint I(x, y) \cdot \Phi_k(x, y) dx dy.$$

Then the probability of this object being present at this point and on this scale is the *measure* of this set of images I. To ask whether this object is present after scaling by σ, we look instead at the measure of the set of I such that

$$|\langle I, S_\sigma \Phi_k \rangle - a_k| \leq \varepsilon,$$

where S_σ is defined by

$$S_\sigma \Phi(x, y) = \sigma^2 \Phi(\sigma x, \sigma y).$$

Notice that the factor σ^2 is there so that the scaled sensor has the *same sensitivity* as the original; that is, so that

$$\langle S_\sigma I, \Phi \rangle = \langle I, S_{\sigma^{-1}} \Phi \rangle.$$

Thus, we may define rigorously what it means for a continuous image model to be scale-invariant.

Definition 6.1. A stationary scale-invariant model for continuous images should be a probability distribution P on the space of images $I : \mathbb{R}^2 \to \mathbb{R}$ such that

$$\forall \sigma > 0, \quad S_{\sigma,*}(P) = P,$$

$$\forall a, b, \quad T_{a,b,*}(P) = P.$$

(Here $T_{a,b} I(x, y) = I(x + a, y + b)$.)

Exactly what space of functions should be used to model images is discussed next.

6.2.2 Experimental Observations

We describe here three pieces of evidence that people have found for the scale-invariance of images.

1. The first fact is that when we examine the power spectrum of random single images, we find that a reasonably good fit to their power spectrum is given by

$$\mathbb{E}_\mathcal{I}(|\widehat{\mathcal{I}}(\xi,\eta)|^2) \simeq \frac{C}{(\xi^2+\eta^2)^{s/2}},$$

where s depends on the type of image. The numbers s range from 1 to 3, but cluster in the neighborhood of 2. If you take a large data base of images and lump together their power spectra, then the fit with $s=2$ becomes more and more convincing. Van Hateren [220, 221] for example, found power law fits with exponents $1.88 \pm .43$ and $2.13 \pm .36$. The power predicted by scale-invariance is 2. To see this, calculate the total expected power of the signal in an annulus in the Fourier plane $A = \{\rho_1 \leq \sqrt{\xi^2+\eta^2} \leq \rho_2\}$:

$$\mathbb{E}\left(\iint_A |\widehat{\mathcal{I}}(\xi,\eta)|^2 d\xi d\eta\right) \approx \iint_A \frac{d\xi \cdot d\eta}{\xi^2+\eta^2} = 2\pi \log\left(\frac{\rho_2}{\rho_1}\right).$$

This is scale-invariant because the power doesn't change if we shrink or expand the annulus because it depends only on the *ratio* of its inner and outer diameter.

Note that this model implies that the total power of the signal must be infinite. Thus, either the model holds only for some limited range of frequencies (as must be the case for discrete images) or, in the continuous limit, we face both "infrared" and "ultraviolet" catastrophes.

This fit to the empirical power spectrum should correspond to a fit of the empirical autocorrelation because the autocorrelation and the power spectrum are Fourier transforms of each other. In fact, the power spectrum rule implies and is implied by the following rule for autocorrelation: fix a vector displacement β between two pixels and let $\mathcal{A}$ denote a random pixel; then the expected squared difference between image values at pixels separated by β is approximated as

$$\mathbb{E}_{\mathcal{I},\mathcal{A}}\left((\mathcal{I}(\mathcal{A}) - \mathcal{I}(\mathcal{A}+\beta))^2\right) \simeq c_1 + c_2 \log(\|\beta\|).$$

This cannot make literal sense for small β because the term on the left is positive and the term on the right is negative; this is another aspect of the ultraviolet catastrophe caused by scale invariance. As we shall see, scale-invariance implies that images do not have pointwise values at all, but in the

6.2. Scale Invariance in the Discrete and Continuous Setting

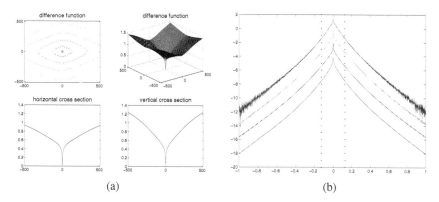

Figure 6.5. (a) Histogram of squared intensity differences for pixel pairs $\alpha, \alpha + \beta$ plotted against β. Note that horizontally and vertically oriented pairs are likely to be more similar than diagonally oriented pairs. For details, see Huang and Mumford. [109]. (b) Histograms of responses of the convolution of the image with a "steerable wavelet" due to Simoncelli plotted for van Hateren's data base, and for the 2×2, 4×4, and 8×8 block-averaged images. The histograms are shifted vertically for legibility; except for increasing noise, they are essentially identical out to six standard deviations. (From [110, Figure 2.20].)

discrete case, small β is meaningless and we can compare this rule with the data. In Figure 6.5(a), this expectation is computed for van Hateren's data base. Note that, because of the geometry of the world, rotational invariance is broken and pixels tend to have more similar values along horizontal and vertical lines than on diagonal lines. Close analysis shows that the log expression fits these curves well for pixel separations of between 2 and 64. The formula $c_1 + c_2 \log(\| \beta \|)$ seems to be valid for values of β up to $\frac{1}{10}$ of the size of the image, but then it seems to change into $c_3 \| \beta \|$, as special facts about our world, such as the sky being on top, seem to have a measurable effect. This points out that in fitting a universal model to low-level features, we must not expect to have a perfect fit. But in the data, the power spectrum fit and the autocorrelation fit are consistent.

2. The second point involves looking at the full histograms of filter responses before and after block averaging. We ask whether

$$\forall \text{ filters } F, \text{ hist}(F * I) = \text{hist}(F * I^{(k)}).$$

This works out astonishingly well; see Figure 6.5(b).

3. Going beyond linear filters altogether, D. Geman and A. Koloydenko [81] have proposed analyzing 3×3 blocks in images by a modified *order statistic*. They first ordered the nine pixel values $a_1 < a_2 < \cdots < a_9$ (assumed

to be in [0,255]) and then mapped them to small numbers by mapping a_1 to 0 and a_k to either the same or one more than the image of a_{k-1}, depending on whether $a_k - a_{k-1} > 16$. The result was a simplified 3×3 block of small numbers, which most often was either all 0s (background blocks with intensity variation less than 16: about 65% of 3×3 blocks) or all 0s and 1s (blocks showing edges or corners with roughly two gray levels present: about 20% of 3×3 blocks). They looked at two statistics: (a) z defined by the range $[0, z]$ of the simplified block, and (b) conditional on $z = 1$ and the block being divided into a connected set of 0s and a connected set of 1s, the number y of pixels in the component not containing the center pixel. They calculated the distributions of z and y for their data base of 80 images and for down-scaled (2×2 block averaged) images. The two histograms appear identical to within experimental fluctuations.

What causes this scale-invariance in images? Certainly one factor is the fact that images are taken at varying distances and this causes their scale to change. We believe a second reason is that "objects" tend to have smaller "objects" on their surfacea and these, in turn, have smaller "objects" on them. (The objects might also be surface markings, or wetness, or reflections, or others; hence, the use of quotes.) We will return to this topic in Section 6.9.

6.3 The Continuous and Discrete Gaussian Pyramids

In Section, 6.2, we defined scale-invariance for discrete images by block averaging, and we showed how this leads to an image pyramid. The pyramid displays simultaneously the same image at all scales. Because images are essentially scale-invariant entities, *we can assume that every part of the image pyramid has the same local properties as any other*. Moreover, a relationship between pixels in an image that is global at one level of the pyramid becomes local higher up, at the point where these pixels are brought together. Thus, pyramids form a very natural data structure for image computations: local analysis at each level plus messages between levels should, in principle, parallelize all image computations. The idea of a multiscale representation of many very different situations arose in the second half of the 20th century and has affected many sciences, such as the theory of turbulence, statistical mechanics, and numerical solution of elliptic problems.

The purpose of this section is to define the most important image pyramid, the *Gaussian pyramid*, in its continuous and discrete form. The 2×2 block-averaged pyramid is a perfectly correct idea, but it is a rather artificial construction biased toward square lattices and subject to aliasing. It corresponds to assuming that the

6.3. The Continuous and Discrete Gaussian Pyramids

sensors of our camera or the receptors of our retina really sample light uniformly over small disjoint squares covering the image plane. This would mean that, if E is the incident photon energy and B is the unit square $\{0 \leq x, y \leq 1\}$, then images are of the form

$$\begin{aligned} I(i,j,\sigma) &= \frac{1}{\sigma^2} \iint_{(i-1)\sigma \leq x \leq i\sigma, (j-1)\sigma \leq y \leq j\sigma} E(x,y) dx dy \\ &= \langle E, \frac{1}{\sigma^2} \mathbb{I}_{\sigma((i,j)-B)} \rangle, \\ &= \left(E * \frac{1}{\sigma^2} \mathbb{I}_{\sigma B} \right)(\sigma i, \sigma j), \end{aligned}$$

where σ is the receptor size (or the camera's resolution) and $\mathbb{I}_S$ is the indicator function of S. If this were true, block averaging would mimic changing the resolution of the camera:

$$\begin{aligned} I(i,j,2\sigma) = \frac{1}{4} \big(& I(2i, 2j, \sigma) + I(2i-1, 2j, \sigma) + I(2i, 2j-1, \sigma) \\ & + I(2i-1, 2j-1, \sigma) \big), \end{aligned}$$

and the discrete pyramid we defined above is just $I(i, j, 2^k \sigma_0)$, a collection of images made with cameras of decreasing resolution.

There are two reasons to reject this. First, no one can make sensors with such sharp cutoffs. The response functions will be much smoother, due to many factors such as (a) restrictions in the placement of the sensors on the image plane, (b) diffraction from the finite size of the lens, and (c) lack of perfect focus due to the varying distances from the lens to different parts of the scene. Second, no one wants them to be like this because sharp cutoffs produce bad aliasing, so that these discrete images are not the best discrete representation of the signal.

What is aliasing? The problem of aliasing already came up in Chapter 2, Section 2.2. If the sensor array spacing is δ and the signal has an oscillation with period $2\delta/3$, what happens is shown in Figure 6.6 (see also Figure 2.4). Note that if the phase is right, two positive lobes and one negative fall on sensor 1, and one positive lobe and two negative on sensor 2. The result is that it records the signal as a weaker oscillation with longer period 2δ. Writing the sensor spacing frequency as $f = 1/\delta$, the true frequency $(3/2)f$ is aliased down to $(1/2)f$ by the sensor spacing frequency f. This is especially important in sound recordings, which are almost all oscillations. Yet another pyramid is that produced by low-pass filtering, cutting off sharply all spatial frequencies above some limit:

$$I(x, y, \sigma) = (\widehat{I} \cdot \mathbb{I}_{\|\xi, \eta\| \leq \sigma^{-1}})\widehat{}.$$

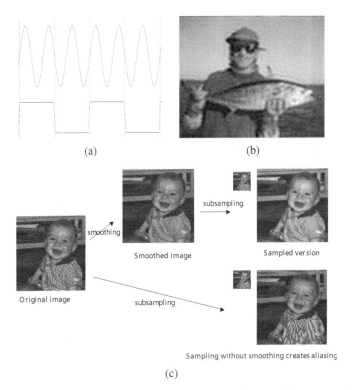

Figure 6.6. (a) Aliasing of a one-dimesional signal: the sine wave is sampled by averaging between the dashedlines. This produces an oscillation at one-third the frequency. (b) The effect of truncating all spatial frequencies in an image above some limit. All sharp edges produce parallel ringing bands. (c) Aliasing in an image. Note how the baby's T-shirt has a completely different pattern. (Photo courtesy of Jean-Michel Morel and Frédéric Guichard.)

(Here (ξ, η) are spatial frequencies, and we recall that the "hat" at the end means taking the inverse Fourier transform of the preceding term.) This produces nasty "ringing" in bands parallel to edges, as shown in Figure 6.6(b) and is usually not suited to images (but see [39]). See also the discussion about the aliasing phenomenon in Chapter 2, Section 2.2 and Section 2.7, Exercise 7 about Shannon-Nyquist sampling theorem.

6.3.1 The Continuous Gaussian Pyramid

Engineers have studied many methods to reduce aliasing, but the simplest mathematical model—which is not far off from engineering practice—is to replace block averaging by Gaussian averaging. Gaussian filtering is much nicer mathematically and gives, in certain senses, the optimal trade-off between small support

6.3. The Continuous and Discrete Gaussian Pyramids

and small aliasing. We simply replace the block sensor $(1/\sigma^2) \cdot \mathbb{I}_{\sigma B}$ by the two-dimensional Gaussian sensor

$$G_\sigma(x,y) = \frac{1}{2\pi\sigma^2} e^{-\frac{x^2+y^2}{2\sigma^2}}.$$

We then associate with an image I the *continuous Gaussian image pyramid*:

$$\begin{aligned}I^{\text{GP}}(x,y,\sigma) &= \iint E(x+u, y+v) G_\sigma(u,v) du dv \\ &= (E * G_\sigma)(x,y),\end{aligned}$$

which gives a simple multiscale description of the image. Recall that we are denoting by $E(x,y)$ the incident photon energy at (x,y), so $I^{GP}(\cdot,\cdot,\sigma)$ represents the light signal smoothed at a scale σ. A beautiful fact about this image pyramid is that it satisfies the heat equation! In fact,

$$\frac{1}{\sigma} \cdot \frac{\partial(I^{GP})}{\partial \sigma} = \Delta(I^{GP}).$$

This relation is simple to prove. First we check that

$$\frac{1}{\sigma} \cdot \frac{\partial G_\sigma}{\partial \sigma} = \Delta(G_\sigma).$$

Second, by substitution, we can rewrite the integrand above as $E(u,v) \cdot G_\sigma(x-u, y-v)$, and then

$$\left(\frac{1}{\sigma}\frac{\partial}{\partial \sigma} - \Delta\right) I^{GP}(x,y) = \iint E(u,v) \left(\frac{1}{\sigma}\frac{\partial}{\partial \sigma} - \Delta\right) G_\sigma(x-u, y-v) du dv$$

$$= 0.$$

Actually, if we change variables, substituting $\tau = \sigma^2/2$, this PDE becomes the ordinary heat equation. Thus, the images at various levels of σ are just smoothed versions of the original using heat flow.

6.3.2 The Burt-Adelson Discrete Pyramid

Remarkably, there is a simple discrete way to approximate the Gaussian pyramid, as discovered by Peter Burt and Ted Adelson [38]. Their construction paved the way to the discovery of wavelets, which provide a vast array of multiscale representations of signals. To describe their construction most simply, we start with the one-dimensional case. Let

$$P^{(n)} = \mathbb{I}_{[0,1]} * \cdots * \mathbb{I}_{[0,1]}$$

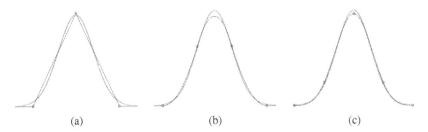

Figure 6.7. The approximations of the Gaussian (dotted) by (a) linear, (b) quadratic, and (c) cubic splines. The knots, or the discontinuities in the first, second, and third derivatives, respectively, are shown with small circles. The Kolmogorov-Smirnov distances between the Gaussian and the spline distributions are about .016, .01, and .007 in the three cases, respectively.

be the result of convolving the indicator function of the unit interval with itself n-times. This defines a series of piecewise polynomial approximations that converge to the Gaussian as $n \to \infty$. The first few P are graphed in Figure 6.7. The function $P^{(2)}$ is just the piecewise linear "tent" function supported on $[0, 2]$ with values 0 at 0, 1 at 1, 0 at 2. Using the fact that $P^{(n)} = P^{(n-1)} * \mathbb{I}_{[0,1]}$, we see that $(P^{(n)})'(x) = P^{(n-1)}(x) - P^{(n-1)}(x-1)$. Then it follows easily by induction that

1. $P^{(n)}$ will be supported on $[0, n]$;

2. It will be given by an $(n-1)$-order polynomial on each interval $[k, k+1]$;

3. It will have $n-2$ continuous derivatives at the "knots," $x = k$.

Such a function is called a *spline* of order $n-1$. Probabilistically, $P^{(1)} = \mathbb{I}_{[0,1]}$ is the distribution of a random variable uniformly distributed between 0 and 1, and $P^{(n)}$ is the distribution of the sum of n such independent variables. In particular, the mean of the probability density $P^{(n)}$ is $n/2$ and its variance is n times that of $\mathbb{I}_{[0,1]}$, which we check to be 1/12. Thus, by the central limit theorem,

$$\forall y \in \mathbb{R}, \quad \lim_{n \to \infty} \int_{-\infty}^{y} P^{(n)}\left(\sqrt{\frac{12}{n}}\left(x - \frac{n}{2}\right)\right) dx = \int_{-\infty}^{y} G_1(x)\, dx.$$

What is most striking is that the quadratic spline $P^{(3)}$ and the cubic spline $P^{(4)}$ are already very close to the Gaussian (see Figure 6.7). Thus, we can approximate the continuous Gaussian pyramid by filtering with these splines at size 2^k instead

6.3. The Continuous and Discrete Gaussian Pyramids

of with Gaussians:

$$I^{DP,k}(i,j) = \left(E * \frac{1}{4^k}P^{(n)}(\frac{x}{2^k})P^{(n)}(\frac{y}{2^k})\right)(2^k i, 2^k j)$$

$$= \iint E(2^k(i-u), 2^k(j-v))P^{(n)}(u).P^{(n)}(v)dudv.$$

This is a pyramid in the sense that its values at level $k+1$ can be computed from those at level k by a procedure called *filtering and decimating*. We define the two-dimensional binomial filter of order n by

$$F_n = \frac{1}{4^n}\sum_{p,q=0}^{n}\binom{n}{p}\binom{n}{q}\delta_{p,q} = \left(\tfrac{1}{4}(\delta_{0,0} + \delta_{0,1} + \delta_{1,0} + \delta_{1,1})\right)^{*n}.$$

and we define *binomial downsampling* D_n by

$$D_n I(i,j) = (I * F_n)(2i, 2j).$$

Then we have Proposition 6.2:

Proposition 6.2.

$$I^{DP,k+1}(i,j) = D_n(I^{DP,k}) = \frac{1}{4^n}\sum_{p,q=0}^{n}\binom{n}{p}\binom{n}{q}I^{DP,k}(2i-p, 2j-q).$$

Proof: Note that

$$\tfrac{1}{2}\mathbb{I}_{[0,2]} = \tfrac{1}{2}(\mathbb{I}_{[0,1]} + \mathbb{I}_{[1,2]}) = \tfrac{1}{2}(\delta_0 + \delta_1) * \mathbb{I}_{[0,1]},$$

and, hence,

$$\frac{1}{2^n}\left(\sum\binom{n}{k}\delta_k\right) * P^{(n)}(x) = \left(\tfrac{1}{2}(\delta_0 + \delta_1) * \mathbb{I}_{[0.1]}\right)^{*n} = \left(\tfrac{1}{2}\mathbb{I}_{[0,2]}\right)^{*n}$$

$$= \tfrac{1}{2}P^{(n)}(\tfrac{x}{2}).$$

Thus,

$$I^{DP,k+1}(i,j) = \left(E * \frac{1}{4^{k+1}}P^{(n)}(\frac{x}{2^{k+1}})P^{(n)}(\frac{y}{2^{k+1}})\right)(2^{k+1}i, 2^{k+1}j)$$

$$= \left(E * \left(\frac{1}{4^n}\sum\binom{n}{p}\binom{n}{q}\delta_{2^k p}(x)\delta_{2^k q}(y)\right)\right.$$

$$\left. * \left(\frac{1}{4^k}P^{(n)}(\frac{x}{2^k})P^{(n)}(\frac{y}{2^k})\right)\right)(2^{k+1}i, 2^{k+1}j)$$

$$= \left(F_n * I^{DP,k}\right)(2i, 2j). \qquad \square$$

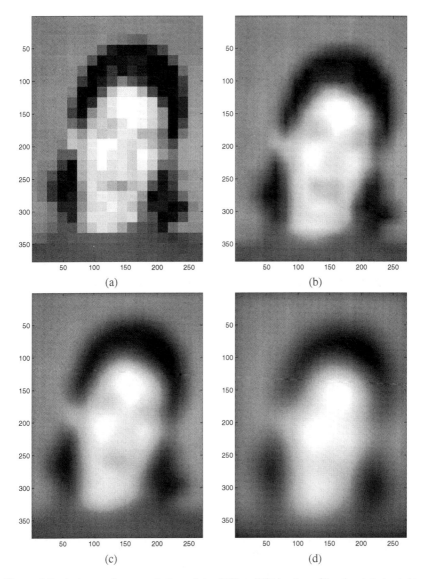

Figure 6.8. An image of a woman's face of size 1152 × 1536 has been filtered and decimated to size 18 × 24 by)a) block averaging D_1, (b) linear filtering D_2, (c) third-order binomial D_3, and (d) eighth-order binomial filters D_8. To view the small image, it has been reexpanded by the transpose operation (see Section 6.4). Note the Mach bands in (b): these are a visual artifact when an image has a discontinuous derivative.

Which n is the best? If $n = 1$, we have block averaging and, as previously mentioned, we get aliasing and unrealistic models of real cameras. For $n > 4$, the discrete images in the resulting pyramid become very smooth, the mutual information between adjacent pixels is very high, and the pixels are highly redundant. The best trade-off between introducing artifacts and over-smoothing seems to be the pyramid with $n = 3$. The original Burt-Adelson pyramid used $n = 4$. Four of these pyramids are compared in Figure 6.8.

These discrete pyramids have another property, a kind of stability, which makes them even more natural. Suppose we have any sensor with compact support and nonzero mean and use it to compute a high-resolution discrete image. Starting here, if we filter and decimate many times by using D_n, *the resulting pyramid converges to the order n discrete pyramid* I^{DP}. This result is stated carefully and proven in the appendix to this chapter, Section 6.10.

6.4 Wavelets and the "Local" Structure of Images

The third major general characteristic of images, as signals, is that they are made up of distinctive local elements. These are the analogs of the phonemes of speech; they are not semantically meaningful parts of the signal but they are the basic reusable building blocks from which larger meaningful elements are constructed. In discussing texture in Section 4.5.1, we referred to the idea that every texture is made up of textons, elementary shapes tat are repeated, regularly or randomly, with variations in the textured area. We have seen that images show statistical scale-invariance, so whatever shapes are used to build up textures ought to be used at larger scales too to build up objects and scenes. The simplest sorts of textons encountered in Chapter 4 were the simple geometric shapes formed by edges, bars, and blobs. *Remarkably, the same basic elements have reappeared in every statistical analysis of natural images as well as in all psychophysical and neurophysiological studies of human and animal vision.* For example, Figure 6.9 displays the results of a typical study showing that the most common structures present in 8×8 image patches are of these types. An intensive study made of 3×3 image patches in very large data bases [43, 137] confirms these conclusions. Edges, bars, and blobs are the phonemes of images.

Whenever we have a class of functions with distinctive local properties, we can seek to represent these functions as sums of elementary functions that are the simplest functions with these properties. Thus, periodic functions are naturally decomposed into sums of sine waves via Fourier analysis. When these properties consist of the function having some stereotyped local form, we can use the simplest functions with such a local form at one point and zero elsewhere. For images made up of edges, bars, and blobs, we can ask for the simplest images, namely those with a *single* edge, bar, or blob and nothing else, that is, zero

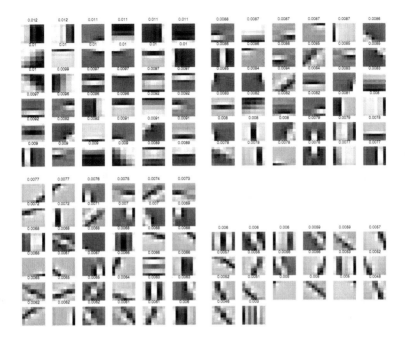

Figure 6.9. A large number of 8 × 8 image patches extracted from the van Hateren data base, each one considered as a point in $\mathbb{R}^{64}$. These patches are the means of clusters found in this cloud of points using k-means. Note that almost all the patches show an edge, bar, or blob. (From [110, Figure 2.49].)

elsewhere (see Figure 6.11). Then we can hope that the whole image decomposes in a natural way into a sum of these simple images. This is the first basic idea behind *wavelets*.

There is a link here with the statistical scale invariance that images exhibit. If phonemes are the atoms of the auditory signal, there is nothing smaller because their parts do not carry any independent information. Nor do we ever find slow phonemes that carry over several seconds. But when we magnify a perfect image, there is always more detail with its own still smaller edges, bars, and blobs. And if we blur an image, we find that large-scale structures in the image are reduced to mere edges, bars, and blobs. This leads to the second basic idea behind wavelets: in addition to moving the wavelet around to different parts of the domain, we should expand and contract the basic wavelet and use this whole collection of functions as a basis for expanding arbitrary signals. This, then, is an additive multiscale representation of the signal.

If $\psi_\alpha(x,y), 1 \leq \alpha \leq A$, are the simple functions, each of which represents one of the possible local properties, these are called *mother wavelets*, and the whole image is decomposed or approximated by a sum of translations and scalings

6.4. Wavelets and the "Local" Structure of Images

of these:
$$I(x,y) \approx \sum_{n,m,k,\alpha} c_{n,m,k,\alpha} \psi_\alpha(2^k x - n, 2^k y - m).$$

This is the fundamental equation of wavelet theory. Note that each term represents the αth basic shape, scaled to size 2^{-k} and translated to the point $(\frac{n}{2^k}, \frac{m}{2^k})$. Here, the wavelets are scaled by powers of 2—dyadic scaling—but any factor $\lambda > 1$ works just as well.

The simplest of these representations is a representation of the image as a sum of Laplacians of Gaussians, to which we turn next.

6.4.1 Frames and the Laplacian Pyramid

The Gaussian pyramid and its approximations are multiscale representations of an image whose values at each level σ represent the image smoothed to that scale. A logical step in analyzing an image as a function is to look at the *difference* of two levels, asking what is added when we pass from a coarse scale to a finer scale. We saw an example in Chapter 0, where we subtracted more smoothed versions of a face image from less smoothed versions. What we found was that these differences represent different levels of structure. At the coarsest scale, they show only the large but blurred shapes; at an intermediate scale, they show the parts of the large shape, which itself has blended into the background; and at the fine scale, they show only sharp edges and textures separated by gray areas. The basic idea is that each difference shows only elements on that scale, which we will call local relative to the scale at which the image has been smoothed.

The differences between scales form what is called the *Laplacian pyramid*. It can be defined in both the continuous and discrete settings. In the continuous case, we fix a coarse resolution σ_0 and a fine resolution σ_1, with $\sigma_1 < \sigma_0$. Then we may write the finer image as a sum of the coarser plus correction terms:

$$\begin{aligned} I^{GP}(x,y,\sigma_1) &= I^{GP}(x,y,\sigma_0) + \int_{\sigma_0}^{\sigma_1} \frac{\partial I^{GP}}{\partial \sigma} d\sigma \\ &= I^{GP}(x,y,\sigma_0) - \int_{\sigma_1}^{\sigma_0} \sigma \triangle I^{GP} d\sigma. \end{aligned}$$

Thus, $I^{LP} = -\sigma \cdot \triangle I^{GP}$ represents what is being added to an image when the resolution is increased; this is why the result is called the Laplacian pyramid. In practice, it is approximated by differences

$$I^{GP}(x,y,\sigma_1) - I^{GP}(x,y,\sigma_0) = (I * (G_{\sigma_1} - G_{\sigma_0}))(x,y),$$

where $G_{\sigma_1} - G_{\sigma_0}$ is called a difference of Gaussians (DOG) filter. An example of this is shown in Figure 6.10.

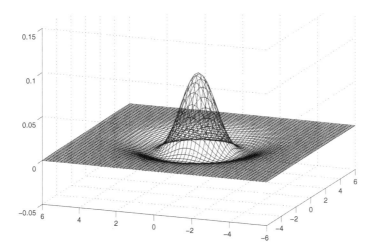

Figure 6.10. The graph of a difference-of-Gaussian (DOG) filter, where the outer negative filter has size $\sqrt{2}$ larger than the inner positive one.

From another point of view, we may think of this filter as a "blob detector." Since the filter is positive in the center and negative in a surrounding annulus, it responds maximally to blobs of the right size against a contrasting surround. If an image has a contrasted edge, the filtered image will have positive responses on one side of the edge and negative responses on the other; hence, it will cross zero on a line that approximates the edge. If there was such a zero-crossing over a range of scales, Marr and Hildreth [151] have proposed that this was a good edge detector. It will have an even larger response all along a bar whose width roughly matches the size of the inner positive circle. Crowley and Parker [53] introduced in 1984 the idea that the local maxima of the absolute value of the Laplacian pyramid were key points in an image. These are isolated points in (x, y, σ) space where a blob in that position and of that scale fits the image optimally. Their idea has been developed by Lowe [143], who named it SIFT (for scale-invariant feature transform) and it is now a very popular preprocessing step for object recognition applications.

Assuming scale-invariance of image statistics, we can work out the expected power at different levels of the Laplacian pyramid. In fact, if we assume $\mathbb{E}_I(|\widehat{I(\xi, \eta)}|^2) \approx C/(\xi^2 + \eta^2)$, then, by using the Fourier transform formula from Section 2.2, we find that $\widehat{G_\sigma}(\xi, \eta) = e^{-2\pi^2\sigma^2(\xi^2+\eta^2)}$; hence,

$$\mathbb{E}_I(|\widehat{I^{LP}(\cdot, \cdot, \sigma)}(\xi, \eta)|^2) = |\widehat{\sigma \triangle G_\sigma}|^2 \cdot \mathbb{E}_I(|\widehat{I(\xi, \eta)}|^2)$$
$$\approx \left|\sigma((2\pi i \xi)^2 + (2\pi i \eta)^2) \cdot \widehat{G_\sigma}(\xi, \eta)\right|^2 \cdot \frac{C}{\xi^2 + \eta^2}$$

6.4. Wavelets and the "Local" Structure of Images

$$= C'\sigma^2(\xi^2 + \eta^2) \cdot e^{-4\pi^2\sigma^2(\xi^2+\eta^2)},$$

where C' is a new constant.

Note that this dies off at both low and high frequencies and, in fact, both the infrared and ultraviolet blow-ups are gone. The total power here, the integral of this over ξ and η, is just C''/σ^2. The power is concentrated in an annulus in the (ξ, η)-plane, peaking at spatial frequency $\sqrt{\xi^2 + \eta^2} = 1/(2\pi\sigma)$. Thus, the Laplacian pyramid roughly separates the different spatial frequency components of the image.

The discrete form of this pyramid, also due to Burt and Adelson, is equally important. It introduced the idea of frames into vision and, even before a systematic theory of wavelets had been worked out, paved the way to apply the theory of wavelets to images. Here, we define an analogous decomposition of a discrete image $I(i, j)$ by using the differences between levels of the discrete Gaussian pyramid $I^{DP,k}(i, j)$. In what follows, we fix the order of the binomial smoothing n, and the binomial downsampling operator D_n will simply be denoted by D. We assume $I(i, j) = I^{DP}(i, j, 0)$ is the level 0 part of the pyramid. *We want to subtract successive levels of the pyramid.* But since they are defined on coarser grids too, to do this we need to interpolate. This is done by the transpose D^t of D:

$$D^t(I)(i, j) = 4 \sum F_n(i - 2k, j - 2l) \cdot I(k, l).$$

This is formally the transpose for the inner products $||I||^2 = 4^k \sum_{i,j} I(i, j)^2$ on an image at level k. The reason for the factor 4^k is that the lattice at level k has 4^{-k} as many pixels as the original lattice. The effect of this factor is that if $I \equiv 1$, then $D^t(I) \equiv 1$.

If the order n of binomial smoothing is one, then D^t just replicates each pixel of the coarser grid four times on a 2×2 block of the finer grid; if the order is two, then D^t puts the pixels of the coarser image at the even grid points and linearly interpolates across the intermediate pixels; and if the order is three or more, it interpolates using several coarser-level values for each grid point. We can iterate D^t half a dozen times and create a large image from a small one, as we saw in Figure 6.8.

Now we take the differences $L(I) = I - D^t(D(I))$. This is a discrete DOG-type filter: $D^t D$ is very close to convolution with the binomial filter B_{2n} of order $2n$, order n going down and order n coming back. The operation $D^t D$ does slightly different things to pixels, depending on whether their coordinates are even or odd with respect to the decimation. But if we average $D^t D$ over the four relative positions, we get exact convolution with B_{2n}. Thus $L(I)$ is essentially $I - B_{2n} * I = (\delta_{0,0} - B_{2n}) * I$. The values $n = 2, 3, 4$ seem to be the most useful.

By using all levels of the discrete Gaussian pyramid I^{DP}, we define the discrete Laplacian pyramid by

$$I^{LP,k} = I^{DP,k} - D^t(I^{DP,k+1}).$$

The key point is that the original image I can be reconstructed from its Laplacian pyramid and one very smoothed image $I^{DP}(\cdot,\cdot,K)$. In fact,

$$\begin{aligned} I &= (1 - D^t D)I + D^t(1 - D^t D)DI + (D^t)^2(1 - D^t D)D^2 I + \cdots \\ &\quad + (D^t)^K D^K I = I^{LP,0} + D^t I^{LP,1} + (D^t)^2 I^{LP,2} + \cdots \\ &\quad + (D^t)^K I^{DP,K}. \end{aligned}$$

Thus, knowledge of I and of the set of successive differences I^{LP} plus one very smoothed image are equivalent.

Is there any gain in this? Specifying I directly requires giving N^2 pixel values, and specifying the terms $I^{LP,k}$ of the Laplacian pyramid requires

$$N^2 + (N/2)^2 + (N/4)^2 + \cdots + (N/2^K N)^2 \approx 4N^2/3$$

pixel values. But what happens in practice is that many of the Laplacian image pixels are nearly zero. Coding the original image requires coding N^2 large numbers, but coding the Laplacian pyramid requires coding 1/3 more pixels, many of which are small. Thus, in practice we use fewer bits. This is called using a *sparse coding*, which is a central idea in wavelet theory. Perceptually, the atomic parts of the image are compactly represented by a relatively small number of Laplacian pyramid pixels. In fact, nature has also "discovered" this fact: the signal from the retina to the cortex in mammals seems to transmit the values of the Laplacian filtered image.[2]

This savings of bits is an easy consequence of the scale-invariance properties of images, as seen in Theorem 6.3.

Theorem 6.3. *Assume we have a probability distribution on discrete $N \times N$ images $\mathcal{I}$, which, for simplicity, we treat cyclically as images on a discrete 2-torus. Assume that the probability distribution is stationary and that the expected power spectrum of the image satisfies the scale-invariant law:*

$$\mathbb{E}_\mathcal{I}\left(|\widehat{\mathcal{I}}(k_1,k_2)|^2\right) = C/(k_1^2 + k_2^2), \quad -N/2 \leq k_1, k_2 \leq N/2, (k_1,k_2) \neq (0,0).$$

---

[2] These values are instantiated by the response properties of certain of the retinal ganglion cells. In primates, these are the cells in the *parvo* pathway.

6.4. Wavelets and the "Local" Structure of Images

Then if $\mathcal{I}$ is a random image and $\mathcal{A}$ is a random pixel,

$$\frac{1}{C}\mathbb{E}_{\mathcal{I},\mathcal{A}}\left(|(1 - B_{2n}) * \mathcal{I}(\mathcal{A})|^2\right) < 8 \text{ for } n \leq 4$$

$$\frac{1}{C}\mathbb{E}_{\mathcal{I},\mathcal{A}}\left(|\mathcal{I}(\mathcal{A}) - \overline{\mathcal{I}}|^2\right) \approx 2\pi \log(N) - 1.08 \approx 34 \text{ (resp. 42) if}$$

$$N = 256 \text{ (resp. } N = 1024\text{)}.$$

Comparing the bound 8 to the value 34 or 42, this shows that the mean squared values of the Laplacian pyramid images are significantly smaller than those of the original image (with mean taken out).

Proof: We again use the discrete Fourier transform and coordinatize the frequency plane by pairs of integers (k_1, k_2) with $-N/2 < k_1, k_2 \leq N/2$, with frequency zero in the middle. Then,

$$N^2 \cdot \mathbb{E}_{\mathcal{I},\mathcal{A}}(|\mathcal{I}(\mathcal{A}) - \overline{\mathcal{I}}|^2) = \mathbb{E}\left(\sum_a |\mathcal{I}(a) - \overline{\mathcal{I}}|^2\right)$$

$$= \mathbb{E}\left(\sum_{k_1,k_2 \neq 0,0} |\widehat{\mathcal{I}}(k_1, k_2)|^2\right)$$

$$= \sum_{k_1,k_2 \neq 0,0} C/(k_1^2 + k_2^2).$$

This sum can be estimated by replacing it with the integral $\iint_D r^{-2} dx dy$, where D is the square $-N/2 \leq x, y \leq N/2$ minus a hole around the pole $(0,0)$. This will create a discretization error (as in the definition of Euler's gamma constant) tending to some constant as $N \to \infty$. This integral is equal to $\iint r^{-1} dr d\theta$, which has value on an annulus with radii r_1, r_2 equal to $2\pi \log(r_2/r_1)$. Rounding the corners of the annulus creates an asymptotically small error, so the main term in the integral is $2\pi \log(N)$. We used MATLAB to estimate the additive constant.

If we apply the filter $(\delta - B_{2n})$, the Fourier transform is multiplied by $1 - \cos(\pi k_1/N)^{2n} \cdot \cos(\pi k_2/N)^{2n}$, which kills all the low-frequency components of $\widehat{\mathcal{I}}$. Then the same calculation as above gives

$$N^2 \cdot \mathbb{E}_{\mathcal{I},\mathcal{A}}|(1 - B_{2n}) * \mathcal{I}(\mathcal{A})|^2$$

$$= \sum_{k_1,k_2 \neq 0,0} C \cdot \frac{\left(1 - \cos(\pi k_1/N)^{2n} \cdot \cos(\pi k_2/N)^{2n}\right)^2}{(k_1^2 + k_2^2)}.$$

The summand is small for $\sqrt{k_1^2 + k_2^2}$ both small and large, and the sum is now bounded for large N and, using MATLAB, we can estimate its value. In fact, what one finds is that $3.55 + 3\log(n)$ is an excellent approximation to this sum. □

The savings in bits using an expansion of an image by an *orthonormal basis of wavelets* is developed in Exercise 5 at the end of this chapter. Although the Laplacian pyramid is not a basis, it is the simplest example of a construction known as *frames*.

Definition 6.4. A countable set of vectors $e_n \in \mathcal{H}$ in a Hilbert space is a frame with frame bounds A and B if, for all $x \in \mathcal{H}$:

$$A \cdot ||x||^2 \leq \sum_n \langle x, e_n \rangle^2 \leq B \cdot ||x||^2.$$

Equivalently, this means that the linear map:

$$R : \mathcal{H} \to \ell^2, R(x) = \{\langle x, e_n \rangle\}$$

is bounded, is injective, and has a closed image $L \subset \ell^2$, and that the inverse from L to $\mathcal{H}$ is also bounded.

In the case of the Laplacian pyramid, we simply take the Hilbert space to be the vector space of images and e_n to be set of images $(D^t)^k(I - D^t D)\delta_i$ for all scales k and pixels i.

A frame differs from a basis in that L may be a proper subset of ℓ^2. In other words, frames always span the space $\mathcal{H}$ but may be dependent. Frames are called *tight* if $A = B$. In the finite-dimensional case, where $\mathcal{H} = \mathbb{R}^N$ and there are M elements in the frame, the linear map R is given by an $N \times M$ matrix. Then A (respectively, B), are just the max (respectively, min) values of $||R(x)||^2/||x||^2$, so that $\sqrt{A}$ and $\sqrt{B}$ are just the maximum and minimum singular values of R. A very simple example of a tight frame is given by the M vertices of a regular M-gon in the plane $e_n = (\cos(2\pi n/M), \sin(2\pi n/M)) \in \mathbb{R}^2$. Tight frames behave like bases in that

$$\langle x, y \rangle = \tfrac{1}{A} \sum_n \langle x, e_n \rangle \langle y, e_n \rangle; \quad \forall x, y;$$

hence,

$$x = \tfrac{1}{A} \sum_n \langle x, e_n \rangle e_n, \quad \forall x$$

The first line follows by polarizing the defining identity for tight frames, and the second line follows because its inner product with any y is true by the first line. If frames are not tight, there are still multiple ways to reconstruct an x from the frame. There is always an optimal L^2 way to reconstruct. Note that $R^* \circ R$ is a self-adjoint bounded map from $\mathcal{H}$ to itself with bounded inverse.

6.4. Wavelets and the "Local" Structure of Images

Define the *dual frame* by $e_k^* = (R^* \circ R)^{-1}(e_k)$. By the definition of R^*, we get $R^* \circ R(x) = \sum_n \langle e_n, x \rangle e_n$. Applying $(R^* \circ R)^{-1}$ to this equality, we get the reconstruction formula:

$$x = \sum_n \langle e_n, x \rangle e_n^*,$$

and by letting $y = R^* \circ R(x)$, we get the dual reconstruction formula:

$$y = \sum_n \langle e_n, (R^* \circ R)^{-1} y \rangle e_n = \sum_n \langle e_n^*, y \rangle e_n.$$

To take advantage of the overcomplete nature of a frame, it is often better not to use this inner–product-based reconstruction formula but to seek directly the most efficient way to represent the signal x in terms of the e_n. Two ways have been proposed. One is the *greedy search method*:

1. Pick the frame element e_n that maximizes $\langle x, e_i \rangle$ over all i.

2. Write $x = \frac{\langle x, e_n \rangle}{\langle e_n, e_n \rangle} e_n + r$, where r is the residue, orthogonal to e_n.

3. Go back to step 1, but apply it to the residue r instead of x.

The other way is to seek to minimize by L^1 techniques, not L^2. That is, to seek the combination of frame elements $\sum a_n e_n$ that solves one of the minimization problems:

(A) $\min \sum_n |a_n|$, under the constraint $x = \sum_n a_n e_n$ or

(B) $\min \left(\|x - \sum_n a_n e_n\|_{L^1} + \lambda \sum_n |a_n| \right)$, λ a Lagrange multiplier.

Here λ controls the trade-off between the accuracy of the approximation and the size of the coefficients. The magic of L^1 minimization is that it often picks out the sparsest approximation—the one with the maximal number of zero coefficients a_n. Moreover, expression (A) is a linear programming problem, so it can be efficiently solved.

Example. Let's look at the one-dimensional Laplacian pyramid for the second-order binomial filter with periodic discrete signals of length $M = 2^m$. This means our signals are sequences $\vec{x} = (x_1, x_2, x_3, \cdots, x_M)$. The smooth-and-decimate mappings are given by

$$D(\vec{x}) = (x_1 + 2x_2 + x_3, x_3 + 2x_4 + x_5, \cdots, x_{M-1} + 2x_M + x_1)/4.$$

$L_1 = 1 - D^t \circ D$ is given by the matrix

$$0.25 \cdot \begin{pmatrix} -1 & 2 & -1 & 0 & 0 & 0 & 0 & 0 & \cdots \\ -.5 & -1 & 3 & -1 & -.5 & 0 & 0 & 0 & \cdots \\ 0 & 0 & -1 & 2 & -1 & 0 & 0 & 0 & \cdots \\ 0 & 0 & -.5 & -1 & 3 & -1 & -.5 & 0 & \cdots \\ \cdots & \cdots & \cdots & \cdots & \cdots & \cdots & & & \end{pmatrix},$$

whose rows form the first part of the frame (the last rows are going to wrap around because the signal is cyclic). Curious coefficients appear! The even and odd rows are different because we are using decimation, but all the even rows are cyclic permutations of each other, as are all the odd rows. Let's carry the pyramid down two levels, so we decompose the signal $\vec{x}$ into $L_1 = (I - D^t D)\vec{x}$, $L_2 = D^t (I - D^t D) D\vec{x}$, and the low-pass part $(D^t)^2 D^2 \vec{x}$. L_2 is computed by multiplication by a matrix whose rows are the linear interpolations of every second row of the previous matrix. That is,

$$0.125 \cdot \begin{pmatrix} -1 & -2 & 1 & 4 & 1 & -2 & -1 & 0 & 0 & 0 & 0 & 0 & \cdots \\ -.5 & -1 & -1.5 & -2 & 2 & 6 & 2 & -2 & -1.5 & -1 & -.5 & 0 & \cdots \\ 0 & 0 & 0 & 0 & -1 & -2 & 1 & 4 & 1 & -2 & -1 & 0 & \cdots \\ 0 & 0 & 0 & 0 & -.5 & -1 & -1.5 & -2 & 2 & 6 & 2 & -2 & \cdots \\ \cdots & \cdots & \cdots & \cdots & \cdots & \cdots & \cdots & & & & & & \end{pmatrix}.$$

Finally, we have the low-pass sequences from two levels down, given by

$$0.0625 \cdot \begin{pmatrix} 1 & 2 & 3 & 4 & 3 & 2 & 1 & 0 & 0 & 0 & 0 & 0 & \cdots \\ 0 & 0 & 0 & 0 & 1 & 2 & 3 & 4 & 3 & 2 & 1 & 0 & \cdots \\ 0 & 0 & 0 & 0 & 0 & 0 & 0 & 0 & 1 & 2 & 3 & 4 & \cdots \\ \cdots & \cdots & \cdots & \cdots & \cdots & \cdots & \cdots & & & & & & \end{pmatrix}.$$

We take as our frame the rows of these three matrices. The constant factors are more or less arbitrary, so we normalize each matrix so that the max norm of its rows is 1. Then we get nearly a tight frame: its frame bounds are approximately $A = 1.04, B = 1.7$, independent of m. Working with frames is a lot like cooking: a pinch of this, a pinch of that, and we get the best balanced creation.

Frames in general are important for the same reason that the Laplacian pyramid is important. If we are given some class of vectors in $\mathcal{H}$ (e.g., as random samples from some probability distribution), then one hopes to find a frame in which these vectors can be well approximated with only a small number of large coefficients.

6.4.2 Wavelets Adapted to Image Structures

DOG filters are a good general-purpose tool for extracting the geometric structures in an image around a certain scale. But, as we know, textons are rather more

6.4. Wavelets and the "Local" Structure of Images

specific shapes—edges, bars, and blobs. To optimally find these and to optimally expand the image with the smallest set of functions, matched filters work better. These are wavelets whose shape is a generic version of the texton itself. As described earlier, we seek an expansion

$$I(x,y) \approx \sum_{n,m,k,\alpha} c_{n,m,k,\alpha} \psi_\alpha(2^k x - n, 2^k y - m),$$

where each ψ_α mimics one type of edge, of bar, or of blob. Of the many books currently avialble on on wavelets, we recommend [54, 113, 145, 205].

One class of such wavelets are the so-called *Gabor functions*. Gabor discovered that among functions $f(\vec{x})$, products of sine and cosine by Gaussians minimize the product of $\int \|\vec{x}\|^2 \cdot |f|^2 / \int |f|^2$ and $\int \|\vec{\xi}\|^2 |\hat{f}|^2 / \int |\hat{f}|^2$. This means they have the most compact support possible *simultaneously* in space and frequency. When Hubel and Wiesel discovered that a class of neurons in the primary visual area of experimental animals respond preferentially to edges, bars, and blobs, many people sought the simplest mathematical model for their response properties [17]. Their response to an image flashed before the animal's eyes was modeled by a linear filter, and the best-fitting linear filter appears to be a particular class of Gabor functions, namely

$$\text{Gab}_{\sigma,\theta,\text{even}} = R_\theta \circ S_\sigma \left((\cos(\kappa.x) - e^{-\kappa^2/2}).e^{-\frac{1}{2}(x^2 + (y/2)^2)} \right)$$

$$\text{Gab}_{\sigma,\theta,\text{odd}} = R_\theta \circ S_\sigma \left(\sin(\kappa.x) e^{-\frac{1}{2}(x^2 + (y/2)^2)} \right),$$

where R_θ, S_σ = rotation by θ and scaling by σ; that is,

$$S_\sigma f(x,y) = f(x/\sigma, y/\sigma)/\sigma.$$

These functions are plotted in Figure 6.11. The term $y/2$ in the exponent makes the Gaussian factor elliptical with eccentricity 2; of course, a wide range of cells are measured and this is an average. Likewise, the sine and cosine factors have a period related to the size of the Gaussian, which can be varied by setting the value of κ, producing more or less "sidelobes" in the response profile. The values $2.5 \leq \kappa \leq 3.5$ represent most cortical simple cells: $\kappa = 2.5$ has small sidelobes and $\kappa = 3.5$ has more developed sidelobes (making the cell more narrowly tuned in spatial frequency along the x-axis). Finally, the term $e^{-\kappa^2/2}$ is a very small constant, which makes the even Gabor function have mean 0; this is needed for wavelet theory. The graphs in Figure 6.11 are for $\kappa = 2.5$.

Gabor filters have been used to construct pyramid-based filter banks and frames by sampling the space of possible translations, scalings, and rotations.

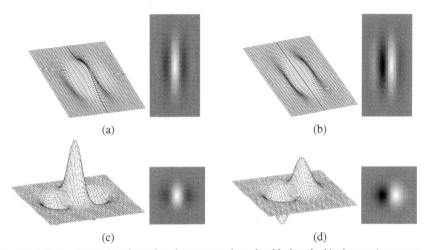

Figure 6.11. Four examples of wavelets shown as graphs and as black-and-white images (gray=zero values): (a) the even Gabor wavelet; (b) the odd Gabor wavelet; bottom left; (c) the even steerable wavelet; (d) the odd steerable wavelet. The Gabor wavelets have eccentricity 2:1 and $\kappa = 2.5$ with a black line to show the section $x = 0$.

T. S. Lee [139] has analyzed many versions of this construction and computer their frame bounds. For example, Lee considers the frame with $\kappa = 2.5$ made up of the functions

$$\text{Gab}_{(2^{-f}, \pi\ell/8, even/odd)}\left(x - \frac{0.8}{2^f}n, y - \frac{0.8}{2^f}m\right), \quad 0 \leq \ell < 8,$$

which has eight oriented directions, spatial frequency 2^f, and spatial locations on grids $(0.8/2^f) \cdot (n, m)$ as getting finer as the frequency increases. He proves that this has frame bounds with $B/A \approx 2$.

In representing images as sums of frames, there is a trade-off between using a larger number of wavelets whose supports are highly elongated, hence whose responses are highly selective in orientation, and a smaller number of wavelets that are not so highly tuned in orientation. In the first case, large portions of edges and bars can be represented by single elements of the frame, but the frame is more over-complete and redundant. In the latter case, the frame is more economical, but each element represents only small parts of edges and bars.

A simple frame of the latter type is the so-called *steerable* frame given by derivatives of an isotropic Gaussian:

$$DG_{\sigma,1} = S_\sigma\left(x \cdot e^{-\frac{1}{2}(x^2+y^2)}\right),$$

$$DG_{\sigma,2} = S_\sigma\left(y \cdot e^{-\frac{1}{2}(x^2+y^2)}\right),$$

6.4. Wavelets and the "Local" Structure of Images

$$DG_{\sigma,3} = S_\sigma \left((1-x^2) \cdot e^{-\frac{1}{2}(x^2+y^2)} \right),$$

$$DG_{\sigma,4} = S_\sigma \left(xy \cdot e^{-\frac{1}{2}(x^2+y^2)} \right),$$

$$DG_{\sigma,5} = S_\sigma \left((1-y^2) \cdot e^{-\frac{1}{2}(x^2+y^2)} \right).$$

This set of functions is called steerable because its linear span is rotation invariant. The functions DG_1 and DG_2 represent small pieces of vertical and horizontal edges, respectively, and a suitable combination of them represents such a edgelet with any orientation. Likewise, DG_3 and DG_5 represent small pieces of vertical and horizontal bars; combining them with DG_4, we get bars with any orientation. The constants in DG_3 and DG_5 are there to make their means 0. Note that $DG_3 + DG_5 = -\triangle G$ is the Laplacian of the Gaussian G used above.

In contrast, so-called *curvelet* frames are crafted to achieve as high an orientation tuning as may be needed to represent straight edges and bars of any length. These are not easy to write; the reader is referred to the paper of Candes and Donoho [39] for details.

The goal of much of wavelet theory for images has been to closely reproduce a natural image I by a wavelet expansion with either the smallest number of nonzero terms or the smallest number of bits. The game is finding how few bits per pixel we can get away with and still have a perceptually acceptable approximation to the original image. This is not merely an intellectual contest because such schemes can be used for commercial image compression (e.g., JPEG2000), and there is real money involved. Figure 6.12 was produced by using MATLAB's implementation of JPEG compression.

Figure 6.12. A 200 × 300 image of an eye: the raw image (left); compressed with JPEG at a 16:1 ratio (1/2 a bit per pixel) (middle); compressed with JPEG at a 40:1 ratio (1/5 bit per pixel) (right).

6.5 Distributions Are Needed

6.5.1 The Problem

We have been talking of constructing stochastic models for continuous images, which means putting probability measures on spaces of images defined on the continuous plane $\mathbb{R}^2$. We would like this measure to be invariant under the action of translations $T_{a,b}$ and scalings S_σ. When we actually begin to worry about building such a measure, we run into an obstacle. *There are no such measures on spaces of measurable functions!* In fact, the measures we find in this section turn out to live on images *mod constant images*, and even here we have a problem if we deal only with images that are functions. This is because of Theorem 6.5.

Theorem 6.5. *There is no translation and scale-invariant probability measure on the vector space $L(\mathbb{R}^n)$ of measurable functions on $\mathbb{R}^n$ or even on $L(\mathbb{R}^n)/\text{constants}$ except for trivial measures supported on constant functions.*

The reason for this theorem is quite simple: as you look more and more closely at some surface point in a scene, more and more detail emerges. Biologists tell us that we have mites living on our eyelashes, and light reflecting off and casting shadows from these mites creates finer and finer patterns of black and white.[3] In other words, the scale-invariance of images means that the light/dark variations, that is, the expected difference $\mathbb{E}(\max(\mathcal{I}) - \min(\mathcal{I}))$, is the same in small windows in an image as it is in the whole image.

We need to measure the size of the local light/dark swings of a measurable function. We consider any function f on $\mathbb{R}^n$. We let $B(x, r)$ be the ball of radius r around a point $x \in \mathbb{R}^n$, and define the following functional of f:

$$V_{a,r}(f, x) = \left|\{y \in B(x, r) \mid |f(x) - f(y)| > a\}\right| / |B(x, r)|,$$

where $|S|$ is the measure of the set S. This V measures the local swings of the function f. For example, if f is continuous at x, then for any a, $V_{a,r}(f, x) = 0$ for r sufficiently small. But for any measurable function f, we still have $\lim_{r \to 0} V_{a,r}(f, x) = 0$ for all a and almost every x. This is an easy consequence of Lebesgue's differentiation theorem (see Folland [74], page 98).

Considering V enables us to make precise our comment that scale-invariance of a probability measure on images implies that random images have just as big swings locally as globally. The function $V_{a,r}$ is a bounded measurable function on $L(\mathbb{R}^n)/\text{constants} \times \mathbb{R}^n$. If we have any probability measure on L or $L/\text{constants}$,

---

[3] The fractal nature of reality was anticipated by the Edward Lear poem:
  Big bugs have little bugs upon their backs to bite em,
  And these little bugs have lesser bugs and so on *ad infinitum*.

6.5. Distributions Are Needed

we can consider the expectation $\mathbb{E}_{\mathcal{F}}(V_{a,r}(\mathcal{F},x))$ of V in the variable $\mathcal{F}$. The proof of Theorem 6.5 is then easy: if the measure is translation-invariant, this expectation is independent of x; if it is scale-invariant, the expectation is independent of r. But as $r \to 0$, V is uniformly bounded (by 1) and goes pointwise to zero, so by the Lebesgue bounded convergence theorem, $\lim_{r \to 0} \mathbb{E}_{\mathcal{F}}(V_{a,r}(\mathcal{F},x)) = 0$. Thus, the expectation of V must be zero for all a, r, x. This means our probability measure is a trivial measure supported on constant images.

As we see in the next section, such probability measures do exist on spaces just barely bigger than spaces of locally integrable functions—in fact, by using the Sobolev spaces (defined in the next section) on

$$\bigcap_{\epsilon > 0} H_{\text{loc}}^{-\epsilon}.$$

Our theory suggests that this is the proper function space for images.

6.5.2 Basics XII: Schwartz Distributions

Let's now look more closely at images defined on the continuum. Why should we believe that images have well-defined values $I(x, y)$ at points? No image is known directly by means of its values at precise points. It has to be measured by an instrument, such as a single device in a charge coupled device (CCD) camera or a simple rod or cone in your retina, and every instrument has sensors of finite size and samples the light via its response function $R(u, v)$. This means it returns not a point value $I(x, y)$ of the light image but an *average*,

$$\iint R(u,v) I(x-u, y-v) du dv,$$

over nearby points.

What all this means is that an image is not necessarily a function but a "thing" whose averages,

$$\langle I, \phi \rangle = \iint I(x,y) \phi(x,y) dx dy,$$

make sense for smooth test functions ϕ. Such a "thing" is called a *Schwartz distribution* or a *generalized function*. The use of distributions allows us to deal effortlessly with the ultraviolet aspect of the power blow-up produced by the power spectrum $1/(\xi^2 + \eta^2)$. A good introduction to the theory of distributions is found in Chapter 9 of Folland [74] with excellent exercises. Here we present the full-blown mathematical definition.

Definition 6.6. Let $\mathcal{D}(\mathbb{R}^n)$ be the vector space of C^∞ functions on $\mathbb{R}^n$ with compact support. Then a *distribution* f is a linear map

$$f : \mathcal{D}(\mathbb{R}^n) \to \mathbb{R},$$

written $f : \phi \mapsto \langle f, \phi \rangle$, which is continuous in the sense that

$$f(\phi_n) \to 0 \text{ as } n \to \infty$$

for any sequence of test functions $\phi_1, \phi_2, \cdots \in \mathcal{D}(\mathbb{R}^n)$, such that

(a) $\bigcup_{n=1}^{\infty} \text{support}(\phi_n) \subset$ some fixed ball B;

(b) For all partial derivatives $D^\alpha = \frac{\partial^{\alpha_1 + \cdots + \alpha_n}}{\partial x_1^{\alpha_1} \cdots \partial x_n^{\alpha_n}}$, $\sup_{\vec{x} \in B} |D^\alpha \phi_n(\vec{x})| \to 0$ as $n \to \infty$.

The set of all such distributions is another vector space, denoted $\mathcal{D}'(\mathbb{R}^n)$. But what are distributions like? Here is a list of six examples and procedures for generating new distributions from old.

1. Ordinary locally integrable functions $f(\vec{x})$ are distributions if we define

$$f(\phi) = \int f(\vec{x}) \phi(\vec{x}) d\vec{x}.$$

Because of this, we often write a distribution suggestively as a function $f(\vec{x})$ and we also write $f(\phi)$ as $\langle f, \phi \rangle$ or as $\int f(\vec{x})\phi(\vec{x})d\vec{x}$ (trying to remember though that these pointwise values $f(\vec{x})$ are meaningless!). Even if a function is not locally integrable, it can sometimes be made into a distribution by either the principal value or the renormalization trick. Thus, $1/x$ is a perfectly good distribution if we define

$$\langle 1/x, \phi \rangle = \lim_{\epsilon \to 0} \int_{\mathbb{R}-(-\epsilon,+\epsilon)} \frac{\phi(x)}{x} dx, \text{ called the } \textit{principal value}.$$

The $1/x^2$ can also be made into a distribution if we define

$$\langle 1/x^2, \phi \rangle = \lim_{\epsilon \to 0} \left(\int_{\mathbb{R}-(-\epsilon,+\epsilon)} \frac{\phi(x)}{x^2} dx - 2\frac{\phi(0)}{\epsilon} \right), \text{ called } \textit{renormalization}.$$

2. Delta functions $\delta_{\vec{a}}$ are distributions via

$$\langle \delta_a, \phi \rangle = \phi(\vec{a}),$$

and, more generally, locally finite signed measures on the real line define distributions.

3. All distributions can be differentiated as often as we want by using the integration-by-parts rule: $f \in \mathcal{D}'$, then $\frac{\partial f}{\partial x_i} \in \mathcal{D}'$ is defined by

$$\langle \frac{\partial f}{\partial x_i}, \phi \rangle = -\langle f, \frac{\partial \phi}{\partial x_i} \rangle$$

6.5. Distributions Are Needed

because, if f were a smooth function, we could use integration by parts:

$$\langle \frac{\partial f}{\partial x_i}, \phi \rangle = \int \frac{\partial f}{\partial x_i} \phi \, d\vec{x} = -\int f \frac{\partial \phi}{\partial x_i} \, d\vec{x} = -\langle f, \frac{\partial \phi}{\partial x_i} \rangle.$$

This gives us, for example, derivatives of $\delta_{\vec{a}}$, whose value on a test function are the derivatives of the test function.

4. Two distributions cannot usually be multiplied, but if $f \in \mathcal{D}', \phi \in \mathcal{D}$, then $f\phi$ makes sense via

$$\langle f\phi, \psi \rangle = \int f(\vec{x}) \phi(\vec{x}) \psi(\vec{x}) d\vec{x} = \langle f, \phi\psi \rangle.$$

5. We can convolve a distribution with a test function, obtaining an infinitely differentiable function (although not with compact support). In fact, this is just the distribution evaluated on translates of the inverted test function

$$(f * \phi)(\vec{x}) = \langle f, T_{\vec{x}}(\phi) \rangle,$$

where $T_{\vec{x}}(\phi)(\vec{y}) = \phi(\vec{x} - \vec{y})$.

6. Finally, with a mild restriction, distributions have Fourier transforms. We need to assume the distribution doesn't blow up totally wildly at ∞. The condition is given in Definition 6.7:

Definition 6.7. A distribution f is *tempered* if its value $\langle f, \phi \rangle$ can be defined for all C^∞ test functions $\phi(\vec{x})$ whose derivatives decrease to 0 faster than all polynomials (i.e., $\forall \alpha, n, \|\vec{x}\|^n |D^\alpha \phi(\vec{x})|$ is bounded on $\mathbb{R}^n$). In other words, we weaken the compact support condition on ϕ to a rapid decrease.

It is easy to check that test functions ϕ decreasing at ∞ like this have Fourier transforms $\hat{\phi}$ that are again C^∞ and decrease at ∞ like this. Then we define the Fourier transform of a tempered distribution f by

$$\langle \hat{f}, \phi \rangle = \langle f, \hat{\phi} \rangle$$

for all test functions ϕ. For instance, 1 and δ_0 are Fourier transforms of each other. Some other useful Fourier transform pairs—not always easy to find—are shown in Table 6.1 Note that if f is even, then f and $\hat{f}$ are each the Fourier transform of the other, whereas if f is odd, then $\hat{\hat{f}} = -f$. To make the symmetry clear, we use x for the variable in both space and frequency.

Function $f(x)$	Fourier transform $g(x)$	Parity
1	$\delta_0(x)$	even
x	$\frac{i}{2\pi}\delta_0'(x)$	odd
x^n	$(\frac{i}{2\pi})^n \delta_0^{(n)}(x)$	alternating even, odd
$\delta_0^{(n)}(x)$	$(2\pi i x)^n$	alternating even, odd
$\text{sign}(x)$	principal value$(\frac{1}{\pi i x})$	odd
$|x|$	$\frac{-1}{2\pi^2 x^2}$ renormalized as above	even
$\frac{1}{x^n}$ renormalized	$(-i\pi)^n \frac{2^{n-1}}{(n-1)!} x^{n-1} \text{sign}(x)$	alternating even, odd
$1/|x|$ renormalized	$-\log(x^2) + c$	even

Table 6.1. Fourier transform pairs.

The final useful thing we need for distributions is a set of reference points to tell us how wild or how nice the distribution is. This is given by the ladder of *Hilbert-Sobolev spaces* $H^n, -\infty < n < \infty$, which sit like this:

$$\mathcal{D} \subset \cdots \subset H^2 \subset H^1 \subset H^0 = L^2 \subset H^{-1} \subset \cdots \subset H^{-n} \subset \cdots \subset \mathcal{D}'$$

These are very handy so we know where we are in this brave new world of "not-quite-functions." They are defined as follows for $k \geq 0$:

$$H^k = \{f \in \mathcal{D}' \,|\, D^\alpha f \in L^2(\mathbb{R}^n) \text{ for all } D^\alpha \text{ where } \alpha_1 + \cdots + \alpha_n \leq k\}$$
$$= \left\{f \in \mathcal{D}' \,\Big|\, \int |\xi^\alpha \hat{f}|^2 d\vec{\xi} < +\infty \right.$$
$$\left. \text{ for all } \xi^\alpha = \xi_1^{\alpha_1} \cdots \xi_n^{\alpha_n}, \alpha_1 + \cdots + \alpha_n \leq k\right\}$$
$$= \left\{f \in \mathcal{D}' \,\Big|\, \int (1+|\vec{\xi}|^2)^k \cdot |\hat{f}|^2 d\vec{\xi} < \infty \right\},$$

For negative exponents,

$$H^{-k} = \left\{f \in \mathcal{D}' \,\Big|\, \int \frac{|\hat{f}|^2}{(1+|\vec{\xi}|^2)^k} d\vec{\xi} < \infty \right\}.$$

In fact, we can define H^α for all $\alpha \in \mathbb{R}$ by the condition

$$\int (1+|\xi|^2)^\alpha |\hat{f}|^2 d\vec{\xi} < +\infty.$$

It is often more useful to separate local regularity of f from bounds at ∞, and so we define

$$H^\alpha_{\text{loc}} = \{f \in \mathcal{D}' \,|\, f\phi \in H^\alpha, \text{ all } \phi \in \mathcal{D}\}$$
$$= \left\{f \in \mathcal{D}' \,\Big|\, \int (1+|\vec{\xi}|^2)^\alpha |\hat{f} * \hat{\phi}|^2 d\vec{\xi} < +\infty \text{ all } \phi \in \mathcal{D}\right\},$$

which imposes only local regularity on f or, equivalently, bounds at ∞ on smoothed versions of $\hat{f}$.

6.6 Basics XIII: Gaussian Measures on Function Spaces

In the second half of this chapter, we actually construct stochastic models for images. We want these to be scale-invariant, so our models will be probability measures on function spaces. To define these, we need some background. The case of Gaussian measures is the only one that has been worked out to any depth, and we explain these with examples in this section.

First, we present the key definitions.

Definition 6.8. A probability measure P on a function space X is Gaussian if, for every continuous linear map $\pi : X \to \mathbb{R}^n$, $\pi_* P$ is a Gaussian measure on $\mathbb{R}^n$; that is,

$$\pi_* P(\vec{x}) = \frac{1}{Z} e^{-(\vec{x}-\overline{x})^t Q(\vec{x}-\overline{x})} (dx_1 \cdots dx_n) \text{ for some } Q, \overline{x}.$$

Definition 6.9. A probability measure P on a function space X has mean $\overline{f} \in X$ if, for all continuous linear functionals $\ell : X \to \mathbb{R}$,

$$\mathbb{E}(\ell) = \ell(\overline{f}).$$

This is the same as saying $\pi(\overline{f}) = $ mean $\overline{x}$ of $\pi_* P$, with π as in Definition 6.8.

Definition 6.10. If X is a space of functions $f(\vec{x})$, then a function $C(\vec{x}, \vec{y})$ of two arguments—a kernel—is the covariance of a probability measure P if for all continuous linear maps ℓ_1, ℓ_2 given by

$$\ell_i(f) = \int f(x) \phi_i(x) d\vec{x}, \quad i = 1, 2,$$

then

$$\mathbb{E}((\ell_1 - \ell_1(\overline{f})) \cdot (\ell_2 - \ell_2(\overline{f}))) = \int \int \phi_1(x) \phi_2(y) C(x, y) \, d\vec{x} \, d\vec{y}.$$

Note that the functionals $\ell : X \to \mathbb{R}$ take the place of the coordinates X_i on $\mathbb{R}^n$ in the finite-dimensional definition

$$\mathbb{E}((x_i - \overline{x})(x_j - \overline{x})) = C_{ij}.$$

The most important fact about Gaussian distributions is that, just as in the finite-dimensional case, *a Gaussian distribution is determined by its mean and covariance.* This is because the mean and covariance determine its marginals on all continuous linear projections to $\mathbb{R}^n$; hence, they determine the value of the measures on a basis for the σ-algebra of all measurable subsets.

6.6.1 Brownian Motion Revisited

In Chapter 3, we introduced Brownian motion as a probability measure on the function space of the continuous function $f : [0, \infty) \to \mathbb{R}$ such that $f(0) = 0$. We defined it by giving its marginals as suitable Gaussian distributions and citing the Kolmogorov Existence Theorem. It is the simplest and most well-studied example of a Gaussian measure on a function space.

In this chapter, we construct our basic random functions by *expansions*—sums of independent terms, each involving a scalar Gaussian random variable. Because any sum of independent scalar Gaussian random variables is Gaussian, the Gaussian property will be immediate. Let's warm up by doing this for a variant of Brownian motion, periodic Brownian motion. The idea is simply to construct a periodic function $f \in L^2(\mathbb{R}/\mathbb{Z})$ by a random Fourier series,

$$\mathcal{F}(x) = \sum_{n \in \mathbb{Z}, n \neq 0} \mathcal{A}_n e^{2\pi i n X}, \quad \overline{\mathcal{A}_n} = \mathcal{A}_{-n}$$

where the $\mathcal{A}_i$ are independent complex numbers with mean 0 and variance $1/n^2$. We have to be a little careful. To be precise, for $n \geq 1$, we let $\mathcal{A}_n = \mathcal{B}_n + i\mathcal{C}_n$ and assume $\mathcal{B}_n, \mathcal{C}_n$ are all independent with mean 0, variance $1/n^2$ and $\mathcal{B}_{-n} = \mathcal{B}_n, \mathcal{C}_{-n} = -\mathcal{C}_n$. Note incidentally that this means that

$$E(\mathcal{A}_n \mathcal{A}_m) = 0, \quad n \neq \pm m$$
$$E(\mathcal{A}_n^2) = E(\mathcal{B}_n^2 - \mathcal{C}_n^2) + 2iE(\mathcal{B}_n\mathcal{C}_n) = 0$$
$$E(\mathcal{A}_n \mathcal{A}_{-n}) = E(|\mathcal{A}_n|^2) = E(\mathcal{B}_n^2 + \mathcal{C}_n^2) = 2/n^2.$$

In these constructions, there is always a question of where the resulting measure is supported. A key point is given in Lemma 6.11.

Lemma 6.11. *If $\{\mathcal{X}_n\}$ are iid Gaussian random variables with mean 0 and variance 1, then for any $\alpha > 1$, the sum*

$$\sum_n n^{-\alpha} |\mathcal{X}_n|^2$$

is almost surely finite.

6.6. Basics XIII: Gaussian Measures on Function Spaces

Proof: In fact,

$$C \cdot \text{Prob}\left(\sum_{n=1}^{N} n^{-\alpha}|\mathcal{X}_n|^2 > C\right) \leq \mathbb{E}\left(\sum_{n=1}^{N} n^{-\alpha}|\mathcal{X}_n|^2\right) = \sum_{n=1}^{N} \frac{1}{n^\alpha} < \infty,$$

so

$$\text{Prob}\left(\sum n^{-\alpha}|\mathcal{X}_n|^2 \text{ diverges}\right) \leq \lim_{C \to \infty} \text{Prob}\left(\sum n^{-\alpha}|\mathcal{X}_n|^2 > C\right) = 0. \quad \square$$

Use the diagram

$$\ell^{\text{arb}} = (\text{Space of all sequences } \{a_n\}, \overline{a_n} = a_{-n})$$
$$\cup$$
$$\ell^2 = (\text{Space of sequences with } \sum |a_n|^2 < \infty) \xrightarrow{\pi} L^2(\mathbb{R}/\mathbb{Z}).$$

On ℓ^{arb}, we have put the product measure p of all the normal distributions on the $\mathcal{B}_n, \mathcal{C}_n$ with variances $1/n^2$. By the lemma, this measure is supported on ℓ^2, and so via π, it induces a measure on $L^2(\mathbb{R}/\mathbb{Z})$.

To see how the mean and covariance work, let's calculate the mean and covariance for the two versions of Brownian motion. We can use this, then, to show that, modulo the so-called Brownian bridge, they are equal. First, using the definition in Chapter 3, the mean path is clearly 0 because the Gaussian distribution on $(f(a_1), \cdots, f(a_n))$ space has 0 mean. Noting that, we may write

$$f(a_i) = \int \delta_{a_i}(x) f(x) dx, \quad \delta_a = \text{delta function at } a.$$

Then

$$C(a_1, a_2) = \iint \delta_{a_1}(x) \delta_{a_2}(y) C(x, y) dx dy = \mathbb{E}(\mathcal{F}(a_1)\mathcal{F}(a_2)).$$

Look at the marginal on the two numbers $f(a_1), f(a_2)$. Assuming $a_1 < a_2$, it is

$$\frac{1}{Z} e^{-\frac{f(a_1)^2}{2a_1} - \frac{(f(a_1)-f(a_2))^2}{2(a_2-a_1)}}$$
$$= \frac{1}{Z} e^{-\frac{1}{2}\left[f(a_1)^2\left(\frac{a_2}{a_1(a_2-a_1)}\right) + 2f(a_1)f(a_2)\left(\frac{1}{a_2-a_1}\right) + f(a_2)^2\left(\frac{1}{a_2-a_1}\right)\right]},$$

from which

$$C(f(a_1), f(a_2)) = Q^{-1} = \begin{pmatrix} \frac{a_2}{a_1(a_2-a_1)} & \frac{1}{a_2-a_1} \\ \frac{1}{a_2-a_1} & \frac{1}{a_2-a_1} \end{pmatrix}^{-1} = \begin{pmatrix} a_1 & a_1 \\ a_1 & a_2 \end{pmatrix}.$$

So finally (dropping the restriction $a_1 < a_2$),

$$\mathbb{E}(\mathcal{F}(a_1)\mathcal{F}(a_2)) = \min(a_1, a_2).$$

This expresses the fact that if $a_2 > a_1$, $\mathcal{F}(a_2)$ is symmetrically distributed around its mean $f(a_1)$, while $\mathcal{F}(a_1)$ has variance a_1.

In contrast, if we use periodic Brownian motion,

$$\mathbb{E}(\mathcal{F}(x)) = \sum \mathbb{E}(\mathcal{A}_n) e^{2\pi i n x} = 0,$$

so the mean is 0. But

$$\mathbb{E}(\mathcal{F}(x_1)\mathcal{F}(x_2)) = \sum_{n,m} \mathbb{E}(\mathcal{A}_n \mathcal{A}_m) e^{2\pi i (n x_1 + m x_2)}.$$

Using the previous remark that $\mathbb{E}(\mathcal{A}_n \mathcal{A}_m) = 0$ unless $n = -m$, we get

$$\mathbb{E}(\mathcal{F}(x_1)\mathcal{F}(x_2)) = \sum_n \mathbb{E}(|\mathcal{A}_n|^2) e^{2\pi i n (x_1 - x_2)} = 2 \sum_{n \neq 0} \frac{1}{n^2} e^{2\pi i n (x_1 - x_2)}.$$

This Fourier series can be summed (especially for someone trained as an electrical engineer), and with some pain, we get

$$\mathbb{E}(\mathcal{F}(x_1)\mathcal{F}(x_2)) = (x_1 - x_2)^2 - |x_1 - x_2| + 1/6.$$

(Why? A hint is that it is a periodic function f of $x_1 - x_2$ with $f'' = -\delta_0$, $\int f = 0$).

These covariances don't seem very close, but actually, they are closer than they look. Let's look at the *Brownian bridge*. Let $\mathcal{F}_F$ be a random sample from the free (nonperiodic) Brownian motion. Let

$$\mathcal{F}_P(x) = \mathcal{F}_F(x) - \left(\int_0^1 \mathcal{F}_F(y) dy\right) - (x - 1/2)\mathcal{F}_F(1), \quad 0 \leq x \leq 1.$$

Then $\mathcal{F}_P$ is a continuous periodic function of mean 0. After a little algebra, we can check that $\mathbb{E}(\mathcal{F}_F(a)) = 0$ and $\mathbb{E}(\mathcal{F}_F(a_1)\mathcal{F}_F(a_2)) = \min(a_1, a_2)$ implies that $\mathbb{E}(\mathcal{F}_P(a)) = 0$ and $\mathbb{E}(\mathcal{F}_P(a_1)\mathcal{F}_P(a_2)) = (a_1 - a_2)^2 - |a_1 - a_2| + 1/6$. Thus, $\mathcal{F}_P$ is a random sample from the periodic Brownian space!

6.6.2 White Noise

The same idea of using an explicit expansion allows us to give an explicit construction of *white noise*. This is the simplest and most fundamental probability model we can put on $\mathcal{D}'(\mathbb{R}^n)$. It can be characterized as the unique Gaussian model of distributions $\mathcal{N} \in \mathcal{D}'(\mathbb{R}^n)$ ($\mathcal{N}$ for "noise") with mean 0 and covariance

$$\mathbb{E}(\langle \mathcal{N}, \phi_1 \rangle \cdot \langle \mathcal{N}, \phi_2 \rangle) = \sigma^2 \int_{\mathbb{R}^n} \phi_1(\vec{x}) \phi_2(\vec{x}) d\vec{x}.$$

6.6. Basics XIII: Gaussian Measures on Function Spaces

In particular, if ϕ_1, ϕ_2 have disjoint supports, $\mathbb{E}(\langle \mathcal{N}, \phi_1 \rangle \cdot \langle \mathcal{N}, \phi_2 \rangle) = 0$. So if U_1, U_2 are disjoint open sets, $\mathcal{N}|_{U_1}$ and $\mathcal{N}|_{U_2}$ are *independent*. This generalizes the definition of discrete white noise as iid Gaussian samples at each point of the discrete space. We write this as

$$C_{\mathcal{N}}(\vec{x}, \vec{y}) = \sigma^2 \delta_\Delta(\vec{x}, \vec{y}),$$

where Δ is the diagonal of $\mathbb{R}^n \times \mathbb{R}^n$. It's easiest to work up to the definition of $\mathcal{N}$ by first defining periodic white noise by a random Fourier series:

$$\mathcal{N}_{\text{per}}(\vec{x}) = \sum_{\vec{n} \in \mathbb{Z}^n} \mathcal{A}_{\vec{n}}\, e^{2\pi i \vec{n} \cdot \vec{x}},$$

where $\mathcal{A}_{\vec{n}}$ are Gaussian independent with mean 0 and such that $\overline{\mathcal{A}_{\vec{n}}} = \mathcal{A}_{-\vec{n}}$, $\text{Var}(\mathcal{A}_0^2) = \sigma^2$, $\text{Var}((\text{Re}\,\mathcal{A}_{\vec{n}})^2) = \text{Var}((\text{Im}\,\mathcal{A}_{\vec{n}})^2) = \sigma^2/2$ if $n \neq 0$, so that $\text{Var}(|\mathcal{A}_{\vec{n}}|^2) = \sigma^2$, for all $\vec{n}$. With this definition, it's mechanical to check

$$\mathbb{E}(\langle \mathcal{N}_{\text{per}}, \phi_1 \rangle \cdot \langle \mathcal{N}_{\text{per}}, \phi_2 \rangle) = \int_{(\mathbb{R}^n/\mathbb{Z}^n)} \sum_{\vec{n}} \phi_1(\vec{x}+\vec{n}) \cdot \sum_{\vec{m}} \phi_2(\vec{x}+\vec{m}) d\vec{x}.$$

Heuristically, we may write the probability measure for periodic white noise as

$$P(d\mathcal{N}_{\text{per}}) = \frac{1}{Z} e^{-\frac{\mathcal{A}_0^2}{2\sigma^2} - \sum_{n \neq 0} \frac{(\text{Re}\,\mathcal{A}_n)^2}{2\sigma^2} + \frac{(\text{Im}\,\mathcal{A}_n)^2}{2\sigma^2}} \cdot P_0(d\mathcal{N}_{\text{per}})$$
$$= \frac{1}{Z} e^{-\sum_{\text{all } n}(|\mathcal{A}_n|^2/2\sigma^2)} \cdot P_0(d\mathcal{N}_{\text{per}}) = \frac{1}{Z} e^{-\int_{\mathbb{R}^n/\mathbb{Z}^n} \mathcal{N}_{\text{per}}(\vec{x})^2 d x} \cdot P_0(d\mathcal{N}_{\text{per}}),$$

where $P_0(d\mathcal{N}_{\text{per}}) = \prod d\,\text{Re}(\mathcal{A}_n) \cdot d\,\text{Im}(\mathcal{A}_n)$. Note that in one dimension, $\mathcal{N}_{\text{per}} = \mathcal{B}'_{\text{per}}$, $\mathcal{B}_{\text{per}}$ being periodic Brownian motion. Another heuristic way to see this in one dimension is to write $\mathcal{B}_{\text{per}}(x) = \sum_n \mathcal{A}_n e^{2i\pi n x}$, where the $\mathcal{A}_n$ are independent and such that $\mathbb{E}(\mathcal{A}_n) = 0$ and $\mathbb{E}(|\mathcal{A}_n|^2) = 2/n^2$. Then, by "differentiating" this, we get $\mathcal{B}'_{\text{per}}(x) = 2i\pi \sum_n n\mathcal{A}_n e^{2i\pi n x}$, where now $\mathbb{E}(n\mathcal{A}_n) = 0$ and $\mathbb{E}(|n\mathcal{A}_n|^2) = 2$ for all n. This the definition of $\mathcal{N}_{\text{per}}$.

How wild is white noise as a distribution? Using Lemma 6.11, we find that one-dimensional white noise $\mathcal{N} \in H_{\text{loc}}^{-1/2-\epsilon}$. In higher dimensions, white noise gets more and more irregular. In fact, in dimension n, $\mathcal{N} \in H_{\text{loc}}^{-n/2-\epsilon}$. This follows from the estimate

$$\mathbb{E}\left(\sum_{\vec{n}} (1+\|\vec{n}\|^2)^{-\frac{n}{2}-\epsilon} |\mathcal{A}_{\vec{n}}|^2 \right) =$$

$$\sigma^2 \sum_{\vec{n}} \frac{1}{(1+\|\vec{n}\|^2)^{n/2+\epsilon}} \approx \sigma^2 \int_1^\infty \frac{r^{n-1} dr}{(1+r^2)^{n/2+\epsilon}} < +\infty.$$

To extend this constructive approach to generate white noise on the whole of $\mathbb{R}^n$ in one fell swoop, we can use a double series, using discrete sampling in both space and frequency space:

$$\mathcal{N}(\vec{x}) = \sum_{\vec{n},\vec{m}} \mathcal{A}_{\vec{n},\vec{m}} e^{2\pi i \vec{n} \cdot \vec{x}} w(\vec{x} - \vec{m}/2),$$

where w is a "window" function. More precisely, we want

(a) $\mathrm{Supp}(w) \cap (\mathrm{Supp}(w) + \vec{m}) = \emptyset$, all $\vec{m} \in \mathbb{Z}^n, \vec{m} \neq 0$

(b) $\sum_{\vec{m} \in \mathbb{Z}^n} w(\vec{x} - \vec{m}/2)^2 \equiv 1$.

This can easily be achieved by starting with some C^∞ function $w_1(x)$ of one variable, with $\mathrm{supp}(w_1) \subset [-1/2, +1/2]$ and $w_1 > 0$ on $[-1/4, +1/4]$. Let

$$w(\vec{x}) = \prod_{i=1}^n \left(w_1(x_i) \bigg/ \sqrt{\sum_m w_1^2(x_i + m/2)} \right).$$

We are thus building white noise from components that "tile" space and frequency simultaneously (see Figure 6.13).

As before, the $\mathcal{A}_{\vec{n},\vec{m}}$ are independent Gaussian except for $\mathcal{A}_{-\vec{n},\vec{m}} = \overline{\mathcal{A}_{\vec{n},\vec{m}}}$. To verify that we indeed have white noise, we just compute covariances. The

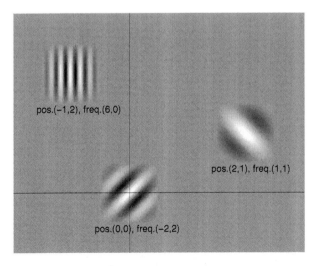

Figure 6.13. The tiling of the four-dimensional image-plane $\times$ frequency-plane by elementary images with positions $\vec{m}$, frequencies $\vec{n}$.

6.6. Basics XIII: Gaussian Measures on Function Spaces

pieces $\mathcal{N}^{(\vec{m})}(\vec{x}) = \sum_{\vec{n}} \mathcal{A}_{\vec{n},\vec{m}} e^{2\pi i \vec{n} \vec{x}}$ of $\mathcal{N}$ are *independent* samples of periodic noise, and

$$\langle \mathcal{N}, \phi_i \rangle = \sum_{\vec{m}} \langle \mathcal{N}^{(\vec{m})}(\vec{x}), w(\vec{x} - \vec{m}/2) \phi_i(\vec{x}) \rangle$$

By using the independence of the $\mathcal{N}^{(\vec{m})}$ and the fact that $w(\vec{x}) w(\vec{x} + \vec{n}) \equiv 0$ if $\vec{n} \neq 0$, we check

$$\mathbb{E}(\langle \mathcal{N}, \phi_1 \rangle \cdot \langle \mathcal{N}, \phi_2 \rangle) = \sigma^2 \int_{\mathbb{R}^n} \phi_1(\vec{x}) \phi_2(\vec{x}) d\vec{x}.$$

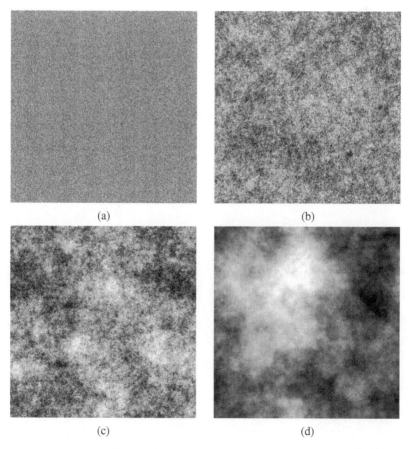

Figure 6.14. Gaussian random images whose coefficients scale as $\mathbb{E}(|\mathcal{A}_{\vec{n}}|^2) = 1, ||n||^{-1.5}, ||n||^{-2}$, and $||n||^{-3}$. For viewing purposes, the real axis has been scaled to [0,1] by using the logistic function $1/(1 + e^{-x/a})$, where a is set to twice the median absolute value.

This is the defining property of Gaussian white noise, which, as previously, we write informally by

$$C_{\mathcal{N}} = \sigma^2 \delta_\Delta(\vec{x}, \vec{y})$$

$$\text{or} \quad P(d\mathcal{N}) = \frac{1}{Z} e^{-\int \mathcal{N}(\vec{x})^2 \, d\vec{x}/2\sigma^2} \cdot \prod_{\vec{x}} d\mathcal{N}(\vec{x}).$$

These two constructions with random Fourier series are special cases of a general technique of using random Gaussian Fourier coefficients with a prescribed rate of decrease of their expected squared absolute value. We may consider

$$\sum_{\vec{n} \in \mathbb{Z}^n, \vec{n} \neq 0} \mathcal{A}_{\vec{n}} e^{2\pi i \vec{n} \cdot \vec{x}},$$

where $\overline{\mathcal{A}_{\vec{n}}} = \mathcal{A}_{-\vec{n}}$, $\mathcal{A}_{\vec{n}}$ Gaussian with $\mathbb{E}(\mathcal{A}_{\vec{n}}) = 0$, and $\text{Var}(\mathcal{A}_{\vec{n}}) = ||\vec{n}||^{-\lambda}$. For $\lambda = 0$, we get periodic white noise; for dimension $n = 1$ and $\lambda = 2$, we get periodic Brownian motion. In Figure 6.14, we show a few images of two-dimensional random functions. These have been popularized by Benoit Mandelbrot in various graphical guises. They are trivial to generate: one simply picks iid Gaussian Fourier coefficients and then scales them down to the desired rate of expected decrease.

6.7 Model I: The Scale-, Rotation- and Translation-Invariant Gaussian Distribution

Can we define a stochastic model for images that has the three fundamental properties discussed in Sections 6.1, 6.2, and 6.4? This is not easy! We begin by addressing scale-invariance.

There is a unique translation-, rotation- and scale-invariant Gaussian probability distribution on the space of generalized functions, or Schwartz distributions, on $\mathbb{R}^n$ (as we saw before, it cannot "live" on a space of actual functions). We now construct it in two dimensions, that is, on images $\mathcal{I}$. The right choice is given by the heuristic formula:

$$P(d\mathcal{I}) = \frac{1}{Z} e^{-\iint ||\nabla \mathcal{I}||^2 dx dy/2\sigma^2} \cdot \prod_{x,y} d\mathcal{I}(x,y),$$

which, purely formally, is scale-invariant due to the elementary fact stated in Proposition 6.12.

6.7. The Scale-, Rotation- and Translation-Invariant Gaussian Distribution

Proposition 6.12. *In dimension 2 (and only in dimension 2), $\iint_{\mathbb{R}^2} \|\nabla I\|^2 dxdy$ is scale-invariant; that is, if we define $I_\sigma(x,y) = I(\sigma x, \sigma y)$, then*

$$\iint_{\mathbb{R}^2} \|\nabla I_\sigma\|^2 dxdy = \iint_{\mathbb{R}^2} \|\nabla I\|^2 dxdy \quad \text{for any smooth function } I.$$

The proof is immediate by substitution. Intuitively, in the expression

$$\left(\left(\frac{\partial I}{\partial x} \right)^2 + \left(\frac{\partial I}{\partial y} \right)^2 \right) dxdy,$$

there are two distances in the numerator and two in the denominator.

But the formula doesn't make any sense literally. It does make sense, however, if we make a finite approximation:

$$P(d\mathcal{I}) = \frac{1}{Z} e^{-\sum_i \sum_j ((\mathcal{I}_{i+1,j} - \mathcal{I}_{i,j})^2 + (\mathcal{I}_{i,j} - \mathcal{I}_{i,j+1})^2)/2\sigma^2} \cdot \prod_{i,j} d\mathcal{I}_{i,j}.$$

If we treat i, j as cyclic variables on $\mathbb{Z}$ modulo N and M, this is a special type of quadratic form and special Gaussian distribution—almost. One problem is that it is also invariant under $I_{ij} \to I_{ij} + \overline{I}$. This makes it an "improper" distribution with an infinite integral. We have to accept this and consider the distribution as a probability measure not on the images themselves but on *images up to an additive constant*, that is,

$$I \in \mathbb{R}^{NM} \Big/ \left(\begin{array}{c} \text{subspace of constant} \\ \text{images } I_{ij} \equiv \overline{I} \end{array} \right).$$

We could define the Gaussian distribution on Schwartz distributions by considering the projections $(\mathcal{D}')^2 \to \mathbb{R}^{(\mathbb{Z}/n\mathbb{Z})^2}$ given by choosing an array of test functions and then taking a suitable limit of the pull-back measures.

Another method is to use the Fourier transform and the fact that $\widehat{(I_x)} = -2\pi i \xi \hat{I}$ and $\widehat{(I_y)} = -2\pi i \eta \hat{I}$. This gives us

$$\iint \|\nabla I\|^2 = \iint I_x^2 + \iint I_y^2 = \iint |\hat{I}_x|^2 + \iint |\hat{I}_y|^2 = 4\pi^2 \iint (\xi^2 + \eta^2) |\hat{I}|^2.$$

Hence, our heuristic formula can be rewritten as

$$P(d\hat{\mathcal{I}}) = \frac{1}{Z} \prod_{\xi, \eta} e^{-4\pi^2 (\xi^2 + \eta^2) |\hat{\mathcal{I}}(\xi,\eta)|^2 / 2\sigma^2} \cdot d\hat{\mathcal{I}}(\xi, \eta).$$

This suggests that we can define a random image $\mathcal{I}$ by multiplying random white noise in the Fourier variables (ξ, η) by the weighting function $(\xi^2 + \eta^2)^{-1/2}$ and taking the inverse Fourier transform.

However, we can also construct the scale-invariant distribution by taking a wavelet-based modification of the construction in the last section of white noise. We present this method in detail here because it extends the analysis of images through filter responses (wavelets) given in Chapter 4. The idea is not to sample (space)×(frequency) by a lattice, but to sample by points $(\vec{x}, \vec{\xi})$ with the property that when $(\vec{x}, \vec{\xi})$ is used, so is $(2\vec{x}, \frac{1}{2}\vec{\xi})$. The key point here is that when the spatial scale is doubled, the corresponding frequency scale is halved. We need to start with an elementary function $w(\vec{x})$, a "mother" wavelet, reasonably well localized in space, around, say, 0, and with no high or low frequencies in its Fourier transform $\hat{w}$. That is, its support $\hat{w}$ in frequency space satisfies

$$\hat{w}(\vec{\xi}) \neq 0 \Rightarrow C^{-1} \leq \|\vec{\xi}\| \leq C$$

for some C. Then we define a random image by the usual wavelet expansion:

$$\mathcal{I}(\vec{x}) = \sum_{k=-\infty}^{+\infty} \left(\sum_{\vec{n} \in \mathbb{Z}^n} \mathcal{A}_{\vec{n},k} w(2^k \vec{x} - \vec{n}) \right).$$

Notice that the $(\vec{n}, k)$th term is spatially located near $(\vec{n}/2^k)$ and has frequencies $\vec{\xi}$ in the range $2^k C^{-1} \leq \|\vec{\xi}\| \leq 2^k C$, or, in other words, the higher frequency terms are more densely sampled in space, the lower frequency terms less densely. Figure 6.15 illustrates the tiling of space-frequency with this expansion and should be compared with Figure 6.13.

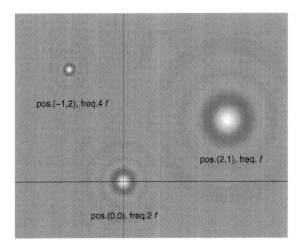

Figure 6.15. The tiling of the four-dimensional image-plane × frequency-plane by elementary wavelet images with positions $\vec{m}$, radial frequency $2^n f$.

6.7. The Scale-, Rotation- and Translation-Invariant Gaussian Distribution

We look at the expansion of $\mathcal{I}$ like this:

$$\mathcal{I}(\vec{x}) = \sum_{k=-\infty}^{+\infty} \mathcal{I}_k(2^k \vec{x})$$

where

$$\mathcal{I}_k(\vec{x}) = \sum_{\vec{n}} \mathcal{A}_{\vec{n},k} w(\vec{x} - \vec{n})$$

are independent samples from the same band-pass distribution. This formulation makes it clear that the statistics of $\mathcal{I}$ are set up precisely to make them invariant under scaling by powers of 2. In fact, we can make them invariant under all scalings.

Right off, though, we encounter a problem. If the above expansion is to make any sense, then $\mathcal{I}_k(0)$ will be independent samples from some Gaussian distribution. Whereas $\mathcal{I}_k$ for $k \gg 0$ will probably oscillate very fast near 0, after convolution with a test function, it can be expected to be very small, and $\mathcal{I}_k$ for $k \ll 0$ is changing very slowly. So the variance of $\sum_{k=0}^{-N} \langle \mathcal{I}_k, \phi \rangle$ will build up linearly in N (ϕ is a test function, which is a simple blip around 0). Thus, the series for $\mathcal{I}$ cannot converge almost all the time. In terms of images, if we had true scale-invariance, this page would be part of infinitely many even larger shapes with random gray levels right here, and this sum wouldn't converge.

But fortunately, large band-passed shapes are very smooth locally, so their effect is very close to adding a constant. Thus we are saved from this "infrared catastrophe" since we are defining a probability distribution not on $\mathcal{D}'(\mathbb{R}^2)$ but on $\mathcal{D}'(\mathbb{R}^2)$/constants. This means that $\langle f, \phi \rangle$ will not be defined for all test functions $\phi \in \mathcal{D}(\mathbb{R}^2)$, but only for those such that $\int \phi = 0$, so that $\langle f + c, \phi \rangle = \langle f, \phi \rangle$ for all constants c. Likewise, the covariance $C(\vec{x}, \vec{y})$ need only integrate against functions ϕ_i with $\int \phi_i = 0$. What should C be? If we want the probability distribution to be translation- and rotation-independent, then C should be a function only of $\|\vec{x} - \vec{y}\|$. If we want it to be scale-invariant, we need $C(\sigma \|\vec{x} - \vec{y}\|) = C(\|\vec{x} - \vec{y}\|)$+constant. The only such function is $C(\vec{x}, \vec{y}) = A + B \cdot \log(\|\vec{x} - \vec{y}\|)$. We can ignore A, and the second constant B just determines the magnitude of the samples $\mathcal{I}$. In fact, we have Theorem 6.13:

Theorem 6.13. *There is a wavelet w that satisfies*
  1. *The support of $\widehat{w}(\xi, \eta)$ is contained in the annulus $0.1 \leq \rho \leq 0.4$, $\rho = \sqrt{\xi^2 + \eta^2}$.*
  2. $\sum_k 4^k \widehat{w}^2(2^k \xi, 2^k \eta) = \rho^{-2}$.

For any such w, the random distribution $\mathcal{I}$ defined above, considered modulo constants, defines a Gaussian distribution on $\mathcal{D}'(\mathbb{R}^2)$/constants with mean 0 and covariance $C(\vec{x}, \vec{y}) = 2\pi \log(1/\|\vec{x} - \vec{y}\|)$. *It is translation-, rotation-, and scale-invariant.*

Proof: First, let's construct such a wavelet w. Simply start with some positive C^∞ function $\omega(\rho)$ supported in $[0.1, 0.4]$. Let

$$\omega^*(\rho) = \rho^2 \sum_k 2^{2k} \omega(2^k \rho), \quad \rho > 0.$$

It is easy to verify that $\omega^*(2\rho) = \omega^*(\rho)$. Take w to be the Fourier transform of $\sqrt{\omega(\rho)/\omega^*(\rho)}$. The two properties are immediate.

We need Lemma 6.14.

Lemma 6.14. *If we make $1/(x^2 + y^2)$ into a distribution by renormalization, that is,*

$$\langle (x^2 + y^2)^{-1}, \phi \rangle = \lim_{\epsilon \to 0} \iint_{\mathbb{R}^2 - B_\epsilon} \frac{\phi(x,y)}{x^2 + y^2} dx dy + 2\pi \log(\epsilon) \phi(0,0),$$

then its Fourier transform, up to a constant, is $-2\pi \log(\xi^2 + \eta^2)$.

This is a standard calculation, proven via the Hankel transform [31] (which is the Fourier transform restricted to radially symmetric functions). Essentially, this comes from the fact that the log term is the Green's function of the two-dimensional Laplacian; hence, its Fourier transform is 1 over the Fourier transform of the Laplacian.

If we now look at $\widehat{\mathcal{I}}$:

$$\widehat{\mathcal{I}}(\xi, \eta) = \sum (\widehat{(\mathcal{I}_k)_{2^k}})(\xi, \eta)$$
$$= \sum 4^{-k} \widehat{\mathcal{I}}_k(2^{-k}\xi, 2^{-k}\eta),$$

where

$$\widehat{\mathcal{I}}_k(\xi, \eta) = \widehat{w}(\xi, \eta) \sum_{n,m} \mathcal{A}_{n,m,k} e^{2\pi i(n\xi + m\eta)}.$$

But $\widehat{w}$ is supported in a circle of diameter 0.8, and the series represents periodic white noise with period 1. Thus their product is the same as $\widehat{w}$ times pure white noise, and we know

$$\mathbb{E}(\langle \widehat{\mathcal{I}}_k, \psi_1 \rangle \cdot \langle \widehat{\mathcal{I}}_k, \psi_2 \rangle) = \iint \widehat{w}(\vec{\xi})^2 \psi_1(\vec{\xi}) \psi_2(\vec{\xi}).$$

Take two test functions $\psi_i(\vec{\xi}$ in Fourier space and assume that $\psi_i(0,0) = 0$ to remove convergence problems). Use the fact that the $\mathcal{I}_k$ are independent and put

6.7. The Scale-, Rotation- and Translation-Invariant Gaussian Distribution

everything together:

$$\begin{aligned}
\mathbb{E}(\langle \widehat{\mathcal{I}}, \psi_1 \rangle . \langle \widehat{\mathcal{I}}, \psi_2 \rangle) &= \sum_k 16^{-k} \mathbb{E}(\langle \widehat{\mathcal{I}_k} \circ 2^{-k}, \psi_1 \rangle . \langle \widehat{\mathcal{I}_k} \circ 2^{-k}, \psi_2 \rangle) \\
&= \sum_k \mathbb{E}(\langle \widehat{\mathcal{I}_k}, \psi_1 \circ 2^k \rangle . \langle \widehat{\mathcal{I}_k}, \psi_2 \circ 2^k \rangle) \\
&= \sum_k \iint \widehat{w}(\vec{\xi}^2) \psi_1(2^k \vec{\xi}) \psi_2(2^k \vec{\xi}) \\
&= \iint \sum_k 4^{-k} \widehat{w}(2^{-k} \vec{\xi}) \psi_1(\vec{\xi}) \psi_2(\vec{\xi}) \\
&= \iint \psi_1(\vec{\xi}) \psi_2(\vec{\xi}) / (\xi^2 + \eta^2).
\end{aligned}$$

So, on test functions that are 0 at (0,0),

$$C_{\widehat{\mathcal{I}}}((\xi, \eta), (\xi', \eta')) = \frac{1}{(\xi^2 + \eta^2)} \delta_\Delta$$

To go back to $\mathcal{I}$, let $\phi_i(x, y)$ be two test functions with $\int \phi_i = 0$, and let $\psi_i = \widehat{\phi_i}$. Then

$$\begin{aligned}
\mathbb{E}(\langle \mathcal{I}, \phi_1 \rangle \langle \mathcal{I}, \phi_2 \rangle) &= \mathbb{E}(\langle \widehat{\mathcal{I}}, \psi_1(-\vec{\xi}) \rangle \cdot \langle \widehat{\mathcal{I}}, \psi_2(-\vec{\xi}) \rangle) \\
&= \iint \frac{1}{\xi^2 + \eta^2} \psi_1(-\xi, -\eta) \psi_2(-\xi, -\eta) d\xi d\eta \\
&= 2\pi \iiiint (-\log \|\vec{x} - \vec{y}\|) \phi_1(\vec{x}) \phi_2(\vec{y}),
\end{aligned}$$

or

$$C_\mathcal{I}(\vec{x}, \vec{y}) = 2\pi \log \left(\frac{1}{\|\vec{x} - \vec{y}\|} \right). \qquad \square$$

A final question is about how close the samples $\mathcal{I}$ of this process are to being functions. The answer is within a hairsbreadth! In fact,

$$\mathcal{I} \in H_{\text{loc}}^{-\epsilon}(\mathbb{R}^2) \quad \text{for almost all samples } \mathcal{I}, \text{ all } \epsilon > 0.$$

To check this, note that locally $\mathcal{I}$ is given by a Fourier series whose coefficients $\mathcal{A}_{n,m}$ satisfy

$$\mathbb{E}(|\mathcal{A}_{n,m}|^2) \approx 1/(n^2 + m^2).$$

Thus,

$$\mathbb{E}(\sum_{n,m} (n^2 + m^2)^{-\epsilon} |\mathcal{A}_{n,m}|^2) \approx \sum \frac{1}{(n^2 + m^2)^{1+\epsilon}} \sim \int_1^\infty \frac{r \cdot dr}{r^{2+2\epsilon}} < \infty.$$

It is not difficult to synthesize these images. When working on finite discrete images, the above construction reduces to synthesizing discrete Fourier series with independent coefficients going to zero like the inverse of frequency squared: so-called "$1/f^2$" noise. This is, in fact, what we did in Section 6.6 (see Figure 6.14(c)). That figure is not exactly a sample from the scale-invariant model because it uses the Fourier expansion, and hence is periodic. Then the low frequency terms will link the top to the bottom and the left to the right side—which would not happen if we had used the wavelet expansion.

What do these images look like? Essentially clouds, with cloudlets and cloud banks on every scale, but no hard edges. This is because the phases of the various coefficients are random and the kurtosis of all filter responses will be 3. To incorporate more of the basic facts on images, we need a non-Gaussian model.

6.8 Model II: Images Made Up of Independent Objects

The previous model, even though it is scale-invariant, has Gaussian marginals with kurtosis 3 and no local structure such as textons. To get a better model, we synthesize a random image as a superposition of elementary images, which will correspond to objects placed at random positions and scales. This section is based on the work of Gidas and Mumford [88]. Before discussing this theory, we need another tool, *infinitely divisible distributions*.

6.8.1 Basics XIV: Infinitely Divisible Distributions

Many signals can be considered, at least to a first approximation, as the sum or superposition of many independent similar signals. Signals of this type are a natural generalization of Gaussian distributed signals but they always have higher kurtosis. To understand the mathematical nature of such models, we study the general theory of random variables, which are a sum of arbitrarily many iid random variables. Definition 6.15 defines infinitely divisible (ID) distributions.

Definition 6.15 (Infinitely divisible distributions). A probability distribution P_1 on a vector space V (defining a random variable $\mathcal{X}^{(1)}$) is infinitely divisible iff for all integers $n > 0$, there exists a probability distribution $P_{1/n}$ such that if $\mathcal{X}_1^{(1/n)}, \ldots, \mathcal{X}_n^{(1/n)}$ are independent random variables following the distribution $P_{1/n}$, then

$$\mathcal{X}^{(1)} \sim \mathcal{X}_1^{(1/n)} + \ldots + \mathcal{X}_n^{(1/n)},$$

where $\sim$ means "having the same probability distribution."

6.8. Model II: Images Made Up of Independent Objects

Note that a Gaussian random variable is infinitely divisible: if $\mathcal{X}^{(1)}$ is Gaussian with mean X_0 and variance σ^2, then we just define the random variables $\mathcal{X}^{(n)}$ to have mean X_0/n and variance σ^2/n. By the additivity of the mean and the variance, $\mathcal{X}^{(1)}$ has the same distribution as the sum of n independent copies of $\mathcal{X}^{(n)}$, as required.

The striking fact is that there is a very elegant complete description of all infinitely divisible distributions, given by the *Levy-Khintchin theorem*. This is treated in all standard probability texts, such as Feller [70] and Breiman [33]. We give here a sketch of the result, emphasizing the motivation. We assume that the random variable $\mathcal{X}$ is real-valued for simplicity. Nothing fundamental changes in the more general case.

1. The Fourier transform of $\mathcal{X}$, that is, its so-called *characteristic function*,

$$F_\mathcal{X}(\xi) = \mathbb{E}(e^{2\pi i \xi \cdot \mathcal{X}}),$$

is nowhere zero. The Fourier transform comes in naturally because whenever $\mathcal{X}$ and $\mathcal{Y}$ are *independent* random variables, then the probability distribution of $\mathcal{X} + \mathcal{Y}$ is the convolution of those of $\mathcal{X}$ and $\mathcal{Y}$; hence, the characteristic functions multiply:

$$F_{\mathcal{X}+\mathcal{Y}}(\xi) = F_\mathcal{X}(\xi) \cdot F_\mathcal{Y}(\xi).$$

Note that the Fourier transform of a *Gaussian* random variable with mean X_0 and variance σ^2 is

$$F_{\text{Gaussian}}(\xi) = \int \frac{1}{\sqrt{2\pi}\sigma} e^{-\frac{(x-X_0)^2}{2\sigma^2}} e^{2i\pi\xi x} dx = e^{2\pi i X_0 \xi - 2\pi^2 \sigma^2 \xi^2}.$$

For any infinitely divisible random variable $\mathcal{X}$ with finite variance σ^2, we have an inequality between its Fourier transform and that of the Gaussian:

$$|F_\mathcal{X}(\xi)| \geq C \cdot e^{-C_2 \sigma^2 \xi^2}.$$

One of the key points of the theory is that there are very interesting infinitely divisible distributions with infinite variance (e.g., the *stable distributions*), but these will not be our focus here. A simple proof of the lower bound for the finite variance case is the following.

Proof: We use the Markov inequality: for any positive random variable $\mathcal{Y}$ and any a, we have $a \cdot \mathbb{P}(\mathcal{Y} \geq a) \leq \mathbb{E}(\mathcal{Y})$. Thus, if $\mathcal{Y}$ has mean 0,

$$|\mathbb{E}(e^{2\pi i \xi \mathcal{Y}} - 1)| \leq \max_{|Y| \leq a} |e^{2\pi i \xi Y} - 1| + 2\mathbb{P}(\mathcal{Y}^2 \geq a^2) \leq 2\pi|\xi|a + 2 \cdot \frac{\text{Var}(\mathcal{Y})}{a^2}.$$

We apply this to $\mathcal{Y} = \mathcal{X}^{(1/n)} - \mathrm{mean}(\mathcal{X}^{(1/n)})$ so that $\mathrm{Var}(\mathcal{Y}) = \mathrm{Var}(\mathcal{X})/n$. Moreover, we take carefully chosen a, n:

$$a = 1/(5\pi|\xi|), n = 125\pi^2\xi^2\mathrm{Var}(\mathcal{X}).$$

Substituting these into the last inequality, we find that the right-hand side is bounded by 4/5. Then it follows readily that

$$|\mathbb{E}(e^{2\pi i \xi \mathcal{X}})| = |\mathbb{E}(e^{2\pi i \xi \mathcal{X}^{1/n}})|^n > (1/5)^n = C_1 \cdot e^{-C_2 \xi^2 \mathrm{Var}(\mathcal{X})}. \qquad \square$$

2. We now have a family of random variables $\mathcal{X}^{(r/s)}$, one for each positive rational number r/s, which we define to be the sum of r independent copies of $\mathcal{X}^{(1/s)}$. We interpolate these random variables and define a continuous family of random variables $X^{(t)}$. First, we let $\psi(\xi) = \log(F_\mathcal{X}(\xi))$, using the fact that $F_\mathcal{X}$ is nowhere zero and taking a continuous branch of the log which is 0 when $\xi = 0$. Thus, $F_{\mathcal{X}^{(1/n)}}(\xi) = e^{\psi(\xi)/n}$. We require that the law of $\mathcal{X}^{(t)}$ be the inverse Fourier transform of $e^{t\psi(\xi)}$. Since this inverse Fourier transform is a positive measure for all rational t, by continuity it is always positive and does define a random variable. Note that

$$\forall t, s > 0, \quad X^{(t+s)} \sim X^{(t)} + X^{(s)}.$$

3. Now, having the variables $\mathcal{X}^{(t)}$, we can go further and put them together into a stochastic process $\{\mathcal{X}_t | t \geq 0\}$ whose values at time t have the distribution $\mathcal{X}^{(t)}$. In fact, we construct the process $\mathcal{X}_t$ to be stationary and Markov, so that for any $s < t$, (a) $\mathcal{X}_t - \mathcal{X}_s$ has the same distribution as $\mathcal{X}^{(t-s)}$, and (b) $\mathcal{X}_t$ is conditionally independent of $\mathcal{X}_{[0,s]}$, given $\mathcal{X}_s$. The construction uses the Kolmogorov existence theorem exactly as in the case of Brownian motion (see Breiman [33], Proposition 14.19). $\mathcal{X}_t$ is *not* necessarily continuous now. It is customary to get rid of measure zero ambiguities by asking that the process be *cadlag* ("continu à droite, limite à gauche," or continuous from the right but with limiting values from the left).

4. Besides Gaussians, which are, in many respects, very special infinitely divisible distributions, where do infinitely divisible distributions come from? The basic example is the following. Let ν be any finite positive measure on the real line and let $\mathcal{S} = \{x_k\}$ be a Poisson process with density ν. Let $\overline{\mathcal{S}} = \sum_k x_k$ be the random *sum* of the process. Then $\overline{\mathcal{S}}$ is an infinitely divisible variable. In fact, we can scale up or down the density to a multiple $t\nu$, and the Poisson process with density $(s+t)\nu$ is just the union of

6.8. Model II: Images Made Up of Independent Objects

independent Poisson processes with densities $s\nu$ and $t\nu$. Thus their sums add up and $\overline{S}$ is infinitely divisible. The process $\mathcal{X}_t$ can be constructed by taking a Poisson process $\{x_k, t_k\}$ on $\mathbb{R}^2$ with density $\nu \times dt$. Then we set

$$\mathcal{X}_t = \sum_{t_k \leq t} x_k.$$

The samples $\mathcal{X}_t$ are *jump functions*: at each t_k; their left and right limits differ by x_k.

The characteristic function of $\overline{S}$ is readily computed by conditioning on the number n of points $\{x_k\}$. The probability of each value of n is $\frac{|\nu|^n}{n!}e^{-|\nu|}$, where $|\nu|$ is the total measure of ν. Thus, the probability distribution of $\overline{S}$ is

$$\sum_n \frac{|\nu|^n}{n!} e^{-|\nu|} \; \overset{n \text{ times}}{\nu * \cdots * \nu},$$

and its characteristic function is

$$F_{\overline{S}}(\xi) = \sum_n \frac{|\nu|^n}{n!} e^{-|\nu|} F_\nu(\xi)^n = e^{|\nu| \cdot (F_\nu(\xi) - 1)}.$$

5. In summary, any infinitely divisible distribution is a sum of a Poisson process and a Gaussian. But there is a little hitch: we can let the measure ν have infinite size $|\nu|$ at the origin if we sum the corresponding Poisson process conditionally. Here there is also a basic example that explains the general story.

We consider a Cauchy random variable defined from a Poisson process. Choose a positive number σ, and choose $\{x_i\} \in \mathbb{R}$ a Poisson process with density $\sigma \cdot dx$. Then define

$$\mathcal{X}^{(\sigma)} = \sum_i \frac{1}{x_i} = \lim_{N \to +\infty} \sum_{|x_i| \leq N} \frac{1}{x_i}.$$

This will converge conditionally almost surely! Notice that $\{y_i\} = \{\frac{1}{x_i}\}$ is a Poisson process with density $\sigma \cdot \frac{dy}{y^2}$. We leave to the reader the exercise to prove that $\mathcal{X}^{(\sigma)}$ is Cauchy distributed, and that, moreover,

$$\mathcal{X}^{(\sigma + \tau)} \sim \mathcal{X}^{(\sigma)} + \mathcal{X}^{(\tau)}.$$

This can be done by calculating the Fourier transforms of both probability densities.

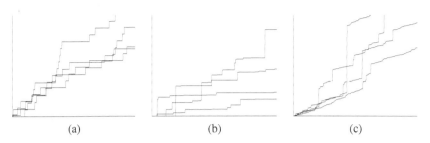

Figure 6.16. Three examples of increasing, stationary, independent-increment jump processes. (a) The Poisson processes are used, $\nu = \delta_1$, and the jumps are discrete and of size 1. (b) The gamma distributions are used, $\nu = \frac{1}{x}e^{-x}dx$, and the jumps are dense and of many sizes. (c) The positive half of the Cauchy distributions are used, $\nu = dx/x^2$, and most jumps are tiny but occasional jumps are huge.

Then again, we can get a bit trickier and just take the positive terms in the above sum, but then compensate with a convergence factor:

$$\mathcal{X}^{(\sigma,+)} = \lim_{N \to +\infty} \left(\sum_{0 \leq x_i \leq N} \frac{1}{x_i} - \log(N) \right).$$

In general, we can take any positive measure ν on the real line for which $\int \min(1, x^2) d\nu < \infty$ and define a sum of the Poisson process associated with ν by using the convergence factor $\int_U x \cdot d\nu$, $U = [-1, 1] - [-\varepsilon, \varepsilon]$.

Figure 6.16 illustrates three simple examples of stationary, independent-increment processes associated with infinitely divisible distributions.

We can now state the final result as Theorem 6.16.

Theorem 6.16 (Levy-Khintchin Representation Theorem). *Every infinitely divisible probability distribution is a sum of a Gaussian distribution and a sum of a Poisson process (possibly with convergence factors) associated with a measure ν, called the Levy measure.*

A very clear treatment of the theorem can be found in Breiman's book, *Probability* [33, Section 9.5].

As mentioned, except in the pure Gaussian case, the kurtosis in ID families is greater than 3:

Lemma 6.17 (Kurtosis in ID Families). *If $\mathcal{X}_t$ is an infinitely divisible family such that $\mathbb{E}(\mathcal{X}_t^4) < \infty$, then there exist constants a, b, c, with $b, c \geq 0$, such*

6.8. Model II: Images Made Up of Independent Objects

that

$$\mathbb{E}(\mathcal{X}_t) = a \cdot t$$
$$\mathbb{E}((\mathcal{X}_t - a \cdot t)^2) = b \cdot t$$
$$\mathbb{E}((\mathcal{X}_t - a \cdot t)^4) = c \cdot t + 3 \cdot (bt)^2;$$

hence,

$$\text{kurtosis}(\mathcal{X}_t) = 3 + \frac{c}{b^2 t} \geq 3.$$

(Remember that 3 is the kurtosis of the Gaussian distributions, so that if $\mathcal{X}_t$ is Gaussian then $c = 0$).

Proof: Since $\mathcal{X}_{s+t} \sim \mathcal{X}_s + \mathcal{X}_t$, it follows that $\mathbb{E}(\mathcal{X}_s)$ is linear, call it $a \cdot t$. Let $Z_t = \mathcal{X}_t - a \cdot t$. Then we get that $\mathbb{E}(\mathcal{X}_s^2)$ is linear, call it $b \cdot t$, and $b \geq 0$ as usual. Furthermore,

$$\mathbb{E}(Z_{s+t}^4) = \mathbb{E}(Z_s^4) + 4\mathbb{E}(Z_s^3)\cdot\mathbb{E}(Z_t) + 6\mathbb{E}(Z_s^2)\cdot\mathbb{E}(Z_t^2) + 4\mathbb{E}(Z_s)\cdot\mathbb{E}(Z_t^3) + \mathbb{E}(Z_t^4)$$
$$= \mathbb{E}(Z_s^4) + 6b^2 s \cdot t + \mathbb{E}(Z_t^4)$$

Therefore, $\mathbb{E}(Z_t^4) - 3\mathbb{E}(Z_t^2)^2$ is also linear, call it $c \cdot t$. But if $c < 0$, then for *small* t, we would get $\mathbb{E}(Z_t^4) < 0$, which is impossible. □

The theory of infinitely divisible distributions and their associated stochastic process $\mathcal{X}_t$ extends easily to vector-valued and even Banach–space-valued random variables (see, for example, the book by Linde [142]). In particular, we may consider infinitely divisible *images*.

6.8.2 Random Wavelet Models

We want to construct here a probability distribution that will reflect the fact that the world is a union of quasi-independent objects; hence, an image of the world is, to some degree, approximated by a sum of independent "imagelets." In a first step, we express this by

$$I(x, y) = \sum_{\text{objects } \alpha} I_\alpha(x, y),$$

where I_α is the image of one "object." As we have seen, if the components are random samples from an auxiliary Levy measure of one-object images, this defines an infinitely divisible random image. We can form spare images by decreasing the number of α, and more cluttered images by increasing it. We can even create a stochastic process image, that is, a random image $\mathcal{I}_t(x, y)$ which starts out at

$t = 0$ as a blank image and gradually accumulates more and more objects in it, asymptotically becoming Gaussian.

To make this model scale- and translation-invariant, the objects α have to be at all positions and scales. We can achieve this if we set

$$\mathcal{I}(x,y) = \sum_k \mathcal{I}_k(e^{\mathcal{S}_k}x - \mathcal{U}_k, e^{\mathcal{S}_k}y - \mathcal{V}_k),$$

where $(\mathcal{U}_k, \mathcal{V}_k, \mathcal{S}_k)$ is a *Poisson process* in $\mathbb{R}^3$ (i.e., a random countable subset of $\mathbb{R}^3$ with given density; see Chapter 2), and the $\mathcal{I}_k$ are chosen independently from some auxiliary probability distribution ν of "normalized one-object images" (e.g., with support in the unit disk and in no smaller disk). Note that the center of the kth object is $(\mathcal{X}_k = e^{-\mathcal{S}_k}\mathcal{U}_k, \mathcal{Y}_k = e^{-\mathcal{S}_k}\mathcal{V}_k)$ and its size is $\mathcal{R}_k = e^{-\mathcal{S}_k}$. We check that $(\mathcal{X}_k, \mathcal{Y}_k, \mathcal{R}_k)$ are a Poisson process from the scale-invariant density $dxdydr/r^3$. The first point is the convergence of $\sum \mathcal{I}_k$, which is proven in [88].

Theorem 6.18 (Convergence). *Assume that the samples from ν are "normalized imagelets" meaning their support is in the unit disk and in no smaller disk. Assume that there is a positive ε such that*

$$\int ||J||_\varepsilon^2 d\nu(J) < \infty.$$

Then the $\sum \mathcal{I}_k$ converges almost surely to a distributional image that is in $H_{\text{loc}}^{-s}/\text{constants}$ for all $s > 0$.

Note that a free parameter in this definition is the density of the Poisson process $(\mathcal{U}_k, \mathcal{V}_k, \mathcal{S}_k)$. A high density will produce scenes that are very cluttered; a low density will produce scenes that are clean and simple.

As a model for generic natural images of the world, various points come out right, but also various points come out wrong. The right qualities include:

- scale-invariance;
- stationarity;
- creates images with distinct objects.

The main wrong point is that this model does not choose a foreground and background when objects overlap, but both act as though they are partly transparent colored glass, because we simply add the images of the different objects to each other. Another point that is wrong is that no grouping laws are incorporated since the imagelets are totally independent and have no desire to have "good continuation" or to be parallel, symmetric, or another assertion. Texture can be incorporated into the Levy measure but does not appear as a consequence of grouping

6.8. Model II: Images Made Up of Independent Objects

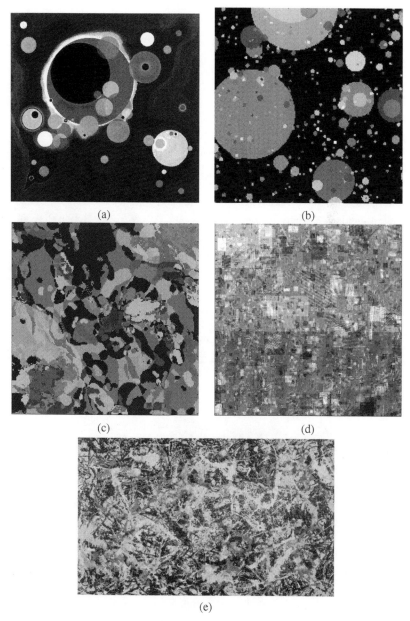

Figure 6.17. A comparison of abstract art and random wavelet models. (a) A 1926 painting entitled "Quelques Cercles" by Wassily Kandinsky (1926, ©2009 Artists Rights Society (ARS), New York / ADAGP, Paris). (e) Jackson Pollock's "Convergence" (1952, ©2009 The Pollock-Krasner Foundation / Artists Rights Society (ARS), New York). (b)–(d) Random wavelet images from the work of Gidas and Mumford: In (b), the Levy measure is concentrated on simple images of disks with random colors; in (c) the Levy measure is concentrated on snake-like shapes with high clutter; in (d), the Levy measure is on textured rectangles. (See Color Plate IX.)

laws acting on the textons. Some examples of random wavelet models are shown in Figure 6.17.

6.9 Further Models

In this section, we sketch possible ways to synthesize random images that are more realistic. The first lack in the random wavelet model was the failure to deal with occlusion: the images of all component objects were added together, which is equivalent to their being transparent. The dead leaves model of Serra and Matheron [192] addresses exactly this issue. The simplest way to define this model is to take the random wavelet model and linearly order the indices k of the summands $\mathcal{I}_k$. Then, instead of adding the terms $\mathcal{I}_k$, we define the value of $\mathcal{I}(x, y)$ to be that of the one summand $\mathcal{I}_k(x, y)$ whose index is least among those summands for which $(x, y) \in \text{support}(\mathcal{I}_k)$.

To make this precise, we define

$$(f \triangleright g)(x, y) = \begin{cases} f(x, y) \text{ if } f(x, y) \neq 0, \\ g(x, y) \text{ if } f(x, y) = 0. \end{cases}$$

Using the setup of the random wavelet model, we now ask that for every k,

$$\mathcal{I}_{t_k, +} = \mathcal{J}_k \triangleright \mathcal{I}_{t_k, -},$$

instead of

$$\mathcal{I}_{t_k, +} = \mathcal{J}_k + \mathcal{I}_{t_k, -},$$

where the subscripts indicate the left and right limits of the process at the point t_k. But there is a problem with this. For any fixed point (x_0, y_0) in the domain of the image and any sample $\mathcal{J}_k$ from ν, this point (x_0, y_0) gets occluded by the new term if and only if $(\mathcal{U}_k, \mathcal{V}_k) \in \text{support}(\mathcal{J}_k) - e^{s_k} \cdot (x_0, y_0)$. This defines a tube in (u, v, s)-space of infinite volume at both ends. What this means is that almost surely for a dense set of times t, a huge object blots out the whole image near (x_0, y_0) and at another dense set of points, microscopic particles of fog blot out every specific point (x_0, y_0).

To make the dead leaves model work we need both infrared and ultraviolet cutoffs, that is, values $s_{\max}$ and $s_{\min}$, which fix maximum and minimum values, respectively, of the radius r of the objects. Then the Poisson process $(\mathcal{U}_k, \mathcal{V}_k, \mathcal{S}_k)$ has only a finite number of values for which the image $\mathcal{J}_k$ affects any bounded area in the image $\mathcal{I}$. Unfortunately, the images we get now have only approximate scale invariance. The effects of letting the cutoffs go to infinity have been studied by A. Lee [135, 135] and Y. Gousseau [28, 89].

6.9. Further Models

We can also define the resulting images by a Markov chain on images $\mathcal{I}$ defined on a finite domain $D \subset \mathbb{R}^2$. We ask that there be transitions from $\mathcal{I}_1$ to $\mathcal{I}_2$ whenever $\mathcal{I}_2 = \mathcal{F} \triangleright \mathcal{I}_1$, where $\mathcal{F}(x,y) = \mathcal{J}(e^{\mathcal{S}}x - \mathcal{U}, e^{\mathcal{S}}y - \mathcal{V})$, $(\mathcal{U}, \mathcal{V}, \mathcal{S})$ is a random point in $D \times [s_{\min}, s_{\max}]$, and $\mathcal{J}$ is a sample from ν. This is the original dead leaves idea: each transition is a new leaf falling off the tree and landing on the ground, occluding everything under it. This version of the model gives a new view of clutter. In the random wavelet model, as the density parameter λ increases, more and more terms are added, and the final image gets closer and closer to a sample from a Gaussian model. But in the dead leaves setup, if we run this Markov chain indefinitely, we approach an equilibrium distribution that is not at all Gaussian. In fact, from time to time, a huge object will obliterate the whole image domain D, and we have to start afresh. Each of the falling leaves covers up the older objects under it and even the accumulation of smaller leaves will eventually cover the whole of D.

In contrast, occlusion is really a three-dimensional phenomenon. This suggests we ought to start with a Poisson process in $\mathbb{R}^4$, that is, in the (x, y, z, r)-space, where (x, y, z) are three-dimensional coordinates and r represents the radius of the object. We give this process the density

$$\lambda \cdot \frac{dxdydzdr}{r^3}$$

for reasons explained below. We can then model a random solid world of three-dimensional opaque objects as a subset of $\mathbb{R}^3$,

$$\bigcup \left[\mathcal{R}_i(\mathrm{Obj}_i) + (\mathcal{X}_i, \mathcal{Y}_i, \mathcal{Z}_i) \right],$$

where the object Obj_i is to be painted with an albedo function $\mathcal{F}_i$. We may use a suitable Levy measure to make both the objects and their surface albedo random. But again, we need an infrared cutoff: the expected density of these opaque objects is $\int \frac{4}{3}\pi r^3 \cdot \lambda/r^3 \cdot dr$, which diverges as $r \to \infty$.

If we now look at this from the origin—from the point of view of projective geometry—we get two-dimensional coordinates:

$$u = \frac{x}{z}, \qquad v = \frac{y}{z}, \qquad r_2 = \frac{r}{z}.$$

We can thus model the resulting two-dimensional image in two different ways. The first way, which corresponds to a "transparent world," is simply

$$\mathcal{I}(u, v) = \sum_i \mathcal{F}_i \left(\frac{u - \mathcal{U}_i}{\mathcal{R}_{2,i}}, \frac{v - \mathcal{V}_i}{\mathcal{R}_{2,i}} \right),$$

which is formally identical to the random wavelet model. In fact,

$$\lambda \cdot \frac{dxdydzdr}{r^3} = \lambda \cdot \frac{dudvdr_2}{r_2^3} \cdot dz.$$

So if we integrate with cutoffs on z instead of r, we get exactly the random wavelet model. The second way, if we have cutoffs on r_2 and z, we can define

$$\mathcal{I}(u,v) = \triangleright_i \mathcal{F}_i\left(\frac{u-\mathcal{U}_i}{\mathcal{R}_{2,i}}, \frac{v-\mathcal{V}_i}{\mathcal{R}_{2,i}}\right),$$

which means the terms are combined using $\triangleright$ applied using the order specified by $\mathcal{Z}_i$. Then we get the dead leaves model.

We don't want to belabor these models because they really are wrong in another respect: the terms with small r produce fog or dust in the three-dimensional scene, which is simply not the way the world is structured. In the real world,

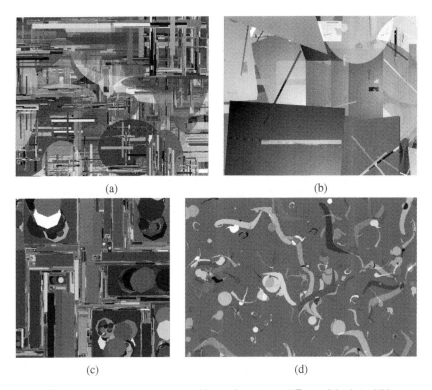

Figure 6.18. Four synthetic images generated by random trees. (a) Trees of depth 4, children constrained to occlude and lie inside parent, top level transparent; (b) top level with uniform sizes but random types and transparency of subobjects; (c) trees of depth 3 with scaling size distribution, disjoint at top level, children lie inside parent; (d) one scale, ν supported at four delta functions, random transparency. (Courtesy of L. Alvarez, Y. Gousseau and J.-M. Morel. See Color Plate X.)

small things cluster on the surface of larger objects: houses sit on the surface of the earth, doorknobs on the surface of doors, and dust itself clusters on floors and in corners. This leads us directly back to the grammatical models of Chapter 3. The component objects of a three-dimensional scene should be generated by a random branching tree, where the child nodes define objects on or near the surface of the parent node. In fact, this allows the model to be more realistic in many ways. Child objects can be constrained to have common characteristics—think of the similarity of the set of leaves of a tree. We can produce coherent textures and implement Gestalt grouping as in Chapter 3. The end result is a view of random geometry that combines the grammatical ideas of Chapter 3 with the scale-invariant models of this chapter. The difference is that the parse tree of Chapter 3 is now situated in scale space rather than being merely an abstract tree. Quite striking random images can be produced by defining suitable random branching trees and seeding scale space by one or more mother trees. Some examples from the work of Alvarez, Gousseau, and Morel are shown in Figure 6.18.

6.10 Appendix: A Stability Property of the Discrete Gaussian Pyramid

We begin by reexamining the discrete Gaussian pyramid. Suppose an image $I(i,j)$ is related to photon energy E by $I(i,j) = (E * R)(i,j)$, where $R(x,y)$ is the response function of the sensor. A natural assumption for the sensor is that

$$\sum_{n,m} R(x+n, y+m) \equiv 1$$

so that light at any point (x,y) creates an image I with the same total energy. We call this the *partition of unity constraint*. Now, filtering an image in any way is the same as convolving the image with some sum of delta functions $F = \sum_{i,j} c_{i,j} \delta_{i,j}$. Then filtering and decimating creates the new image:

$$(E * R * F)(2i, 2j).$$

Repeating, we get

$$(E * R * F * F_2)(4i, 4j), \quad \text{where } F_2 = \sum c_{i,j} \delta_{2i, 2j}.$$

Repeating this n times, we get

$$(E * R * F * F_2 * \cdots * F_{2^{n-1}})(2^n i, 2^n j), \quad \text{where } F_t = \sum c_{i,j} \delta_{ti, tj}.$$

Then the result we want to prove is given in Theorem 6.19.

Theorem 6.19. *If $E(x,y)$ is bounded, $R(x,y)$ satisfies the partition of unity condition and has compact support, and F is the two-dimensional binomial filter of order m, then for some C,*

$$\sup_{x,y}\left|\left(E*R*F*\cdots*F_{2^{n-1}}\right)-\left(E*\frac{1}{4^n}P^{(m)}(\frac{x}{2^n})P^{(m)}(\frac{y}{2^n})\right)\right|\leq\frac{C}{2^n}.$$

This theorem expresses the idea that the discrete pyramid construction converges to the same values regardless of what sensor is used to create a digital image. The term on the left is the result of using any sensor R and following it with Burt-Adelson smoothing. The term on the right is straight Gaussian-spline smoothing, and the result is that they are very close.

Proof: The first step is to work out $F*F_2*\cdots*F_{2^{n-1}}$ for the first order filter $F = \frac{1}{4}(\delta_{0,0}+\delta_{0,1}+\delta_{1,0}+\delta_{1,1})$. We can check easily that

$$F*F_2*\cdots*F_{2^{n-1}} = \frac{1}{4^n}\sum_{i,j=0}^{2^n-1}\delta_{i,j}.$$

Call this filter D_n. Note that $(D_n)^{*m}$ equals the filter $(F'*F'_2*\cdots*F'_{2^n-1})$, for F' equal to the binomial filter of order m. Also we let $\chi = 4^{-n}\mathbb{I}_{[0,2^n]\times[0,2^n]}$, so that $\mathbb{I}^{*m}$ is the filter $4^{-n}P^{(m)}(x/2^n)P^{(m)}(y/2^n)$. Then, because $\|f*g\|_\infty \leq \|f\|_\infty \cdot \|g\|_1$, we can drop E from the theorem and it will suffice to verify that

$$\|R*(D_n)^{*m} - \mathbb{I}^{*m}\|_1 \leq C/2^n.$$

We prove this first for $m=1$. For this case, it says

$$\iint\left|\frac{1}{4^n}\sum_{i,j=0}^{2^n-1}R(x+i,y+j)-\mathbb{I}(x,y)\right|dxdy \leq \frac{4L}{2^n},$$

where L is the width of the smallest square containing the support of R. The reason is that, due to the partition of unity assumption, the term inside the integral is *zero* except in a strip around the perimeter of the square $[0,2^n]\times[0,2^n]$ of width equal to L, where it is bounded by 4^{-n}.

In general, we can write

$$R*(D_n)^{*m}-\mathbb{I}^{*m} = (R*D_n-\mathbb{I})*(D_n)^{*m-1}+(D_n-\mathbb{I})*(D_n)^{*m-2}*\mathbb{I}$$
$$+\cdots+(D_n-\mathbb{I})*\mathbb{I}^{m-1}.$$

Next, check in one dimension what

$$\left(2^{-n}\sum_{k=0}^{2^n-1}\delta_k\right)*\left(2^{-n}\mathbb{I}_{[0,2^n]}\right)-4^{-n}\mathbb{I}_{[0,2^n]}^{*2}$$

looks like. The term on the right is a "tent" function with a base of size 2^{n+1}, rising to a height 2^{-n} (hence area 1). The term on the left is a step-function approximation to this tent with jumps at each integer, so the maximum difference is 4^{-n} and the L^1 norm of the difference is at most 2^{-n+1}. Applying

$$\iint |f(x)f(y) - g(x)g(y)| \leq \max(\|f\|_1, \|g\|_1) \cdot \|f - g\|_1$$

to the terms on the left and right, we get the two-dimensional bound

$$\|D_{2^n} * \mathbb{I} - \mathbb{I}^{*2}\|_2 \leq 2^{-n+1}.$$

We put this together by using the inequality $\|f * g\|_1 \leq \|f\|_1 \cdot \|g\|_1$, which holds also for measures g such as D_n. □

6.11 Exercises

1. Analyzing the Statistics of Your Favorite Set of Images

Take any data base of images I. If these photos are true calibrated photos, replace I by $\log(I)$; otherwise, assume they have been gamma corrected or whatever (see Section 6.1) and leave them alone—there is no way to make them better.

1. Look at the decay of their power in high frequencies. It is tempting to take their fast Fourier transform (FFT; in MATLAB fft). If we do this, and then plot the log of the absolute value of $\widehat{I}$, we will certainly see bright lines along the horizontal and vertical axes. *This is caused by treating the image as periodic.* We get discontinuities by wrapping around, abutting the top and bottom, left and right sides. To avoid this, we use the discrete cosine transform (DDT; dct in MATLAB), instead. One way to think of this is that dct replaces an $n \times m$ image I by an image I' of size $2n \times 2m$ obtained by reflecting the image across its edges. Then I' really is periodic and its FFT is the DCT of I. Alternately, we can expand an image $I(i,j), 1 \leq i \leq N, 1 \leq j \leq M$ in terms given by the products of cosines

$$\frac{2}{\sqrt{NM}} \cos((i - \tfrac{1}{2})\xi\pi/N) \cdot \cos((j - \tfrac{1}{2})\eta\pi/M),$$

$$0 \leq \xi \leq N - 1, 0 \leq \eta \leq M - 1,$$

which have zero normal derivatives at the "frame" around the image given by $i = \tfrac{1}{2}$ or $N + \tfrac{1}{2}$. If we divide the $\xi = 0$ and $\eta = 0$ terms by an additional $\sqrt{2}$, we can easily check that this is an orthogonal basis for the space of functions I. In any case, we can do this with the MATLAB dct function, getting the coefficients $\widehat{I}(\xi, \eta)$ in this expansion. Let $\rho = \sqrt{\xi^2 + \eta^2}$ and bin the range of ρ (omitting the largest values where the full circle is not represented in the discrete data). Average $|\widehat{I}(\xi,\eta)|^2$ over the annuli defined by these bins. Now fit this with a power law (i.e., regress the logs of these averages against $\log(\rho)$). Also, plot the fit and see if it is convincing. Hopefully, you will confirm the results of Section 6.2!

2. Choose your favorite filter. Filtering all the images with this (carefully omitting filter responses that need values beyond the edges of the images), make a histogram of the responses. Plot the log of the frequencies against the filter values; what sort of curve do you find? Then take all the images and take 2×2 block averages, getting images of half the size. Again, histogram the filter responses and plot their logs. How close are the two curves? Repeat this three or four times. Do you get something like Figure 6.5? If not, let us know—that's interesting.

2. A Guessing Game

To gain some insight into the Gaussian pyramid of an image and do a little psychophysical exploration of how humans do object recognition, we recommend the following parlor guessing game—which also can be used to advantage if you have to give a talk to a lay audience. We take your favorite image—in color and with some content that is not completely common. Then, we prepare a series of Gaussian blurs and present them, blurriest to sharpest to our "guinea-pig" and try to figure out what clues they pick up at what resolution. The following is some code due to Mumford (used with an image of his daughter-in-law bicycling past an alligator in the Florida Keys!):

```
y = imread('YourFavoriteImage');
[i,j,k] = size(y);                    % k should be 3
a = .1; kmax = 10; reschange = 2;     % These can be adjusted
for c = 1:3
    yhat(:,:,c) = double(dct2(y(:,:,c)));
end
z = zeros(i,j,3);
for k = 1:kmax
    freq=((0:i-1).^2)'*ones(1,j) + ones(i,1)*((0:j-1).^2);
    factor = exp(-a*freq);
    for c=1:3
        z(:,:,c) = uint8(idct2(yhat(:,:,c).*factor));
    end
    a = a/reschange;
% Plot and pause and/or save this image z
end
```

Another option is to make an interactive tool in which the subject can move a window of fixed resolution anywhere in the pyramid. This means that they can make a small window in the high-resolution image or a big window in the low-resolution image, or anything in between. To what extent can people work out the content of an image, substituting their memories of the content of small windows for a simultaneous viewing of the whole image at high resolution?

3. Warm-Ups in Learning Distributions

1. Here's a problem using the definition from scratch: show that any distribution f on $\mathbb{R}$ that satisfies $f' = f$ equals Ce^x for some constant C. (*Hint*: show that f annihilates all $\phi \in \mathcal{D}$ of the form $\psi' + \psi$, $\psi \in \mathcal{D}$.) Figure out with basic ODE theory what the space of such ϕ is.

6.11. Exercises

2. If f, g are two distributions and $U \subset \mathbb{R}^n$ is an open set, define $f = g$ on U to mean $\langle f, \phi \rangle = \langle g, \phi \rangle$ for all ϕ supported by U. Show that if $\{U_i\}$ is an open cover of $\mathbb{R}^n$ and $f = g$ on each open set U_i, then $f = g$.

3. Note that if $r = \sqrt{x_1^2 + \ldots + x_n^2}$, then $\log(r)$ is an integrable function at 0 in any dimension, and hence defines a distribution. Show from first principles, in dimension 2, that the distribution $\triangle \log(r)$ equals δ_0.

4. Haar Wavelets

The set of Haar wavelets is the first simple example of a wavelet orthonormal basis of $L^2(\mathbb{R})$. This basis is constructed from a single mother wavelet ψ and its translations and dilations.

Let ψ be the function defined on $\mathbb{R}$ by

$$\psi(x) = \begin{cases} 1 & \text{if } 0 \leq x < 1/2, \\ -1 & \text{if } 1/2 \leq x < 1, \\ 0 & \text{otherwise.} \end{cases}$$

Consider the family of functions $\{\sqrt{2^{-j}}\psi(2^{-j}x - k)\}_{(j,k)\in\mathbb{Z}^2}$; then this family is an orthonormal basis of $L^2(\mathbb{R})$. The aim of this exercise is to prove this result.

1. For $(j,k) \in \mathbb{Z}^2$, let $\psi_{j,k}$ denote the function $\sqrt{2^{-j}}\psi(2^{-j}x - k)$. What is the support of $\psi_{j,k}$? Prove that the family of functions $(\psi_{j,k})_{(j,k)\in\mathbb{Z}^2}$ is an orthonormal family in $L^2(\mathbb{R})$, which means that

$$\forall (j,k) \text{ and } (j',k') \in \mathbb{Z}^2, \quad \int_{\mathbb{R}} \psi_{j,k}(x)\psi_{j',k'}(x)\,dx = \delta(j-j')\,\delta(k-k'),$$

where $\delta(n)$ has value 1 when $n = 0$, and value 0 otherwise.

2. We now need to show that any function of $L^2(\mathbb{R})$ can be approximated (in the L^2 sense) by a linear combination of functions $\psi_{j,k}$. To see how this works, we first start with the characteristic function of the interval $[0,1[$. For $n \geq 0$, let f_n denote the function that has value 1 on $[0,2^n[$, and 0 elsewhere. To approximate f_0 with Haar functions, we recursively apply a "double support-averaging trick" (see the book of I. Daubechies [54]). We write f_0 as the sum of its average on $[0,2]$ and a Haar function: $f_0 = \frac{1}{2}f_1 + \frac{1}{\sqrt{2}}\psi_{1,0}$. Then we do this for f_1, then $f_2, \ldots$ and so on. Prove

$$f_0 = \frac{1}{2^n}f_n + \sum_{j=1}^{n}\frac{1}{\sqrt{2^j}}\psi_{j,0}, \quad \text{and then} \quad \| f_0 - \sum_{j=1}^{n}\frac{1}{\sqrt{2^j}}\psi_{j,0} \|_{L^2} = \frac{1}{\sqrt{2^n}}.$$

3. Now let f be any function in $L^2(\mathbb{R})$, and let $\varepsilon > 0$. Approximate f by a compact support and piecewise constant function on intervals of the form $[k2^{-j}, (k+1)2^{-j}]$, and apply the same averaging trick as above to prove that there exist coefficients $c_{j,k}$ such that

$$\| f - \sum_{j,k} c_{j,k}\psi_{j,k} \|_{L^2} \leq \varepsilon.$$

5. Image Compression with Orthonormal Wavelets

Clearly, the $1/f^2$ power falloff in high frequencies means that natural scene images can be coded with far fewer bits than can white noise images. In this exercise, we want to measure this fact by using orthonormal wavelets.

1. Assume we have a finite set of mother wavelets $\psi^{(\alpha)}(i,j)$ and an expansion factor λ such that translates and expansions of the mother wavelets form an orthonormal basis for the vector space of images. To be precise, assume we are dealing with discrete images of size $\lambda^s \times \lambda^s$, where λ and s are integers, and that the wavelets $\psi^{(\alpha)}(i,j), 1 \leq \alpha < \lambda^2$ are orthonormal with mean 0 and support in $1 \leq i,j \leq \lambda$. Then show that the translated and expanded set of functions:

$$e_{\alpha,t,k,l}(i,j) = \lambda^{-t}\psi^{(\alpha)}\left(\lceil \lambda^{-t}i \rceil - \lambda k, \lceil \lambda^{-t}j \rceil - \lambda l\right)$$

for $0 \leq t \leq s-1, 0 \leq k,l \leq \lambda^{s-1-t} - 1, 1 \leq i,j \leq \lambda^s$ are an orthonormal basis of functions $I(i,j), 1 \leq i,j \leq \lambda^s$ with mean zero. Write the expansion of a mean 0 image I as

$$I(i,j) = \sum_{\alpha,t,k,l} \widehat{I}(\alpha,t,k,l) e_{\alpha,t,k,l}(i,j).$$

The triple of two-dimensional Haar wavelets is an example of this with $\lambda = 2$,

$$\psi^{(1)} = \tfrac{1}{2}\begin{pmatrix} 1 & 1 \\ -1 & -1 \end{pmatrix}, \psi^{(2)} = \tfrac{1}{2}\begin{pmatrix} 1 & -1 \\ 1 & -1 \end{pmatrix}, \psi^{(3)} = \tfrac{1}{2}\begin{pmatrix} 1 & -1 \\ -1 & 1 \end{pmatrix}.$$

We now assume we have some probability distribution p on images whose marginals on all pixels (i,j) are equal and which is scale-invariant in the sense of Section 6.2. Let

$$\sigma_\alpha^2 = \mathbb{E}(\widehat{I}(\alpha,0,0,0)^2).$$

Show that:

(a) $\mathbb{E}(I(i,j)) = \mathbb{E}(\widehat{I}(\alpha,t,k,l)) = 0$, for all i,j,α,t,k,l

(b) $\mathbb{E}\left(\widehat{I}(\alpha,t,k,l)^2\right) = \lambda^{2t}\sigma_\alpha^2$

(c) $\mathbb{E}(I(i,j)^2) = s\lambda^{-2}\sum_\alpha \sigma_\alpha^2$

2. Now we introduce differential entropy, as in Chapter 2. Suppose that $\{e_\beta\}, 1 \leq \beta \leq \lambda^{2s}$ is any orthonormal basis for the vector space of mean 0 images I, that $I = \sum_\beta \widehat{I}(\beta)e_\beta$, and that $\sigma_\beta^2 = \mathbb{E}(\widehat{I}(\beta)^2)$. Consider the following two Gaussian models for the images I: p_{wavelet} in which the transform coefficients $\widehat{I}(\beta)$ are independent with mean 0 and variance σ_β^2, and p_{white}, in which the pixel values $I(i,j)$ are independent with mean 0 and variance $\lambda^{-2s}\sum_\beta \sigma_\beta^2$. Show that

$$D(p\|p_{\text{white}}) - D(p\|p_{\text{wavelet}}) = D(p_{\text{wavelet}}\|p_{rmwhite})$$
$$= H_d(p_{\text{white}}) - H_d(p_{\text{wavelet}})$$
$$= \frac{1}{2}\left[\lambda^{2s}\log_2\left(\frac{\sum_\beta \sigma_\beta^2}{\lambda^{2s}}\right) - \sum_\beta \log_2(\sigma_\beta^2)\right] \geq 0$$

6.II. Exercises

with equality if and only if all the σ_β^2 are equal. This is the number of "bits saved" by using the encoding in the basis $\{e_\beta\}$.

3. Combine the previous two parts and, by using the scale-invariant wavelets $\psi_{\alpha,t,k,l}$, show that

   (a) $\sum_{\alpha,t,k,l} \log_2(\mathbb{E}(\widehat{I}(\alpha,t,k,l)^2)) = \frac{\lambda^{2s}-1}{\lambda^2-1}\left(\log_2(\lambda^2) + \sum_\alpha \log_2(\sigma_\alpha^2)\right)$
   $-s\log_2(\lambda^2);$

   (b) By ignoring terms with λ^{-2s}, we finally get for the *per pixel* bits saved:

   $$\frac{D(p_{\text{wavelet}} \| p_{\text{white}})}{\lambda^{2s}} \approx \frac{1}{2}\log_2(s) + \frac{1}{2}\left[\log_2\left(\sum_\alpha \sigma_\alpha^2\right) - \frac{\sum_\alpha \log_2(\sigma_\alpha^2)}{\lambda^2-1}\right]$$
   $$-\frac{\lambda^2}{\lambda^2-1}\log_2(\lambda).$$

   Note that the scale for image intensity cancels out here—only the ratio of the σ_α's matters.

4. Look at some examples. First, for Haar wavelets and 512×512 images, we have $\lambda = 2, s = 9$, and, from experimental data (or the $1/f^2$ power law), the ratio of the variances are $\sigma_1^2 = \sigma_2^2 = 2\sigma_3^2$. Show that Haar encoding gives a savings of about 1.1 bits per pixel. This is not very impressive, but if we add to the picture the high kurtosis of the wavelet responses, this increases a lot.

   Second, we want to look at JPEG compression. The original 1992 JPEG image compression [176] incorporates the essentials of this analysis. JPEG works like this: the image is broken up into disjoint 8×8 blocks of pixels. On each of these blocks, JPEG then takes the DCT of the image as described in Exercise 1. This gives, for each block, a *real* 8×8 transform $\widehat{I}(\xi,\eta)$ depending on pairs of spatial frequencies $0 \leq \xi, \eta < 8$. Unlike the Fourier transform, $(7,7)$ is the highest frequency JPEG encoding gains by throwing away the high frequency coefficients in the DCT when they are too small. The $1/f^2$ power law predicts that $\mathbb{E}(|\widehat{I}(\xi,\eta)|^2) \approx C/(\xi^2+\eta^2)$. This fits the data quite well (we can check this against the data in [110], pages 71–72).

   We can use our machinery now with $\lambda = 8, s = 3$ to encode 512×512 images as a 64×64 collection of 8×8 blocks with zero means, then the means can be encoded by 64 higher 8×8 blocks and their means in turn by one super 8×8 block. There are now 63 α given by $(\xi,\eta), 0 \leq \xi, \eta < 8$ (excluding $(0,0)$) and $\sigma_{(\xi,\eta)}^2 = C/(\xi^2+\eta^2)$. If we plug all this in, we now find this saves about 1.3 bits per pixel. Again, the high kurtosis is key to more compression: many 8×8 blocks have very small DCT coefficients in higher frequencies.

6. Mandelbrot's Fractal Landscapes

It is a remarkable fact that many structures in nature exhibit statistical self-similarity when suitably scaled. This means that when we zoom in on them by a factor λ, they have similar structures up to some rescaling of the "size" of their structures. We have seen the simplest examples in Section 6.6: random Fourier series whose coefficients are sampled from Gaussian distributions with variances being a power of their frequency:

$$f(x) = \sum_n a_n e^{2\pi i n x},$$

where $a_n = \overline{a_{-n}}$, $\text{Re}(a_n)$, $\text{Im}(a_n)$ are independent Gaussian random variables with mean 0 and variance $|n|^{-d}$. Mandelbrot had the idea of doing this for functions $f(x,y)$ and using this as the elevation, $z = f(x,y)$. This is very simple to play with in MATLAB, and we immediately get scenes quite reminiscent of real-world landscapes. The following code helps us to experiment with this. Read Mandelbrot's book [147] for many ideas of this type.

```
% or .85 or 1.5 or 2
n=256; d=1;
coeff = complex(randn(n,n), randn(n,n));
freq = [0 1:n/2 ((n/2)-1):-1:1];
filt = 1./((freq'.^2)*ones(1,n)+ones(n,1)*freq.^2);
       filt(1,1)=0;
landscape = real(ifft2(coeff.*(filt.^d)));
surf(max(landscape,0), 'EdgeColor', 'none',
      'AmbientStrength', .4, ... 'FaceLighting', 'Phong' ),
% sea, rocks, forest shading darker, rocks,snowy peaks
colormap([0 .5 1; .5 .5 .5; zeros(17,1) (1:-.05:.2)'
     zeros(17,1); ... .5 .5 .5; ones(5,3)]);
% Can adjust heights of peaks
axis([0 257 0 257 0 max(landscape(:)*2]), axis off
% A bluish sky
set(gca, 'AmbientLightColor', [.3 .3 1])
% An orange sun low on horizon
light('Position',[1 0 .15], 'Style', 'infinite',
     'Color', [1 .75 0]);
```

7. Some Computations on White Noise Images

Let $(x,y) \to \mathcal{N}(x,y)$ be white noise of variance σ^2 defined on $\mathbb{R}^2$. We recall that according to the discussion in Section 6.6.2, which can be extended to functions in L^2 (see, for instance, [105] for more details about this), the main properties of white noise are:

- For every $f \in L^2(\mathbb{R}^2)$, the random variable $\int f(x,y)\mathcal{N}(x,y)\,dx\,dy$ is a Gaussian random variable with mean 0 and variance $\sigma^2 \int f(x,y)^2\,dx\,dy$.

- For every f and g in $L^2(\mathbb{R}^2)$, the random vector

$$\left(\int f(x,y)\mathcal{N}(x,y)\,dx\,dy, \int g(x,y)\mathcal{N}(x,y)\,dx\,dy \right)$$

is Gaussian with cross-covariance $\sigma^2 \int f(x,y)g(x,y)\,dx\,dy$.

6.11. Exercises

The aim of this exercise is to compute the law of the derivatives of a smoothed white noise image. We start with white noise $\mathcal{N}$ defined on $\mathbb{R}^2$. We then smooth it by convolution with, for instance, a Gaussian kernel of standard deviation r. We thus obtain a smooth (random) image denoted by $\mathcal{I}_r$ and defined by

$$\forall (u,v) \in \mathbb{R}^2 \quad \mathcal{I}_r(u,v) = \frac{1}{2\pi r^2} \int_{\mathbb{R}^2} e^{-((x-u)^2+(y-v)^2)/2r^2} \mathcal{N}(x,y)\, dxdy.$$

1. Prove that $(\frac{\partial \mathcal{I}_r}{\partial x}(u,v), \frac{\partial \mathcal{I}_r}{\partial y}(u,v), \Delta \mathcal{I}_r(u,v))$ is a Gaussian random vector with mean 0 and covariance matrix

$$\begin{pmatrix} \frac{\sigma^2}{8\pi r^4} & 0 & 0 \\ 0 & \frac{\sigma^2}{8\pi r^4} & 0 \\ 0 & 0 & \frac{\sigma^2}{2\pi r^6} \end{pmatrix}.$$

Notice that this implies that the gradient and the Laplacian of $\mathcal{I}_r$ are independent.

2. Prove that the law of $\|\nabla \mathcal{I}_r\|^2$ is $(\sigma^2/8\pi r^4)\chi^2(2)$, where $\chi^2(2)$ denotes the χ^2-law with two degrees of freedom. In other words, prove that

$$\mathbb{P}(\|\nabla \mathcal{I}_r\| \geq t) = \exp\left(-\frac{8\pi r^4 t^2}{2\sigma^2}\right).$$

Bibliography

[1] Milton Abramowitz and Irene Stegun. *Handbook of Mathematical Functions*. New York: Dover, 1965.

[2] Yael Adini, Yael Moses, and Shimon Ullman. "Face Recognition: The Problem of Compensating for Changes in Illumination Direction." *IEEE Transactions on Pattern Analysis and Machine Intelligence* 19 (1997), 721–732.

[3] Adrian Akmajian, Richard Demers, and Robert Harnish. *Linguistics*. Cambridge, MA: MIT Press, 1984.

[4] David Aldous and Persi Diaconis. "Shuffling Cards and Stopping Times." *The American Mathematical Monthly* 93:5 (1986), 333–348.

[5] Luigi Ambrosio and Vincenzo M. Tortorelli. "Approximation of Functionals Depending on Jumps by Elliptic Functionals via Γ-Convergence." *Communications on Pure and Applied Mathematics* 43:8 (1990), 999–1036.

[6] Yali Amit and Donald Geman. "Shape Quantization and Recognition with Randomized Trees." *Neural Computation* 9 (1997), 1545–1588.

[7] Yali Amit and Donald Geman. "A Computational Model for Visual Selection." *Neural Computation* 11 (1999), 1691–1715.

[8] Yali Amit, Donald Geman, and Bruno Jedynak. "Efficient Focusing and Face Detection." In *Face Recognition: From Theory to Applications*, ASI Series F, NATO, edited by H. Wechsler et al., pp. 157–173. New York: Springer, 1998.

[9] Vladimir Arnold. "Sur la géométrie différentielle des groupes de Lie de dimension infinie et ses applications à l'hydrodynamique des fluides parfaits." *Annnales de l'Institut Fourier (Grenoble)* 16:1 (1966), 319–361.

[10] Krishna Athreya and Peter Ney. *Branching Processes*. New York: Springer, 1972.

[11] Jean-Francois Aujol, Guy Gilboa, Tony Chan, and Stanley Osher. "Structure-Texture Image Decomposition: Modeling, Algorithms, and Parameter Selection." *International Journal of Computer Vision* 67:1 (2006), 111–136.

[12] Adrian Barbu and Song-Chun Zhu. "Generalizing Swendsen-Wang to Sampling Arbitrary Posterior Probabilities." *IEEE Transactions on Pattern Analysis and Machine Intelligence* 27 (2005), 1239–1253.

[13] Horace Barlow. "Possible Principles Underlying the Transformation of Sensory Messages." In *Sensory Communication*, edited by W. A. Rosenblith, pp. 217–234. Cambridge, MA: MIT Press, 1961.

[14] Leonard Baum. "An Inequality and Associated Maximization Technique in Statistical Estimation of Probabilistic Functions of Finite State Markov Chain." In *Inequalities*, Vol. 3, edited by Oved Shisha, pp. 1–8. New York: Academic Press, 1972.

[15] Rodney Baxter. *Exactly Solved Models in Statistical Mechanics*. New York: Dover, 2008.

[16] Dave Bayer and Persi Diaconis. "Trailing the Dovetail Shuffle to Its Lair." *The Annals of Applied Probability* 2:2 (1992), 294–313.

[17] Mark Bear, Barry Connors, and Michael Paradiso. *Neuroscience: Exploring the Brain*, Third edition. Philadelphia, PA: Lippincott, Williams, & Wilkins, 2006.

[18] Faisal Beg, Michael Miller, Alain Trouvé, and Laurent Younes. "Computing Large Deformation Metric Mappings via Geodesic Flows of Diffeomorphisms." *International Journal of Computer Vision* 61 (2005), 139–157.

[19] Richard Bellman. *Dynamic Programming*. Princeton, NJ: Princeton University Press, 1957. Dover paperback edition, 2003.

[20] Hans Bethe. "Statistical Theory of Superlattices." *Proc. Roy. Soc. London A* 150:871 (1935), 552–575.

[21] Peter Bickel and Kjell Doksum. *Mathematical Statistics*. Upper Saddle River, NJ: Prentice-Hall, 2006.

[22] Elie Bienenstock, Stuart Geman, and Daniel Potter. "Compositionality, MDL Piors, and Object Recognition." In *Advances in Neural Information Processing Systems*, 9, edited by M. Mozer, M. Jordan, and T. Petsche, 9, pp. 838–844. Cambridge, MA: MIT Press, 1997.

[23] Patrick Billingsley. *Probability and Measure*, Third edition. Wiley Series in Probability and Mathematical Statistics, New York: John Wiley & Sons Inc., 1995.

[24] Andrew Blake and Andrew Zisserman. *Visual Reconstruction*. Cambridge, MA: MIT Press, 1987.

[25] Harry Blum. "Biological Shape and Visual Science." *Journal of Theoretical Biology* 38 (1973), 205–287.

[26] Béla Bollobás. *Graph Theory*. New York: Springer, 1979.

[27] Fred Bookstein. *Morphometric Tools for Landmark Data*. Cambridge, UK: Cambridge University Press, 1991.

[28] Charles Bordenave, Yann Gousseau, and François Roueff. "The Dead Leaves Model: An Example of a General Tessellation." *Advances in Applied Probability* 38 (2006), 31–46.

[29] Eran Borenstein, Eitan Sharon, and Shimon Ullman. "Combining Top-down and Bottom-up Segmentation." In *IEEE Conference on Computer Vision and Pattern Recognition*, Vol. 4, p. 46. Los Alamitos, CA: IEEE Computer Society, 2004.

[30] Yuri Boykov, Olga Veksler, and Ramin Zabih. "Fast Approximate Energy Minimization via Graph Cuts." *IEEE Transactions on Pattern Analysis and Machine Intelligence* 23 (2001), 1222–1239.

[31] Ronald Bracewell. *The Fourier Transform and its Applications*. New York: McGraw-Hill, 1999.

[32] Leo Breiman, Jerome Friedman, Charles Stone, and R. A. Olshen. *Classification and Regression Trees*. Boca Raton, FL: Chapman & Hall, 1984.

[33] Leo Breiman. *Probability*. Philadelphia, PA: SIAM, 1992.

Bibliography

[34] Michael Brent. "An Efficient, Probabilistically Sound Algorithm for Segmentation and Word Discovery." *Machine Learning* 34:1 (1999), 71–105.

[35] Phil Brodatz. *Textures: A Photographic Album for Artists and Designers.* New York: Dover Publications, 1966.

[36] Peter F. Brown, John Cocke, Stephen Della Pietra, Vincent J. Della Pietra, Frederick Jelinek, John D. Lafferty, Robert L. Mercer, and Paul S. Roossin. "A Statistical Approach to Machine Translation." *Computational Linguistics* 16:2 (1990), 79–85.

[37] Peter Brown, Vincent J. Della Pietra, Stephen Della Pietra, and Robert Mercer. "The Mathematics of Statistical Machine Translation: Parameter Estimation." *Computational Linguistics* 19:2 (1993), 263–311.

[38] Peter Burt and Ted Adelson. "The Laplacian Pyramid as a Compact Image Code." *IEEE Transactions on Communications* 31 (1983), 532–540.

[39] Emmanuel Candes and David Donoho. "Curvelets: A Surprisingly Effective Nonadaptive Representation for Objects with Edges." In *Curves and Surfaces*, edited by L. L. Schumaker et al., pp. 69–119. Nashville, TN: Vanderbilt University Press, 1999.

[40] John Canny. "A Computational Approach to Edge Detection." *IEEE Transactions on Pattern Analysis and Machine Intelligence* 8 (1986), 679–698.

[41] Frédéric Cao. "Good Continuations in Digital Image Level Lines." In *IEEE International Conference on Computer Vision*, pp. 440–447. Los Alamitos, CA: IEEE Computer Society, 2003.

[42] Frédéric Cao. "Application of the Gestalt Principles to the Detection of Good Continuations and Corners in Image Level Lines." *Computing and Visualisation in Science* 7 (2004), 3–13.

[43] Gunnar Carlsson, Tigran Ishkhanov, Vin de Silva, and Afra Zomorodian. "On the Local Behavior of Spaces of Natural Images." *International Journal of Computer Vision* 76:1 (2008), 1–12.

[44] Manfredo Do Carmo. *Differential Geometry of Curves and Surfaces.* Englewood Cliffs, NJ: Prentice-Hall, 1976.

[45] Vicent Caselles, Ron Kimmel, and Guillermo Sapiro. "Geodesic Active Contours." *International Journal of Computer Vision* 1:22 (1997), 61–79.

[46] Olivier Catoni. "Simulated Annealing Algorithms and Markov Chains with Rate Transitions." In *Seminaire de Probabilites XXXIII,* Lecture Notes in Mathematics 709, edited by J. Azema, M. Emery, M. Ledoux, and M. Yor, pp. 69–119. New York: Springer, 1999.

[47] Tony Chan and Luminita Vese. "Active Contours without Edges." *IEEE Transactions on Image Processing* 10:2 (2001), 266–277.

[48] Hong Chen, Zi Jian Xu, Zi Qiang Liu, and Song-Chun Zhu. "Composite Templates for Cloth Modeling and Sketching." In *Proceedings of the 2006 IEEE Conference on Computer Vision and Pattern Recognition*, Vol. 1, pp. 943–950. Los Alamitos, CA: IEEE Computer Society, 2006.

[49] Hyeong In Choi, Sung Woo Choi, and Hwan Pyo Moon. "Mathematical Theory of Medial Axis Transform." *Pacific Journal of Mathematics* 181 (1997), 57–88.

[50] Thomas Cormen, Charles Leiserson, Ronald Rivest, and Clifford Stein. *Introduction to Algorithms*. New York: McGraw-Hill, 2003.

[51] Thomas Cover and Joy Thomas. *Elements of Information Theory*, Second edition. New York: Wiley-Interscience, 2006.

[52] William Croft. *Syntactic Categories and Grammatical Relations*. Chicago: University of Chicago Press, 1991.

[53] James Crowley and Alice Parker. "A Representation for Shape Based on Peaks and Ridges in the Difference of Low Pass Transform." *IEEE Transactions on Pattern Analysis and Machine Intelligence* 6 (1984), 156–170.

[54] Ingrid Daubechies. *Ten Lectures on Wavelets*. Philadelphia, PA: SIAM, 1992.

[55] Arthur Dempster, Nan Laird, and Donald Rubin. "Maximum Likelihood from Incomplete Data via the EM Algorithm." *Journal of the Royal Statistical Society—Series B* 39 (1977), 1–38.

[56] Agnès Desolneux, Lionel Moisan, and Jean-Michel Morel. "Meaningful Alignments." *International Journal of Computer Vision* 40:1 (2000), 7–23.

[57] Agnès Desolneux, Lionel Moisan, and Jean-Michel Morel. "Edge Detection by Helmholtz Principle." *Journal of Mathematical Imaging and Vision* 14:3 (2001), 271–284.

[58] Agnès Desolneux, Lionel Moisan, and Jean-Michel Morel. "Variational Snake Theory." In *Geometric Level Set Methods in Imaging, Vision and Graphics*, edited by S. Osher and N. Paragios, pp. 79–99. New York: Springer, 2003.

[59] Agnès Desolneux, Lionel Moisan, and Jean-Michel Morel. *From Gestalt Theory to Image Analysis: A Probabilistic Approach*, Interdisciplinary Applied Mathematics, 34. New York: Springer, 2008.

[60] Persi Diaconis, Ronald Graham, and William Kantor. "The Mathematics of Perfect Shuffles." *Advances in Applied Mathematics* 4 (1983), 175–196.

[61] Persi Diaconis, James A. Fill, and Jim Pitman. "Analysis of Top to Random Shuffles." *Combinatorics, Probability and Computing* 1:2 (1992), 135–155.

[62] Persi Diaconis. "From Shuffling Cards to Walking around the Building: An Introduction to Modern Markov Chain Theory." In *Proceedings of the International Congress of Mathematicians: Documenta Mathematica* 1 (1998), 187–204.

[63] Ian Dryden and Kanti Mardia. *Statistical Shape Analysis*. New York: Wiley, 1998.

[64] Richard Duda and Peter Hart. *Pattern Classification and Scene Analysis*. New York: Wiley, 1973.

[65] Harry Dym and Henry McKean. *Fourier Series and Integrals*. New York: Academic Press, 1972.

[66] Gareth Edwards, Timothy Cootes, and Christopher Taylor. "Face Recognition Using Active Appearance Models." In *Proceedings of the European Conference on Computer Vision*, Vol. 2, pp. 581–595. London: Springer-Verlag, 1998.

[67] Russell Epstein, Peter Hallinan, and Alan Yuille. "5 ± 2 Eigenimages Suffice: An Empirical Investigation of Low-Dimensional Lighting Models." In *IEEE Workshop on Physics-Based Modeling in Computer Vision*, pp. 108–116. Los Alamitos, CA: IEEE Computer Society, 1995.

Bibliography

[68] Leonard Euler. *Methodus inveniendi lineas curvas maximi minimive proprietate gaudentes sive solutio problematis isoperimetrici latissimo sensu accepti.* Lausanne, 1744.

[69] Lawrence Evans and Ronald Gariepy. *Measure Theory and Fine Properties of Functions.* Boca Raton, FL: CRC Press, 1992.

[70] William Feller. *Introduction to Probability Theorey and Its Applications*, 2. New York: Wiley, 1971.

[71] Pedro Felzenszwalb and Dan Huttenlocher. "Pictorial Structures for Object Recognition." *International Journal of Computer Vision* 61 (2005), 55–79.

[72] David Field. "Relations between the Statistics of Natural Images and the Response Properties of Cortical Cells." *Journal of the Optical Society of America A* 4 (1987), 2379–2394.

[73] Martin Fischler and Robert Elschlager. "The Representation and Matching of Pictorial Structures." *IEEE Transactions on Computers* 22:1 (1973), 67–92.

[74] Gerald Folland. *Real Analysis.* New York: Wiley-Interscience, 1999.

[75] Philippe Di Francesco, Pierre Mathieu, and David Sénéchal. *Conformal Field Theory.* New York: Springer, 1999.

[76] Pascal Fua and Yvan Leclerc. "Model-Driven Edge Detection." *Machine Vision and Applications* 3 (1990), 45–56.

[77] Michael Gage and Richard Hamilton. "The Heat Equation Shrinking Convex Plane Curves." *Journal of Differential Geometry* 23 (1986), 69–96.

[78] Mark Gales and Steve Young. "The Application of Hidden Markov Models in Speech Recognition." *Foundations and Trends in Signal Processing* 1 (2007), 195–304.

[79] Meirav Galun, Eitan Sharon, Ronen Basri, and Achi Brandt. "Texture Segmentation by Multiscale Aggregation of Filter Responses and Shape Elements." In *IEEE International Conference on Computer Vision*, p. 716. Los Alamitos, CA: IEEE Computer Society, 2003.

[80] Donald Geman and Stuart Geman. "Stochastic Relaxation, Gibbs Distribution, and the Bayesian Restoration of Images." *IEEE Transactions on Pattern Analysis and Machine Intelligence* 6 (1984), 721–741.

[81] Donald Geman and Alexey Koloydenko. "Invariant Statistics and Coding of Natural Microimages." In *Workshop on Statistical and Computational Theories of Vision.* IEEE, 1999. Available online (http://citeseer.ist.psu.edu/425098.html).

[82] Donald Geman and Chengda Yang. "Nonlinear Image Recovery and Half-Quadratic Regularization." *IEEE Transactions on Image Processing* 4 (1995), 932–946.

[83] Donald Geman, Stuart Geman, Christine Graffigne, and Ping Dong. "Boundary Detection by Constrained Optimization." *IEEE Transactions on Pattern Analysis and Machine Intelligence* 12 (1990), 609–628.

[84] Stuart Geman, Daniel Potter, and Zhiyi Chi. "Composition Systems." *Quarterly of Applied Mathematics* 60 (2002), 707–736.

[85] Hans-Otto Georgii. *Gibbs Measures and Phase Transitions.* Berlin: Walter de Gruyter, 1988.

[86] Zhubin Ghahramani and Michael Jordan. "Factorial Hidden Markov Models." *Machine Learning* 29 (1997), 245–273.

[87] James Gibson. *The Ecological Approach to Visual Perception*. Boston: Houghton-Mifflin, 1979.

[88] Basilis Gidas and David Mumford. "Stochastic Models for Generic Images." *Quarterly of Applied Mathematics* 59 (2001), 85–111.

[89] Yann Gousseau and François Roueff. "Modeling Occlusion and Scaling in Natural Images." *SIAM Journal of Multiscale Modeling and Simulation* 6 (2007), 105–134.

[90] Matthew Grayson. "The Heat Equation Shrinks Embedded Plane Curves to Round Points." *Journal of Differential Geometry* 26 (1987), 285.

[91] D. M. Greig, B. T. Porteous, and A. H. Seheult. "Exact Maximum A Posteriori Estimation for Binary Images." *J. R. Statist. Soc. B* 51:2 (1989), 271–279.

[92] Ulf Grenander and Anuj Srivastava. "Probability Models for Clutter in Natural Images." *IEEE Transactions on Pattern Analysis and Machine Intelligence* 23 (2001), 424–429.

[93] Ulf Grenander. *General Pattern Theory: A Mathematical Study of Regular Structures*. Oxford: Oxford University Press, 1994.

[94] Ulf Grenander. *Elements of Pattern Theory*. Baltimore, MD: Johns Hopkins University Press, 1996.

[95] Geoffrey Grimmett and David Stirzaker. *Probability and Random Processes*, Third edition. Oxford: Oxford University Press, 2001.

[96] Morton Gurtin. *An Introduction to Continuum Mechanics*. New York: Academic Press, 1981.

[97] Peter Hallinan, Gaile Gordon, Alan Yuille, Peter Giblin, and David Mumford. *Two- and Three-Dimensional Patterns of the Face*. Natick, MA: A K Peters, Ltd., 1999.

[98] Peter Hallinan. "A Low-Dimensional Representation of Human Faces for Arbitrary Lighting Conditions." In *IEEE Conference on Computer Vision and Pattern Recognition*, pp. 995–999. Los Alamitos, CA: IEEE Computer Society, 1994.

[99] Peter Hallinan. "A Deformable Model for the Recognition of Human Faces under Arbitrary Illumination." Ph.D. thesis, Harvard University, Cambridge, MA, 1995.

[100] Feng Han and Song-Chun Zhu. "Bottom-Up/Top-Down Image Parsing by Attribute Graph Grammar." In *Proceedings of the Tenth IEEE International Conference on Computer Vision*, Vol. 2, pp. 1778–1785. Los Alamitos, CA: IEEE Computer Society, 2005.

[101] Theodore Harris. *The Theory of Branching Processes*. Berlin: Springer-Verlag, 1963.

[102] Keith Hastings. "Monte Carlo Sampling Methods Using Markov Chains and Their Applications." *Biometrika* 57 (1970), 97–109.

[103] David J. Heeger and James R. Bergen. "Pyramid-based texture analysis/synthesis." In *Proceedings of SIGGRAPH*, pp. 229–238. New York: ACM Press, 1995.

Bibliography

[104] John Hertz, Anders Krogh, and Richard Palmer. *Introduction to the Theory of Neural Computation.* Reading, MA: Addison-Wesley, 1991.

[105] Takeyuki Hida, Hui-Hsiung Kuo, Jürgen Potthoff, and Ludwig Streit. *White Noise. An Infinite Dimensional Calculus.* Mathematics and Its Applications, 253, Dordrecth, Netherlands: Kluwer Academic Publishers Group, 1993.

[106] Yosef Hochberg and Ajit Tamhane. *Multiple Comparison Procedures.* New York: John Wiley & Sons, Inc., 1987.

[107] Paul Hoel, Sidney Port, and Charles Stone. *Introduction to Stochastic Processes.* Boston: Houghton-Mifflin, 1972.

[108] B. K. P. Horn. "The Curve of Least Energy." *ACM Transactions on Mathematical Software* 9 (1983), 441–460.

[109] Jinggang Huang and David Mumford. "Statistics of Natural Images and Models." In *IEEE Conference on Computer Vision and Pattern Recognition.* Los Alamitos, CA: IEEE Computer Society, 1999.

[110] Jinggang Huang. "Statistics of Natural Images and Models." Ph.D. thesis, Div. of Applied Math., Brown University, Providence, RI, 2000. Available at http://www.dam.brown.edu/people/mumford/Papers/Huangthesis.pdf.

[111] Michael Isard and Andrew Blake. "Contour Tracking by Stochastic Propagation of Conditional Density." In *Proceedings of the 4th European Conference on Computer Vision*, Vol. 1, pp. 343–356. London: Springer-Verlag, 1996.

[112] Michael Isard and Andrew Blake. "CONDENSATION: Conditional Density Propagation for Visual Tracking." *International Journal of Computer Vision* 29:1 (1998), 5–28.

[113] Stéphane Jaffard, Yves Meyer, and Robert Ryan. *Wavelets, Tools for Science and Technology.* Philadelphia, PA: SIAM, 2001.

[114] Ya Jin and Stuart Geman. "Context and Hierarchy in a Probabilistic Image Model." In *IEEE Conference on Computer Vision and Pattern Recognition*, pp. 2145–2152, 2006.

[115] Michael Jones and Paul Viola. "Robust Real-Time Face Detection." *International Journal of Computer Vision* 57:2 (2004), 137–154.

[116] Jürgen Jost. *Riemannian Geometry and Geometric Analysis*, Third edition. New York: Springer, 2002.

[117] Bela Julesz. *Foundations of Cyclopean Perception.* Cambridge, MA: MIT Press, 1971.

[118] Daniel Jurafsky and James Martin. *Speech and Language Processing*, Second edition. Upper Saddle River, NJ: Prentice Hall, 2008.

[119] Gaetano Kanizsa. *Organization in Vision.* Austin, TX: Holt, Rinehart & Winston, 1979.

[120] Gaetano Kanizsa. *Grammatica del Vedere.* Bologna, Italy: Il Mulino, 1980. Also published in French as *La Grammaire du Voir* by Éditions Diderot in 1997.

[121] Gaetano Kanizsa. *Vedere e Pensare.* Bologna, Italy: Il Mulino, 1991.

[122] Michael Kass, Andrew Witkin, and Demetri Terzopoulos. "Snakes: Active Contour Models." In *Proceedings of the IEEE International Conference on Computer Vision*, pp. 259–268. Los Alamitos, CA: IEEE Computer Society, 1987.

[123] Ross Kindermann and Laurie Snell. *Markov Random Fields and Their Applications*. Providence, RI: American Mathematical Society, 1980.

[124] John Kingman. *Poisson Processes*. Oxford Studies in Probability, 3, Oxford, UK: Oxford University Press, 1993.

[125] Michael Kirby and Lawrence Sirovich. "Application of the Karhunen-Loeve Procedure for the Characterization of Human Faces." *IEEE Trans. Pattern Anal. Mach. Intell.* 12:1 (1990), 103–108.

[126] Georges Koepfler, Jean-Michel Morel, and Sergio Solimini. "Segmentation by Minimizing Functionals and the Merging Methods." In *Actes du 13eme Colloque GRETSI, Juan-les-Pins*, pp. 1033–1036. Hawthorne, NJ: Walter de Gruyter, 1991.

[127] Georges Koepfler, Christian Lopez, and Jean-Michel Morel. "A Multiscale Algorithm for Image Segmentation by Variational Methods." *SIAM Journal of Numerical Analysis* 31 (1994), 282–299.

[128] Andreas Kriegl and Peter Michor. *The Convenient Setting of Global Analysis*, Mathematical Surveys and Monographs, 53. Providence, RI: American Mathematical Society, 1997.

[129] Joseph Kruskal and Myron Wish. *Multidimensional Scaling*. Thousand Oaks, CA: Sage Publications, 1978.

[130] Sanjeev Kulkarni, Sanjoy Mitter, Thomas Richardson, and John Tsitsiklis. "Local versus Nonlocal Computation of Length of Digitized Curves." *IEEE Trans. Pattern Anal. Mach. Intell.* 16:7 (1994), 711–718.

[131] Xianyang Lan, Stefan Roth, Dan Huttenlocher, and Michael Black. "Efficient Belief Propagation with Learned Higher-Order Markov Random Fields." In *European Conference on Computer Vision 2006, part II*, Lecture Notes on Computer Science 3952, pp. 269–282. New York: Springer, 2006.

[132] Serge Lang. *Fundamentals of Differential Geometry*. New York: Springer, 1999.

[133] Pierre Laplace. *Essai philosophique sur les probabilités*. Paris: Ed. C. Bourgeois, 1986.

[134] Yann LeCun, L. Bottou, Y. Bengio, and P. Haffner. "Gradient-Based Learning Applied to Document Recognition." *Proceedings of the IEEE* 86 (1998), 2278–2324.

[135] Ann Lee and David Mumford. "An Occlusion Model Generating Scale-Invariant Images." In *Workshop on Statistical and Computational Theories of Images*. Los Alamitos, CA: IEEE Press, 1999.

[136] Tai-Sing Lee, David Mumford, and Alan Yuille. "Texture Segmentation by Minimizing Vector-Valued Energy Functionals: The Coupled-Membrane Model." In *Proceedings of the Second European Conference on Computer Vision*, Vol. 588, pp. 165–173. New York: Springer, 1992.

[137] Ann Lee, Kim Pedersen, and David Mumford. "The Nonlinear Statistics of High-Contrast Patches in Natural Images." *International Journal of Computer Vision* 54 (2003), 83–103.

Bibliography

[138] Tai Sing Lee. "Surface Inference by Minimizing Energy Functionals: A Computational Framework for the Visual Cortex." Ph.D. thesis, Harvard University, 1993.

[139] Tai Sing Lee. "Image Representation Using 2D Gabor Wavelets." *IEEE Transactions on Pattern Analysis and Machine Intelligence* 18 (1996), 959–971.

[140] John Lee. *Riemannian Manifolds: An Introduction to Curvature.* New York: Springer, 1997.

[141] Kathryn Leonard. "Efficient Shape Modeling: Epsilon-Entropy, Adaptive Coding, and Boundary Curves vs. Blum's Medial Axis." *International Journal of Computer Vision* 74 (2007), 183–199.

[142] Werner Linde. *Infinitely Divisible and Stable Measures on Banach Spaces.* New York: Wiley, 1986.

[143] David Lowe. "Distinctive Image Features from Scale-Invariant Keypoints." *International Journal of Computer Vision* 60 (2004), 91–110.

[144] Jitendra Malik, Serge Belongie, Thomas K. Leung, and Jianbo Shi. "Contour and Texture Analysis for Image Segmentation." *International Journal of Computer Vision* 43:1 (2001), 7–27.

[145] Stéphane Mallat. *A Wavelet Tour of Signal Processing.* New York: Academic Press, 1999.

[146] Benoit Mandelbrot. "The Variation of Certain Speculative Prices." *Journal of Business* 36 (1963), 394–419.

[147] Benoit Mandelbrot. *The Fractal Geometry of Nature.* San Francisco: W. H. Freeman, 1982.

[148] Albert Marden. *Outer Circles.* Cambridge, UK: Cambridge University Press, 2007.

[149] Kevin Mark, Michael Miller, and Ulf Grenender. "Constrained Stochastic Language Models." In *Image Models and Their Speech Model Cousins*, edited by Stephen E. Levinson and Larry Shepp, pp. 131–140. New York: Springer, 1994.

[150] Andrei A. Markov. "An Example of a Statistical Investigation in the Text of "Eugene Onyegin,' Illustrating the Coupling of "Tests' in Chains." *Proc. Acad. Sci. St. Petersburg* 7 (1913), 153–162.

[151] David Marr and Ellen C. Hildreth. "Theory of Edge Detection." *Proceedings of the Royal Society of London* B-207 (1980), 187–217.

[152] David Marr. *Vision: A Computational Investigation into the Human Representation and Processing of Visual Information.* San Francisco: W. H. Freeman, 1982.

[153] Larisa Matejic. "Group Cascades for Representing Biological Variability." Ph.D. thesis, Brown University, Providence, RI, 1997.

[154] Geoffrey J. Mclachlan and Thriyambakam Krishnan. *The EM Algorithm and Extensions.* New York: Wiley-Interscience, 1996.

[155] Nicholas Metropolis, Arianna Rosenbluth, Marshall Rosenbluth, Augusta Teller, and Edward Teller. "Equations of State Calculations by Fast Computing Machines." *Journal of Chemical Physics* 21 (1953), 1087–1092.

[156] Wolfgang Metzger. *Gesetze des Sehens.* Frankfurt: Waldeman Kramer, 1975.

[157] Yves Meyer. *Oscillating Patterns in Image Processing and Nonlinear Evolution Equations: The Fifteenth Dean Jacqueline B. Lewis Memorial Lectures.* Providence, RI: American Mathematical Society, 2001.

[158] Peter Michor and David Mumford. "Vanishing Geodesic Distance on Spaces of Ssubmanifolds and Diffeomorphisms." *Documenta Mathematica* 10 (2005), 217–245.

[159] Peter Michor and David Mumford. "Riemannian Geometries on Spaces of Plane Curves." *Journal of the European Math. Society* 8 (2006), 1–48.

[160] Peter Michor and David Mumford. "An Overview of the Riemannian Metrics on Spaces of Curves using the Hamiltonian Approach." *Applied and Computational Harmonic Analysis* 23:1 (2007), 74–113.

[161] Michael Miller and Laurent Younes. "Group Actions, Homeomorphisms, and Matching: A General Framework." *International Journal of Computer Vision* 41 (2001), 61–84.

[162] Michael Miller, Alain Trouvé, and Laurent Younes. "On the Metrics and Euler-Lagrange Equations of Computational Anatomy." *Annual Review of Biomedical Engineering* 4 (2002), 375–405.

[163] Melanie Mitchell. *An Introduction to Genetic Algorithms.* Cambridge, MA: MIT Press, 1996.

[164] David Mumford and Jayant Shah. "Optimal Approximations by Piecewise Smooth Functions and Associated Variational Problems." *Communications on Pure and Applied Math.* 42 (1989), 577–685.

[165] David Mumford. "On the Computational Architecture of the Neocortex, I: The Role of the Thalamo-Cortical Loop." *Biological Cybernetics* 65 (1991), 135–145.

[166] David Mumford. "On the Computational Architecture of the Neocortex, II: The Role of Cortico-Cortical Loops." *Biological Cybernetics* 66 (1991), 241–251.

[167] David Mumford. "Elastica and Computer Vision." In *Algebraic Geometry and its Applications*, edited by C. Bajaj, pp. 507–518. New York: Springer, 1993.

[168] Mark Nitzberg, David Mumford, and Takahiro Shiota. *Filtering, Segmentation and Depth*, Lecture Notes in Computer Science 662, New York: Springer, 1993.

[169] Ken Ohara. *One.* Cologne, Germany: Taschen Verlag, 1997.

[170] Bernt Øksendal. *Stochastic Differential Equations: An Introduction with Applications*, Sixth edition, Universitext. Berlin: Springer-Verlag, 2003.

[171] Lars Onsager. "Crystal Statistics I. A Two-Dimensional Model with an Order-Disorder Transition." *Physical Reviews* 65 (1944), 117–149.

[172] Joseph O'Rourke. *Computational Geometry in C.* Cambridge, UK: Cambridge University Press, 2001.

[173] Stephen Palmer. *Vision Science.* Cambridge, MA: MIT Press, 1999.

[174] Judea Pearl. *Probabilistic Reasoning in Intelligent Systems.* San Francisco: Morgan-Kaufmann, 1988.

[175] Judea Pearl. *Causality.* Cambridge, UK: Cambridge University Press, 2000.

[176] William Pennebaker and Joan Mitchell. *JPEG Still Image Data Compression Standard*. London, UK: Chapman & Hall, 1993.

[177] Pietro Perona and Jitendra Malik. "Scale-Space and Edge Detection Using Anisotropic Diffusion." In *Proceedings of IEEE Computer Society Workshop on Computer Vision*, pp. 16–22, Los Alamitos, CA: IEEE Computer Society, 1987.

[178] Thomas Pock, Antonin Chambolle, Daniel Cremers, and Horst Bischof. "A Convex Relaxation Approach for Computing Minimal Partitions." In *IEEE Conference on Computer Vision and Pattern Recognition*, pp. 810–817. Los Alamitos, CA: IEEE Computer Society, 2009.

[179] Jake Porway, Z.-Y. Yao, and Song-Chun Zhu. "Learning an And-Or Graph for Modeling and Recognizing Object Categories." Technical report, UCLA Department of Statistics, Los Angeles, CA, 2007.

[180] Franco Preparata and Michael Shamos. *Computational Geometry*. New York: Springer, 1993.

[181] William Press, Saul Teukolsky, William Vetterling, and Brian Flannery. *Numerical Recipes in C*. Cambridge, UK: Cambridge University Press, 1992.

[182] James G. Propp and David B. Wilson. "Exact Sampling with Coupled Markov Chains and Applications to Statistical Mechanics." *Random Structures and Algorithms* 9 (1996), 223–252.

[183] Jan Puzicha, Thomas Hofmann, and Joachim Buhmann. "A Theory of Proximity Based Clustering." *Pattern Recognition* 33 (2000), 617–634.

[184] Lawrence Rabiner and Biing-Hwang Juang. *Fundamentals of Speech Recognition*. Englewood Cliffs, NJ: Prentice-Hall, 1993.

[185] Thomas Richardson and Sanjoy Mitter. "Approximation, Computation, and Distortion in the Variational Formulation." In *Geometry-Driven Diffusion in Computer Vision, Computational Imaging and Vision*, edited by Bart M. Haar Romeny, pp. 169–190. Norwell, MA: Kluwer Academic Publishers, 1994.

[186] Thomas Richardson. "Scale Independent Piecewise Smooth Segmentation of Images via Variational Methods." Ph.D. thesis, Dept. of Electrical Engineering and Computer Science, MIT, Cambridge, MA, 1989.

[187] Jorma Rissanen. *Stochastic Complexity in Statistical Inquiry*. River Edge, NJ: World Scientific Publishing, 1989.

[188] Jorma Rissanen. *Information and Complexity in Statistical Modeling*. New York: Springer, 2007.

[189] Leonid Rudin, Stanley Osher, and Emad Fatemi. "Nonlinear Total Variation Based Noise Removal Algorithms." *Physica D* 60:1-4 (1992), 259–268.

[190] Erez Sali and Shimon Ullman. "Combining Class-Specific Fragments for Object Classification." In *Proceedings of the 10th British Machine Vision Conference*, Vol. 1, edited by Tony P. Pridmore and Dave Elliman, pp. 203–213. Nottingham: British Machine Vision Association, 1999.

[191] Thomas Sebastian, Philip Klein, and Ben Kimia. "Recognition of Shapes by Editing their Shock Graphs." *IEEE Transactions on Pattern Analysis and Machine Intelligence* 26 (2004), 551–571.

[192] Jean Serra. *Image Analysis and Mathematical Morphology*. New York: Academic Press, 1982.

[193] Jean-Pierre Serre. *Lie Algebras and Lie Groups*, Second edition. New York: Springer, 2005.

[194] Claude Shannon. "A Mathematical Theory of Communication." *Bell System Technical Journal* 27 (1948), 379–423, 623–656.

[195] Claude Shannon. "Prediction and Entropy of Printed English." *Bell System Technical Journal* 30 (1951), 50–64.

[196] Jianbo Shi and Jitendra Malik. "Normalized Cuts and Image Segmentation." *IEEE Transactions on Pattern Analysis and Machine Intelligence* 22:8 (2000), 888–905.

[197] Stuart Shieber. *Constraint-Based Grammar Formalisms*. Cambridge, MA: MIT Press, 1992.

[198] Kaleem Siddiqi and Ben Kimia. "Parts of Visual Form: Computational Aspects." *IEEE Transactions on Pattern Analysis and Machine Intelligence* 17 (1995), 239–251.

[199] Kaleem Siddiqi and Stephen Pizer, editors. *Medial Representations: Mathematics, Algorithms and Applications*. New York: Springer, 2008.

[200] Lawrence Sirovich and Michael Kirby. "Low-Dimensional Procedure for the Characterization of Human Faces." *Journal of the Optical Society of America A* 4 (1987), 519–524.

[201] A. Smith, Arnaud Doucet, Nando de Freitas, and Neil Gordon. *Sequential Monte Carlo Methods in Practice*. New York: Springer, 2001.

[202] Michael Spivak. *A Comprehensive Introduction to Differential Geometry*, Third edition. Houston, TX: Publish or Perish, 1999.

[203] Anuj Srivastava, Xiuwen Liu, and Ulf Grenander. "Universal Analytical Forms for Modeling Image Probabilities." *IEEE Transactions on Pattern Analysis and Machine Intelligence* 24 (2002), 1200–1214.

[204] Ivar Stakgold. *Green's Functions and Boundary Value Problems*. New York: Wiley-Interscience, 1998.

[205] Gil Strang and Truong Nguyen. *Wavelets and Filter Banks*. Wellesley, MA: Wellesley-Cambridge Press, 1997.

[206] Gilbert Strang. "Maximum Flow through a Domain." *Mathematical Programming* 26 (1983), 123–143.

[207] Walter Strauss. *Partial Differential Equations, an Introduction*. New York: Wiley, 1992.

[208] Robert Swendsen and Jian-Sheng Wang. "Nonuniversal Critical Dynamics in Monte Carlo Simulations." *Physical Review Letters* 58 (1987), 86–88.

[209] Richard Szeliski, Ramin Zabih, Daniel Sharstein, Olga Veksler, Vladimir Kolmogorov, Aseem Agarwala, Marshall Tappen, and Carsten Rother. "A Comparative Study of Energy Minimization Methods for Markov Random Fields." In *European Conference on Computer Vision 2006, part II*, Lecture Notes on Computer Science 3952, pp. 16–26. New York: Springer, 2006.

Bibliography

[210] Steven Tanimoto and Theo Pavlidis. "A Hierarchical Data Structure for Picture Processing." *Computer Graphics and Image Processing* 4 (1975), 104–119.

[211] Sekhar Tatikonda and Michael Jordan. "Loopy Belief Propagation and Gibbs Measures." In *Proceedings of the Eighteenth Conference on Uncertainty in Artificial Intelligence*, edited by D. Koller and A. Darwiche, pp. 493–500. San Francisco: Morgan-Kaufmann, 2002.

[212] Howard Taylor and Samuel Karlin. *An Introduction to Stochastic Modeling*. New York: Academic Press, 1984.

[213] David Tolliver and Gary Miller. "Graph Partitioning by Spectral Rounding: Applications in Image Segmentation and Clustering." In *IEEE Conference on Computer Vision and Pattern Recognition*, Vol. I, pp. 1053–1060. Los Alamitos, CA: IEEE Computer Society, 2006.

[214] Anne Treisman. "Preattentive Processing in Vision." *Computer Vision, Graphics and Image Processing* 31 (1985), 156–177.

[215] Alain Trouvé and Laurent Younes. "Local Geometry of Deformable Templates." *SIAM Journal on Mathematical Analysis* 37:1 (2005), 17–59.

[216] Alain Trouvé and Laurent Younes. "Metamorphoses through Lie Group Action." *Found. Comput. Math.* 5:2 (2005), 173–198.

[217] Alain Trouvé. "Infinite Dimensional Group Actions and Pattern Recognition." Technical report, DMI, Ecole Normale Superieure, Paris, 1995.

[218] Zhuowen Tu and Song-Chun Zhu. "Image Segmentation by Data-Driven Markov Chain Monte Carlo." *IEEE Transactions on Pattern Analysis and Machine Intelligence* 24:5 (2002), 657–673.

[219] Leonard Uhr. "Layered 'Recognition Cone' Networks." *IEEE Trans. Computers C* 21 (1972), 758–768.

[220] J. Hans van der Hateren. "Theoretical Predictions of Spatio-Temporal Receptive Fields and Experimental Validation." *J. of Comparative Physiology A* 171 (1992), 157–170.

[221] Arjen van der Schaff and J. Hans van der Hateren. "Modeling the Power Spectra of Natural Images." *Vision Resarch* 36 (1996), 2759–2770.

[222] Luminita Vese and Tony Chan. "A Multiphase Level Set Framework for Image Segmentation Using the Mumford and Shah Model." *International Journal of Computer Vision* 50:3 (2002), 271–293.

[223] Paul Viola and Michael Jones. "Rapid Object Detection Using a Boosted Cascade of Simple Features." *IEEE Conference on Computer Vision and Pattern Recognition* 1 (2001), 511–518.

[224] Paul Viola and Michael Jones. "Robust Real-Time Face Detection." *International Journal of Computer Vision* 57:2 (2004), 137–154.

[225] Yair Weiss and William Freeman. "On the Optimality of Solutions of the Max-Product Belief Propagation Algorithm in Arbitrary Graphs." *IEEE Transactions on Information Theory* 47:2 (2001), 723–735.

[226] Yair Weiss. "Belief Propagation and Revision in Networks with Loops." Technical Report AI Memo 1616 (CBCL Paper 155), MIT, 1997.

[227] Max Wertheimer. "Untersuchungen zur Lehre der Gestalt, II." *Psychologische Forschung* 4 (1923), 301–350.

[228] Bernard Widrow. "The 'Rubber-Mask' Technique—I and II." *Pattern Recognition* 5:3 (1973), 175–211.

[229] Jonathan Yedidia, William Freeman, and Yair Weiss. "Bethe Free Energy, Kikuchi Approximations and Belief Propagation Algorithms." Technical Report TR2001-16, Mitsubishi Electric Research Laboratories, Cambridge, MA.

[230] Laurent Younes. "Computable Elastic Distances between Shapes." *SIAM J. Appl. Math.* 58:2 (1998), 565–586 (electronic).

[231] Laurent Younes. "Optimal Matching between Shapes via Elastic Deformations." *Image and Vision Computing* 17:5/6 (1999), 381–389.

[232] Stella Yu and Jianbo Shi. "Multiclass Spectral Clustering." In *Proceedings of the Ninth International Congress on Computer Vision*, Vol. 2, pp. 313–319. Washington, DC: IEEE Computer Society, 2003.

[233] Song-Chun Zhu and David Mumford. "Prior Learning and Gibbs Reaction-Diffusion." *IEEE Transactions on Pattern Analysis and Machine Intelligence* 19 (1997), 1236–1250.

[234] Song-Chun Zhu and David Mumford. "A Stochastic Grammar of Images." *Foundations and Trends in Computer Graphics and Vision* 2 (2007), 259–362.

[235] Song-Chun Zhu and Alan Yuille. "FORMS: A Flexible Object Recognition and Modelling System." *International Journal of Computer Vision* 20:3 (1996), 187–212.

[236] Song-Chun Zhu and Alan Yuille. "Region Competition: Unifying Snake/Balloon, Region Growing and Bayes/MDL/Energy for Multi-Band Image Segmentation." *IEEE Transactions on Pattern Analysis and Machine Intelligence* 18:9 (1996), 884–900.

[237] Song-Chun Zhu, Yingnian Wu, and David Mumford. "Minimax Entropy Principle and Its Application to Texture Modeling." *Neural Computation* 9:8 (1997), 1627–1660.

[238] Song-Chun Zhu, Yingnian Wu, and David Mumford. "Filters Random Fields and Maximum Entropy (FRAME): To a Unified Theory for Texture Modeling." *International Journal of Computer Vision* 27 (1998), 1–20.

[239] Song-Chun Zhu. "Embedding Gestalt Laws in Markov Random Fields." *IEEE Transactions on Pattern Analysis and Machine Intelligence* 21 (1999), 1170–1189.

[240] Song-Chun Zhu. "Stochastic Jump-Diffusion process for computing Medial Axes in Markov Random Fields." *IEEE Transactions on Pattern Analysis and Machine Intelligence* 21 (1999), 1158–1169.

[241] Jacob Ziv and Abraham Lempel. "Compression of Individual Sequences via Variable-Rate Coding." *IEEE Transactions on Information Theory* 24:5 (1978), 530–536.

Index

a contrario methods, 114, 118
acoustic signal, 2
active contour model, 131
adjoint representation, 268, 301
aliasing, 71, 107, 329
alphabet, 19
amodal contour, 145
angular velocity, 303
Arnold's description of geodesics, 301
atlas, 263, 273
attribute, 153, 155, 161
autocorrelation, 69, 217, 326

Banach space, 273
band-pass filter, 246
banded matrix, 73
baseball, 106
Bayes' theorem, 10, 48, 130
BBP, *see* belief propagation algorithm
belief propagation algorithm, 206
Bernoulli distribution, 52
Bessel K-function, 284, 313, 322
Bethe approximation, 204
Bethe free energy, 209
Bhattaracharya metric, 54
bias of discretization, 190, 242
binding function, 40, 155, 161, 233
binomial distribution, 32, 52, 116
binomial downsampling, 333
binomial filter, 339
blob detector, 338
block averaging, 323, 329, 380
blue sky effect, 321
Bonferroni correction, 117
bottom-up architecture, 13
boundary of $\mathcal{G}$, 287
Box-Muller algorithm, 101
branching process, 149, 171, 377
Brownian bridge, 356

Brownian motion, 123, 354
 construction by its marginals, 123
Burger's equation, 280
Burt-Adelson discrete pyramid, 331

$C^\infty(\mathbb{R}^n, \mathbb{R}^m)$, 273
$C_c^\infty(\mathbb{R}^n, \mathbb{R}^m)$, 273
calibrated image, 320
Canny's edge detector, 164
card shuffling, 30
Cauchy distribution, 101, 369
cells of visual cortex, 219, 345
central limit theorem, 62, 332
Cesaro theorem, 35
character recognition, 112, 146, 159
characteristic function, 367
charts of a manifold, 263, 268, 273
checkerboard pattern, 210
χ^2-law, 385
χ^2-statistic, 232
circulant matrix, 103
clique, 161, 183
clutter, 321, 371, 375
 removal of, 226
co-area formula, 191
co-vector, 275
coding, 21
 coding a continuous random variable, 66
 code book, 21, 96
 code length, 21, 44, 144, 158
 code word, 21, 96
 coding a positive integer, 54
 shape encoding, 142
collapse of L^2 metrics, 310
color images, 187
colored noise, 74, 102
compressible fluid, 286
compression wave, 310

conservation of momentum, 301, 304
Constantinople, 158
contours, 111
    completion of, 115, 132, 145
    detection of, 120, 130, 164
    in an image, 113, 227
    straight, 115
convex hull, 140, 169
coordinates
    material coordinates, 260
    spatial coordinates, 260
coupling from the past, 239
curvature of a manifold, 272, 307
curvature of a plane curve, 275
curvelets, 347

data-driven Monte Carlo Markov chain, 198
dead leaves model, 374
Delaunay triangulation, 139, 169, 170
delta function, 283, 311, 350
delta signal, 258
determinants, computation of, 103, 166
Diff($\mathcal{D}$), 277
Diff($\mathbb{R}^n$), 276, 304
diffeomorphism, 252, 276
    volume-preserving diffeomorphism, 281, 286
difference of Gaussians filter, 337, 339
discontinuities, 79, 86, 318
distance on a manifold, 265
distributions, *see* Schwartz distribution
divergence, *see* Kullback-Leibler distance
DNA sequence, 56
DOG, *see* difference of Gaussians filter
dyadic scaling, 337
dynamic programming, 45, 59, 91

edges in an image, *see* contours
eigenfaces, 250, 254, 294
eigenimages, 313
eigensignals, 259
eigenwarps, 292
elastica model, 125
elasticity theory, 259
    elastic energy, 260, 285, 290, 309
    linear elasticity theory, 261

EM, *see* expectation-maximization algorithm
entropy, 21
    conditional entropy, 22, 28, 35
    differential entropy, 65, 78, 224
    entropy of the source, 27
    entropy rate, 35
EPDiff, 280, 305
epipolar constraint, 310
Euler's equation for inviscid, incompressible fluid, 281
Euler's equations in mechanics, 304
Eulerian equation, 260
expansions, 354
expectation-maximization algorithm, 98, 106
exponential distribution, 52
exponential map, 266–268
exponential models, 221, 229

face recognition, 253
false alarms, number of, 117
filtering and decimating procedure, 333
filters, 217, 225, 246, 319, 327, 380
Ford-Fulkerson theorem, 212, 236
Fourier transform, 68, 219, 351, 367
    discrete Fourier transform, 68, 71
    Fourier basis, 73
    Fourier coefficients, 68
    Fourier series, 68, 354
    image border artifacts, 244, 379
    inverse Fourier transform, 68, 368
    properties and examples, 68, 351
frame, 342, 346
    dual frame, 343
    frame bounds, 342, 344, 346
    tight frame, 342
frame indifference property, 260
fundamental group of a graph, 204

$\mathcal{G}$, 276, 304
Gabor filter, 218, 226, 232, 345
gamma correction, 320
gamma distribution, 52
Gaussian distribution, 62, 353, 360, 367
    covariance, 63, 254, 353
    kurtosis, 81

Index

mean, 63, 353
sampling from a, 101
stationary, 73
Gaussian filter, 246, 300, 331, 385
Gaussian model, 76, 219, 253
    piecewise Gaussian model, 89, 91, 104, 108, 228
Gaussian pyramid, 328, 331, 337, 339, 340
generalized function, *see* Schwartz distribution
genetic algorithms, 199
geodesic, 265, 270, 272
    geodesic equation on $\mathcal{L}^N$, 285, 305
    geodesic equation on Diff, 278, 304
    geodesic equation on $SO(n)$, 303
    geodesic general equation, 302
geometric distribution, 53
geometric heat equation, 142, 275
geometric pattern, 7, 89, 257
Gestalt grouping principles, 120, 142, 154, 377
Gibbs distribution, 183
Gibbs field, 183
Gibbs sampler, 198, 234, 239
gradient descent algorithms, 295
    conjugate gradient, 297
    hierarchical approach, 300
    steepest gradient descent, 295
    with line search, 295
gradient of a discrete image, 114
gradient of length of curves, 275
gradient on manifolds, 274
Gram-Schmidt orthogonalization procedure, 297
grammar, 8, 147
    compositional grammar, 156
    probabilistic context-free grammar, 151
    probabilistic context-sensitive grammar, 153
graph cuts algorithms, 212
greedy search method, 343
Green's functions, 283, 305, 311–313
Greig-Porteous-Seheult method, 211, 236
Grenander's approach, 1
grid of landmark points, 292
group of permutations, 31

H^α_{loc}, 352
H^n, 352
Haar wavelets, 381
Hadamard strain energies, 262
Hammersley-Clifford theorem, 183
harmonic, 76
    fundamental, 78
    resonant frequency, 90
    second and third harmonics, 78
heat equation, 331
Hessian matrix, 296
hidden Markov model, 51, 95, 98
hidden variables, 2, 44, 61, 97, 229, 251
Hilbert transform pair, 218, 230
Hilbert-Sobolev spaces, 352
HMM, *see* hidden Markov model
hyperelastic, 260
hypothesis testing, 114, 120
hysteresis, 164, 165

ID, *see* infinitely divisible
illumination, *see* lighting
image patches, 335
image pyramid, 323, 328
incompressible fluid, 286
inertia matrix, 304
infinitely divisible distribution, 366
infinitely divisible images, 371
infinitesimal deformation, 276
information, 21, 53
    bits of information, 21, 38, 65
infrared or ultraviolet catastrophes, 326, 339, 363, 374
inner product, 264, 275, 277, 295
International Phonetic Alphabet, 17
interpolation, 339
Ising model, 161, 176, 197, 210, 234, 236

JPEG, 382
JPEG compression, 347
Julesz textons, 215
jump function, 369
jumps in a signal, *see* discontinuities

Kanizsa triangle, 145
Karcher mean, 271
Karhunen-Loeve expansions, 253

kinetic energy, 286
Kolmogorov existence theorem, 124, 354, 368
Kullback-Leibler distance, 22, 54, 66, 204, 223
kurtosis, 80, 124, 221, 319, 370

$\mathcal{L}^N(\mathbb{R}^n)$, 282, 305
labeled branching process, 171
labeled tree, 148, 151
Lagrangian equation, 260
landmark point method, 252, 291, 314
landmark point space, 282, 305, 313
Langevin equation, 198
language, 19
Laplace distribution, 81, 322
Laplacian of a Gaussian, 165, 217
Laplacian pyramid, 337, 340
law of large numbers, 101
left (or right) invariant metric, 269
length of a discrete curve, 242
level lines of an image, 118
level sets of an image, 118
Levy measure, 370, 371, 373, 414
Levy-Khintchin theorem, 367, 370
license plate, 110, 162
Lie algebra, 266, 276, 301
Lie bracket, 267, 302
Lie group, 266, 301
lighting angles, 257
lighting coefficients, 257, 294
lighting variations, 249, 255, 294, 313
likelihood function, 10
likelihood ratio, 40, 132, 156, 158
line process, 194
linear group, 268
linear programming, 181, 212, 236, 343
linearization technique, 271, 281
local minima, problem of, 210, 298
low-pass filtering, 329

machine translation, 48
manifold
    finite-dimensional, 263
    infinite-dimensional, 273
MAP, *see maximum a posteriori* estimate
Markov chain, 29, 196, 240, 375

aperiodic, 30, 196
detailed balance, 196, 241
equilibrium distribution, 30, 196, 241
homogeneous, 29
information regular, 38
irreducible, 30, 196
period, 30
primitive, *see* irreducible
steady-state distribution, *see* equilibrium distribution
transition matrix, 29, 35, 196
Markov inequality, 367
Markov property, 29, 33, 183
Markov random field, 183, 190, 193
    marginals, 205, 207
    mode, 199, 203, 212
    on trees, 205
    samples, 195
Markov stochastic process, 368
Marr's zero-crossing, 164
matching problem, 310
max-flow algorithm, 212, 236
maximal disk of a shape, 135
maximum a posteriori estimate, 12, 48, 206, 236
maximum likelihood estimate, 43, 47
MCMC, *see* Monte Carlo Markov chain
MDL, *see* minimum description length
mean field approximation, 204
medial axis, 134, 141, 168, 169
    bifurcation points, 136
    endpoints, 136
medical image analysis, 281, 294
metamorphosis model, 294
Metropolis algorithm, 195, 199
Metropolis-Hastings algorithm, 198
Metropolis transition matrix, 196, 201
min-cut algorithm, 212, 232, 236
minimum description length, 12, 141, 144
minimum length path problem, 46
mixture distribution, 98
ML estimate, *see* maximum likelihood estimate
Möbius inversion formula, 184
modal contour, 145
moduli of elasticity, 262
Monte Carlo Markov chain, 196, 225, 240

Index

Mooney faces, 11
Mooney-Rivlin strain energies, 262
motion parallax, 173
MRF, *see* Markov random field
MS energy, *see* Mumford-Shah model
multidimensional scaling, 270
multiscale decomposition, 7, 323, 336
Mumford-Shah model, 95, 187, 191, 193
musical note, 75
musical score, 61
    tempered scale, 90
    tempo, 90
mutual information, 36, 40

n-gram, 18
    frequency tables, 18, 28, 55
    n-gram model, 19, 32, 59
Newton's law, 260, 286
Newton's method, 296
Nitzberg-Shiota's edge detector, 165
nonlinear data analysis, 271
normal distribution, *see* Gaussian distribution
normalized cut algorithm, 233
normalizing constant, 133, 177, 183, 222
null hypothesis, 115, 119

occlusion, 374, 375
optimal expansion algorithm, 189, 213, 238, 411
order statistic, 327
Ornstein-Uhlenbeck process, 128, 130, 169
orthogonal group, 268, 303

parse graph, 154, 156
parse tree, 113, 148, 171, 377
particle filter algorithms, 198
partition function, *see* normalizing constant
path on a manifold
    energy, 265, 278
    length, 264, 278
pattern recognition, 1
pattern synthesis, 5
pattern theory, 1
PCA, *see* principal components analysis
periodic Brownian motion, 354, 357, 360
periodic white noise, 357, 360

phase space, 183, 203
phase transition phenomenon, 209
phoneme, 17, 96, 335, 336
phones, 2, 96
Poisson distribution, 52, 86
Poisson process, 86, 88, 368, 372, 374
pop-out phenomenon, 215
potential energy, 286
potential on cliques, 183, 206
Potts model, 187, 231
power spectrum, 69, 77, 80, 217, 326, 338
prairie fire equation, 136
prefix code, 21, 24, 67
principal components analysis, 253, 257, 292
principle of least action, 286
prior model, 11
production rule, 151, 154, 156, 161, 171
projective geometry, 375
Propp and Wilson algorithm, 239

quotient metric, 284

random Fourier series, 354, 360
random phase technique, 219, 245
random wavelet expansion, 362
random wavelet models, 371, 373, 414
region adjacency graph, 154
region merging algorithm, 202
renormalization fixed point measure, 324
reproducing kernel property, 283
Riemann curvature tensor, 271
Riemannian energy, 285
Riemannian metric, 264, 269, 282, 302, 309
Riemannian structure, 264
Riemannian submersion, 283
Rissanen's coding, 54

sampling
    a cosine function, 106
    dense sampling of a function, 71
    from n-gram models, 19
    from a continuous distribution, 100
    from a discrete distribution, 52
    from a Potts model, 188
    from the Ising model, 178, 239

sampling (*continued*)
  Gibbs fields, 195
  inversion method, 52, 101
  rejection method, 101
  Shannon-Nyquist sampling theorem, 107
  sparse sampling of a function, 71
  the set of partitions of an image, 240
saturation, 320
scale invariance, 322, 324, 338, 340, 348, 360
Schwartz distribution, 349, 360, 380
  examples, 350
  Fourier transform, 351
  tempered distribution, 351
second-order statistics, 217, 219
segmentation hypothesis, 174
Shannon's optimal coding theorem, 21
shock wave, 310
shock-wave-producing equation, 281
significant edge, 120
  maximally significant, 120
significant straight contour, 117
  maximally significant, 117
simulated annealing, 199, 231, 235
skeleton, *see* medial axis
skew symmetric matrix, 269, 303
slice of an image, 108
snake model, 129, 155
Sobolev norm, 277, 288, 312
space of smooth curves, 273
sparse representation, 340, 343
specific heat, 181, 202
spectrogram, 96
speech recognition, 2, 96, 106
spline function, 332
state space, 183
stationary distribution, 217, 325
statistics of images, 317, 318, 322, 327, 379
statistics on manifolds, 270
  covariance, 271
  mean, 271
steerable frame, 346
stereographic projection, 136, 168
stereoscopic vision, 174, 310
stochastic differential equation, 128

strain energy, 260, 294, 309
strain matrix, 259
stress, 260
submersion, 305
Swendsen-Wang-Barbu-Zhu algorithm, 240
synthesizing
  random images, 188, 365, 374, 377
  random music, 103
  random plane curves, 125, 228
  random shapes, 169
  texture, *see* texture synthesis

t-distribution, 63
T-junction, 143, 165
tail of a distribution, 63, 83, 319
tangent bundle, 264
tangent space, 263, 276
temperature, 177, 181, 199, 202
textons, 215, 335, 344
texture, 214
  texture discrimination, 246
  texture segmentation, 228, 229, 231
  texture synthesis, 220, 225, 231, 245
tiling of space-frequency, 358, 362
top-down architecture, 13
topographic map, 118
total variation distance, 32
translations on a group, 266
transparency, 372, 375
triangle inequality, 286, 290

uncertainty principle, 70
universal covering tree, 203, 209
$u + v$ model, 95, 186

value pattern, 6, 89, 257
variance accounted for (VAF), 255
vector field, 266
  left- or right- invariant, 267
velocity, 279, 288
Venn diagram, 37
Voronoi decomposition, 139

warping, 250, 294
wavelets, 336, 344
  mother wavelet, 336, 362, 381

weak string model, 94, 95
weather forecasting, 58
white noise, 74, 93, 356, 384
windowed Fourier transform, 69, 79
words, 39
    spelling map, 43

word boundaries, 39, 56
word frequencies, 43

Younes' metric, 289, 309

Zipf's law, 43